Research Methods for Computing and Informatics

Patrick Ngulube
https://orcid.org/0000-0002-7676-3931
University of South Africa, South Africa

IGI Global
Scientific Publishing
Publishing Tomorrow's Research Today

Vice President of Editorial Melissa Wagner
Director of Acquisitions Mikaela Felty
Director of Book Development Jocelynn Hessler
Production Manager Mike Brehm
Cover Design Jose Rosado

Published in the United States of America by
IGI Global Scientific Publishing
701 East Chocolate Avenue
Hershey, PA, 17033, USA
Tel: 717-533-8845 | Fax: 717-533-7115
Website: https://www.igi-global.com E-mail: cust@igi-global.com

Library of Congress Cataloging-in-Publication Data

LCCN: 2025034565 (CIP Data Pending)
ISBN13: 9798369340189
Isbn13Softcover: 9798369351987
EISBN13: 9798369340196
British Cataloguing in Publication Data
A Cataloguing in Publication record for this book is available from the British Library.

Editorial Advisory Board

Table of Contents

Detailed Table of Contents

reflexive process that adapts to technological and contextual shifts throughout the research lifecycle. Built on five core principles, that is, triple methodological integration, proactive ethics, dynamic adaptability, enhanced transparency, and structured interdisciplinary collaboration. The AIRM ensures scientific rigour while remaining responsive to real-world complexity.

The latest complexities of Big Data and Gen-AI are posing novel difficulties to Information Systems (IS) and Analytics researchers. The conventional mono-methods are inadequate to address this complexity. This chapter argues for methodological pluralism and suggests a synthesized approach, a combination of Mixed-Methods Research (MMR) and Computational Social Science (CSS). The chapter is grounded in a pragmatic philosophy. This approach leverages the interpretive richness of qualitative methods, the scale of the quantitative analysis, as well as the strength of computational research methods. The chapter identifies the MMR designs and emerging types of computational data, and provides examples of their combination, such as computational ethnography. The synthesized approach enables a more holistic and robust understanding of complex socio-technical phenomena in informatics, providing a direction for rigorous and relevant research.

The integration of data analytics in higher education underscores the need to examine the methodological approaches guiding this research domain. While applications of data analytics have gained extensive attention, their methodological foundations remain underexplored. This review investigates the prevalence, characteristics, and contributions of mixed methods research in data analytics. Following PRISMA guidelines, 22 empirical studies from five databases across were synthesised. Results show that MMR effectively captures the complexities of pedagogy, assessment, and implementation through epistemological complementarity rather than mere methodological combination. Methodological plurality enhances explanatory depth and produces triangulated insights into multifaceted educational phenomena. The review identifies four principles for effective MMR: parallel design integration, sequential design connectivity, analytical triangulation, and methodological

This narrative review aimed to track the adoption of the document analysis approach and to understand the extent to which it has been used as a robust stand-alone qualitative research method in the context of computing and Information Systems (IS). Although document analysis research method is seen as a valuable and beneficial research approach, a thematic synthesis of the studies revealed limited or scant use and adoption of this method in the field of computing and Information Systems (IS). This study is intended to bridge the disparity in literature by focusing on how document analysis can be used as a robust stand-alone approach through determining the importance of conducting qualitative document analysis in computing and IS research, the criteria to effectively select the relevant documents for qualitative research as well as the strengths and limitations of using document analysis as stand-alone qualitative research in the context of Computing and IS. The chapter thus mapped and provided insights into the use of document analysis as an independent qualitative research method in order to understand the deeper meaning and underlying patterns within the documents.

Phenomenology is a philosophical approach that focuses on the lived experiences of people in different situations and circumstances. Phenomenology provides a window for conducting qualitative research that explores human experiences by engaging with study participants in detail to collect rich data about phenomena. Phenomenology can be applied in various fields of study, including information science and informatics, to examine the experiences of different individuals and inform theoretical, policy, and practical dimensions. In this chapter, we discuss the definition of phenomenology, its history, the various types of phenomenology, and the justification for adopting phenomenological approaches. We also examine sampling, data collection, and data analysis in phenomenological studies, as well as the limitations of the approach.

Theoretical and conceptual frameworks in computing and informatics offer organised methods for analysing and interpreting phenomena. They help researchers understand complex systems and relationships. In intricate, technologically advanced settings, these frameworks guide the development of research questions, influence methodological decisions, and shape the interpretation of results and findings. This chapter examines the differences and connections between theoretical and conceptual frameworks, focusing on how they support innovation while firmly establishing research. Frameworks offer coherence, support methodological rigour, and align research goals with results and findings. By combining theory and practice, researchers can produce work that is both socially meaningful and scientifically robust. By integrating these elements, they can produce research that is not only scientifically rigorous but also relevant to societal needs. This chapter highlights the role of frameworks in enhancing scholarly inquiry and contributing to the evolving digital landscape.

Hypothesis testing plays a crucial role in validating research claims in the domains of computing and informatics. It provides a systematic framework for data-driven decision-making. This chapter provides a comprehensive and accessible guide to hypothesis testing, specifically tailored for researchers and practitioners in computing and informatics. The chapter begins by introducing the fundamental concepts, including null and alternative hypotheses, significance levels, p-values, and types of statistical errors. It then presents both parametric and non-parametric tests commonly used in empirical computing research, such as t-tests, ANOVA, and the Chi-square test. Real-world case studies—ranging from A/B testing in web applications to algorithm benchmarking—are used to contextualise each technique. The chapter also highlights the integration of hypothesis testing in machine learning model evaluation and provides hands-on guidance using Python and R. By connecting

Preface

The rapid growth of artificial intelligence (AI), data-intensive research, and computational systems has fundamentally altered the epistemological underpinnings of computing and informatics. As the area grows in size and complexity, the need for research methods that are methodologically and philosophically sound and contextually relevant also increases. Conventional distinctions among qualitative, quantitative, and computational methodologies have become increasingly inadequate for analysing contemporary socio-technical phenomena, which are influenced by automation, algorithmic decision-making, hybrid human–machine agency, and the ethical challenges posed by pervasive digital systems.

Research Methods for Computing and Informatics addresses this changing environment. This edited volume brings together papers that illuminate the intersection of current methodological debates with advancements in AI-driven research. The chapters together endorse methodological pluralism and the creation of frameworks, paradigms, and analytical instruments adept at navigating the evolving dimensions of research in computing and informatics. The book advances scholarly discourse by delineating methodological evolution, advocating integrative approaches, enhancing theoretical and sample strategies, and emphasising the necessity for ethical, contextually aware, and scientifically robust research designs.

The volume begins with a comprehensive examination of methodological trends at the intersection of computing, informatics, and organisational research. This analysis highlights the growing importance of both qualitative methodologies—such as ethnography and discourse analysis—and quantitative and computational techniques, including experimental design, statistical modelling, co-citation analysis, and co-word analytics. The chapter strongly advocates for methodological diversity and illustrates how advancements in AI ethics, reinforcement learning, and quantum computing are transforming the intellectual landscape. It emphasises that bridging positivist and constructivist paradigms is not only beneficial but essential for producing research capable of addressing the complexities of the real world.

The second chapter introduces the Adaptive Integrative Research Methodology (AIRM), which contributes to this historical context. AIRM was specifically designed for computing and informatics. It addresses a significant challenge, that is, balancing generalizable methodological principles with the field's unique technical and epistemological demands. By integrating quantitative, qualitative, and computational methods in a dynamic, reflexive manner, AIRM offers a comprehensive research paradigm that is scientifically sound, ethically responsible, and adaptable to rapidly evolving technologies. Its core concepts, such as proactive ethics and structured interdisciplinary collaboration, position AIRM as a key player in current discussions on evolving research methodologies.

Chapters three through six explore the growing importance of mixed methods research (MMR) and computational social science in addressing contemporary challenges. Chapter three argues that traditional mono-methodological approaches are becoming inadequate in a landscape shaped by Big Data, generative AI, and socio-technical interactions. It proposes a synergistic strategy that merges MMR with computational methods such as algorithmic text mining and computational ethnography. Grounded in pragmatism, the chapter demonstrates how this combination enhances the reliability, relevance, and depth of explanations in studies of complex information systems.

The fourth chapter continues the exploration of methods used in analytics for higher education. It provides a thorough evaluation of empirical research, demonstrating how mixed methods research (MMR) offers deeper insights into learning environments, assessment methods, and digital teaching practices. The chapter frames MMR not simply as a combination of different approaches but as a commitment to complementarity. Key principles such as design integration, connectivity across sequential phases, analytical triangulation, and methodological transparency have been shown to strengthen educational research.

Chapters five and six explore mixed methodologies from both contextual and procedural perspectives in greater depth. Chapter five highlights the growing significance of mixed methods research (MMR) in African and indigenous research contexts. It discusses the benefits of methodological pluralism for studying culturally specific and context-driven phenomena. The chapter outlines key MMR concepts, designs, sampling strategies, and data analysis tools, offering valuable insights for conducting ethical, locally relevant research in informatics.

Chapter six focuses on the sampling process, which is fundamental but often misunderstood in mixed-method research. It addresses the unique challenges of sampling in computational and informatics research, where datasets can be extensive, diverse, or dynamically generated. This chapter enhances methodological rigour by offering practical strategies for integrating sampling techniques across qualitative and quantitative elements, ultimately promoting valid, generalizable, and context-sensitive findings.

While computational and data-driven approaches are prevalent, qualitative research remains crucial for understanding how people experience phenomena, the meanings they attach to these experiences, and how organisations function in digital contexts. Chapters seven and eight highlight this importance.

Chapter seven focuses on document analysis as an independent qualitative method, noting its underutilization in computing and informatics research. It demonstrates how document analysis can provide deep insights into patterns, discourses, institutional practices, and historical trends found in textual artefacts by synthesising existing applications and identifying methodological shortcomings. The chapter also presents systematic criteria for selecting documents and discusses the advantages and disadvantages of this methodology.

Chapter eight introduces phenomenology as both a philosophical and methodological framework for exploring lived experiences in digital environments. By concentrating on meaning-making, embodiment, and subjective engagement, phenomenology offers a unique lens for analysing human interaction with technology. This chapter outlines various types of phenomenology, sample methodologies, data collection procedures, and analytical strategies, and addresses common challenges such as interpretive bias and researcher positionality.

Chapters nine and ten concentrate on the epistemological and quantitative foundations of diligent research.

Chapter nine examines the differences between theoretical and conceptual frameworks and discusses how these frameworks help researchers navigate complex digital ecosystems. Frameworks ensure that research remains substantively valuable and aligned with broader scientific and societal goals by providing coherence, guiding methodological choices, and shaping data interpretation. The chapter stresses the importance of grounding computer science and informatics research in clear theoretical ideas to encourage both new ideas and scholarly rigour.

The tenth chapter provides a comprehensive guide to testing hypotheses in computer and information science research. It starts with basic ideas such as significance levels, p-values, null and alternative hypotheses, and types of errors. Then it shows how these ideas can be applied in real life, with examples such as A/B testing, algorithm performance evaluation, and model benchmarking. The chapter also discusses how hypothesis testing fits into machine learning assessment and provides examples in Python and R. This combination of theory and experience demonstrates the importance of quantitative analysis for proving research claims in a field that is becoming increasingly computerised.

This book offers a thorough overview of recent research in computing and informatics, particularly in an era where AI-driven changes are reshaping both the nature of inquiries and the tools available to researchers. The chapters aim to foster discussions on how research in this field can remain responsive, ethically informed, and scientifically robust. This can be achieved through diverse methodologies, reflective philosophical inquiry, computational integration, and a solid quantitative foundation.

Research Methods for Computing and Informatics aspires to serve as both a reference guide and a source of inspiration for future innovation. It brings together various perspectives and new approaches to the field. We believe that this book will assist researchers, educators, and professionals in understanding the complexities of today's digital ecosystems with intellectual clarity, methodological rigour, and a commitment to advancing knowledge that is both cutting-edge and socially beneficial.

Patrick Ngulube

Acknowledgement

A great deal of cooperation, intellectual generosity, and steadfast support from several people and organisations have enabled the creation of *Research Methods for Computing and Informatics*. Thanks to everyone who helped make this volume a reality. I sincerely appreciate your contributions.

I extend sincere gratitude to all the chapter authors whose expertise, dedication, and scholarly rigour made this volume possible. Their diverse perspectives and commitment to methodological excellence have created a rich and relevant resource for the computing and informatics research community.

I also want to thank the editorial and production teams for their professionalism and meticulous attention to detail throughout the publication process, as well as the peer reviewers whose helpful feedback improved the chapters' coherence and clarity. Many of the studies provided here were influenced by the experiences of students, practitioners, and community members, for which I am thankful. Their willingness to share their opinions and views enhances the study's calibre and significance. I would like to express my sincere gratitude to the organisations and colleagues who helped this initiative through conversation, encouragement, and cooperation.

I want to express my gratitude to my family and close friends for their unwavering support and patience throughout the development of this volume.

Chapter 1
A Bibliometric Exploration of Methodological Trends, Gaps, and Future Directions in Qualitative and Quantitative Research:
Navigating AI, Sustainability, and Ethical Challenges Through Methodological Innovation

Sharneet Singh Jagirdar
http://orcid.org/0000-0001-5716-6368
Apex Institute of Management, Chandigarh University, India

Pradeep Kumar Gupta
http://orcid.org/0000-0001-5680-4612
L M Thapar School of Management, Thapar Institute of Engineering and Technology, India

ABSTRACT

This study analyses five decades of research at the intersection of computing, informatics, and organisational studies to trace trends in qualitative (e.g., discourse

DOI: 10.4018/979-8-3693-4018-9.ch001

analysis, ethnography) and quantitative (e.g., experimental design, statistical modelling) methods. Emphasizing methodological pluralism, it bridges positivist and constructivist paradigms using a pragmatism-triangulation-bibliometrics framework (co-citation, co-word analysis) to map the evolution of computational research. It identifies emerging domains, such as AI ethics, reinforcement learning, and experimental designs for new technologies. Findings highlight how interdisciplinary methods are shaped by innovations from machine learning to quantum computing. Recommendations emphasise sustainable technology, ethical AI, and hybrid methods that balance both academic rigour and real-world relevance.

INTRODUCTION

Given its rapid technological changes and growing global interconnections, management scholarship presides over an enormous burden to develop strong, innovative, and interdisciplinary research methods. These methods must be appreciated and uplifted in order to respond to the complexities that arise in organizational settings, the social environment and economic situation. The current study presents a bibliometric analysis that examines the trends, gaps and future orientations of qualitative versus quantitative research methods in management. Through bibliometric analysis, this chapter aims to provide practical insights into the evolution of methodological practices over time, enabling researchers and policymakers to easily identify evidence within this ever-evolving area.

This study addresses the following questions:

1. How have qualitative and quantitative methodological trends evolved in management research?
2. What are the prominent gaps and emerging themes at the intersection of methodology, technology, and ethics?
3. How can methodological pluralism and epistemological integration advance future scholarship?

Management research, throughout its history, has used a very diverse array of methodological traditions ranging from qualitative approaches like discourse analysis, ethnography, and grounded theory to quantitative techniques like cross-cultural methods, experimental research design, and statistical modelling. Each of these methods plays an important role in uncovering subtle insights, validating constructs, and building theories that are crucial for advancing knowledge. As the field matured, a systematic inquiry was required into the extent of methodological novelty. Central to this effort are questions about whether traditional approaches still suffice to deal with the issues of the contemporary world or if there are gaps where new ones can be established or links forged between different fields.

This study builds on the premise that management scholarship can have a better understanding of methodological foundations. The research focuses on key areas such as measurement techniques, analytic approaches and philosophy of science, providing not only an overview of the progress made but, more importantly, identifying still underexplored topics, emerging trends, and methodological gaps. Furthermore, it emphasizes that critical debates on research design, construct validity, and writing for scholars are of prime importance in making the management research quality and impact better.

By using a bibliometric analysis of publications from the Scopus database, this chapter aims to advance and disseminate techniques for collecting, evaluating, and interpreting management scholarship. The findings suggest that they create a framework for researchers, practitioners, and those assigning work in terms of hypertrophy in study methods, which will facilitate a more comprehensive understanding of management techniques. Consequently, the investigation serves as a wake-up call for management researchers to consider adopting different methods related to growing businesses, just as it reflects the global context in which companies are compelled to act smarter.

The next sections of the chapter continue to discuss in detail the thematic development of research methods in management, evaluate trends and gaps in qualitative and quantitative approaches, and finally provide future directions that guide the way methodological developments should go.

LITERATURE REVIEW

The prolific growth of scholarly literature across different disciplines has greatly increased the requirement for formidable methods of literature review to combine knowledge, find gaps, and promote theoretical progress. Conversely, scholarly output had always been at 250% per decade from 190 to 202 (Thelwall & Sud, 2022). This tool has emerged to meet the challenges as a systematic and replicable technique for the assessment of intellectual structure and thematic evolution of academic fields (Zupic & Čater, 2015; Donthu et al., 2021).

From simple citation metrics, bibliometric methodologies have gained much in their development. These include co-citation analysis, bibliographic coupling, and co-word analysis. Co-citation analysis identifies foundation works by clustering frequently cited publications, whereas bibliographic coupling links contemporary research by shared references. Co-word analysis, however, discovers emergent themes by analyzing keyword co-occurrence patterns (Zupic & Čater, 2015; Fan et al., 2022). These approaches enable scholars to map complex academic landscapes, revealing the dynamic interplay of ideas and the evolution of research trends.

In management research, bibliometric analysis has played a key role in the understanding of the structure of academic domains, thus giving quantitative insight into connected studies, influential authors, and thematic clusters. Bibliometric analysis is a strength, possibly for the reason that it complements a regular literature review. While narrative reviews are known for their depth and theorization, the latter grants replicability and transparency with a macro-level perspective. For example, in management studies, bibliometric analyses can be used for both "miner" approaches (deep diving) and "prospector" approaches (wider scan) in order to carry out a well-balanced review (Breslin & Gatrell, 2023). This helps understand the ways in which fields of inquiry evolved over time, thereby figuring out where potential innovations could be introduced because it involves retracing their origins gradually by using indicators of connections between inputs and outputs or trajectories like technological innovation cycles from patents, etc. Through the integration of qualitative and quantitative methodologies, bibliometric analysis captures not only the current state of knowledge but also charts pathways for future inquiry (Paul & Criado, 2020).

Moreover, through bibliometric methods, many gaps in research have been identified, and new theoretical frames have been proposed. In very fast-moving fields such as technology, finance, and environmental studies, this impact is very clearly evident. For example, Fan et al. (2022) used bibliometric analysis to identify neglected topics that were further explored into new uncharted perspectives and advanced theoretical discourse. Given that bibliometric methods can uncover gaps as well as burgeoning fields, it is no mean feat to influence innovation or to make theoretical contributions.

Incorporating SLRs into this form of analysis makes it more effective despite its weaknesses. Further, although they offer quantitative rigor and wide coverage, their integration may lack an appropriate level of background understanding for meaningful theorizing. To obtain a thorough comprehension of research scenarios and for purposes of identifying workable voids, combining qualitative insights with bibliometric tools is essential (Zupic & Čater, 2015). This wide open access to VOSviewer features has also helped promote the use of bibliometrics as an analytic tool for teaching collaborative and transparent scholarship in higher education institutions and beyond (van Eck & Waltman, 2010; UNESCO, 2022).

Over the years, bibliometrics have provided powerful synthesis tools that facilitate advancement within scholarly fields and domains. For managers and related disciplines, the practice of bibliometric analysis provides that critical bridge between empirical rigour and innovative theory, offering insight into methodological trends, gaps, and opportunities for interdisciplinary collaboration. It provides a thorough foundation for exploring complex, dynamic academic fields and drives meaningful contributions to scholarly advancement.

While bibliometrics effectively maps intellectual landscapes, its integration with methodological pluralism remains underexplored. This study bridges that gap by employing bibliometric tools to trace how qualitative and quantitative methods coexist, compete, and converge within management scholarship.

Methodology

This study uses a bibliometric approach to analyse the existing literature systemically with regard to methodological trends and gaps in management scholarship, focusing on qualitative as well as quantitative research methods. Bibliometric analysis is one of the widely recognized and highly structured, quantitative approaches in synthesizing scholarly work; these analyses identify research trends, create thematic developments, and establish unexplored gaps (Donthu et al., 2021; Paul et al., 2021). Following the SPAR4SLR framework, a framework that organizes systematic literature review into three stages: assembling, organizing, and assessing data, for a rigorous study (Alonso-Garcia et al., 2021).

The gathering stage entailed the extraction of relevant academic articles addressing crucial research questions related to methodological trends and intellectual development in management studies. The specific themes that were involved during this stage included discourse analysis, grounded theory, causal mapping, statistical modelling, and action research. The Scopus database was the primary data source used in the study owing to its extensive coverage and indexing standards, which made it possible for only high-quality literature to be included (Fahimnia et al., 2015; Andersen, 2021).

Articles were screened based on title, abstract, and methodological relevance. Inter-rater reliability was assessed during screening to ensure consistency. Although Scopus provides extensive coverage, future replications could include Web of Science and PubMed to enhance comprehensiveness. A comprehensive Boolean search string (TITLE-ABS-KEY("management scholarship" OR "qualitative research methods" OR "quantitative research methods" OR "methodological trends" OR "causal mapping" OR "survey research" OR "experimental research design" OR "cross-cultural methods" OR "grounded theory" OR "textual analysis" OR "statistical modeling" OR "construct validation" OR "measurement development" OR "epistemology" OR "action research" OR "theory development")) \ AND

TITLE-ABS-KEY ("bibliometric analysis" OR "research trends" OR "systematic development" OR "publication analysis" OR "research gaps" OR "research methodology" OR "analytic approaches") was developed that would retrieve all studies dealing with qualitative and quantitative research methodologies, their applications, as well as the corresponding analytical techniques. 6481 articles were returned in this search.

The retrieved documents in the organizing phase were refined and filtered to ensure coherence with the study's thematic and methodological focus. Only articles in English, in the form of journal articles or review papers, were included; studies from different disciplines, such as medicine and engineering, were not considered. For every document, the alignment of each core theme of the study, such as qualitative research methods including ethnography, textual analysis, quantitative approaches including survey design and experimental methods, statistical methodologies, which involve construct validation, and causal mapping, was reviewed. This led to a focused dataset for detailed bibliometric analysis (Lopes et al., 2021).

Advanced bibliometric techniques were applied during the assessing stage, for instance, co-word analysis, citation analysis, and bibliographic coupling to map out the intellectual landscape and thematic evolution of methodology trends in management research. Co-word analysis captured those recurrent topics and rising issues, whereas citation analysis picked out landmark pieces of work as well as influential authors. Bibliographic coupling aided in identifying research clusters and networks for scholarly collaboration.

This work utilised Biblioshiny in R, which involved data visualisation as well as analytics that produced thematic maps, three-field plots, and citation networks. It allows the comprehensive representation of an academic landscape, focusing on where most research activity is and areas that are less represented in scientific literature (Aria & Cuccurullo, 2017).

Though these bibliometric methods bring robust insights, a set of limitations, which would include the dependence on just one database (Scopus) and the keyword search limitation, is acknowledged. A limitation is that relevant works are probably excluded from this research where they are not indexed on Scopus or in alternate terminologies. However, methodologically, the study does present robust analyses which are representative and meaningful enough regarding trends of management research methodologies.

For the promotion of knowledge in management, this bibliometric research is instrumental in delineating the intellectual configuration and exposing gaps that exist in the area as far as themes are concerned. This discovery is useful to academicians, corporate heads and those who make policies since it creates an enabling platform for fresh ideas that would propel growth within qualitative and quantitative research regarding management.

RESULTS

The dataset for this bibliometric study spans from 1967 to 2025, capturing over five decades of scholarly contributions in management research. The corpus encompasses 3,755 sources from journals, books, and conference proceedings with 6,481 documents. The annual growth rate of 1.2% indicates that this field has consistently been expanded moderately, and the average age of a document being 8.59 years reveals a young and maturing domain. Each document is cited on average 21.48 times, demonstrating strong academic engagement, and the dataset draws on a vast pool of references, 283,864 citations, which testifies to the richness and depth of intellectual contributions.

The content analysis is thematically rich with 11,538 "Keywords Plus" and 15,570 "Author's Keywords," which reflect intense methodological exploration and thematic development (see Table 1). These keywords cover classical qualitative and quantitative approaches as well as contemporary trends, such as interdisciplinary research and data-driven methodologies. The dataset also includes contributions from 15,850 authors, showing significant collaboration. Although 1,832 documents are single-authored, most have an average of 2.74 co-authors, which underscores the collaborative nature of the field. Notably, international co-authorship accounts for 15.58% of the total, highlighting meaningful global partnerships that enrich the transdisciplinary and cross-cultural character of management research.

Table 1. Descriptive Analysis – Main Information

Description	Results
MAIN INFORMATION ABOUT DATA	
Timespan	1967:2025
Sources (Journals, Books, etc)	3755
Documents	6481
Annual Growth Rate %	1.2
Document Average Age	8.59
Average citations per doc	21.48
References	283864
DOCUMENT CONTENTS	
Keywords Plus (ID)	11538
Author's Keywords (DE)	15570
AUTHORS	
Authors	15850

continued on following page

Table 1. Continued

Description	Results
Authors of single-authored docs	1832
AUTHORS COLLABORATION	
Single-authored docs	1929
Co-Authors per Doc	2.74
International co-authorships %	15.58
DOCUMENT TYPES	
article	4775
book	146
book chapter	428
conference paper	613
conference review	12
data paper	2
editorial	21
erratum	6
letter	1
note	20
review	449
short survey	8

Analysis of document types indicates that journal articles make up the majority of the corpus, at 4,775, which underlines the role of journal articles as the primary medium for publishing methodological insights. Additionally, 428 book chapters and 146 books provide in-depth theoretical explorations, while 613 conference papers highlight ongoing dialogues and innovative ideas. Reviews (449) play a crucial role in critically assessing existing literature and setting research agendas. Other document types, including editorials and short surveys, add to the field's intellectual diversity, albeit in smaller numbers.

Figure 1. Annual Scientific Production

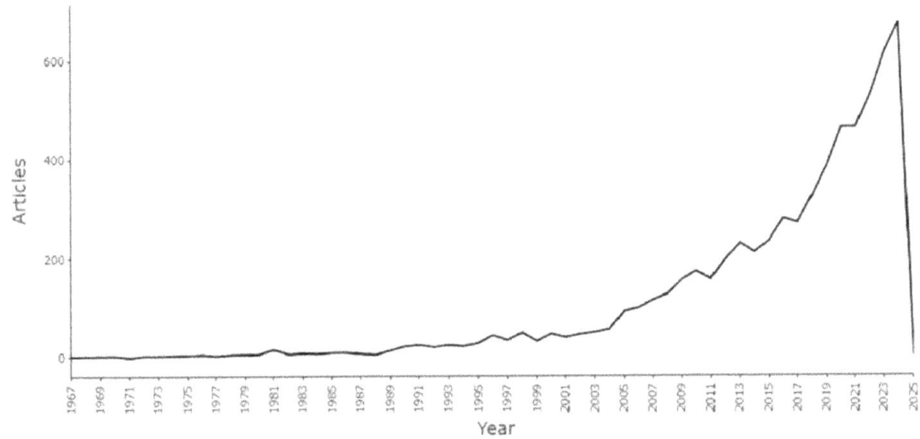

The annual publication production (Figure 1) shows growth and changes over almost six decades. There was not much happening during the early years, which included the 1960s and 1970s, except for publishing 2–6 articles per year. This indicates the infancy of methodological scholarship in management, which also had low production since this was the early stages of setting foundations in research methods.

There was a good trend of increases in annual production from the 1980s, especially sharply in the year 1981: that is, with 15 articles compared with previous ones. The increase was almost smooth in the 1990s because annual production passed 20 articles during the year 1990, and finally, reached peak values of 49 during 1998. This shows that growing interest in new methodological ideas and spreading systematic methodology towards management studies during this decade has been growing.

The pace of publication accelerated further in the early 2000s, with a sharp increase in 2005 (93 articles) and a steady rise thereafter. Annual production had crossed 150 articles by 2010, reaching 173 that year. The next decade, 2010–2020, saw exponential growth, with annual outputs crossing 200 articles in 2012 and reaching 400 articles by 2020. This period reflects the field's maturation and the increasing complexity of research questions, driven by advancements in data science, analytics, and interdisciplinary approaches. The most recent years (2020–2024) showcase a remarkable surge, with annual publications growing from 465 in 2020 to 677 in 2024. The rapid expansion is proof of the rise in the breadth of management research with new technologies, like AI, as well as machine learning and global interest in bedside mannerism to improve controlled trials within organizations.

Figure 2. Average Citations Per Year

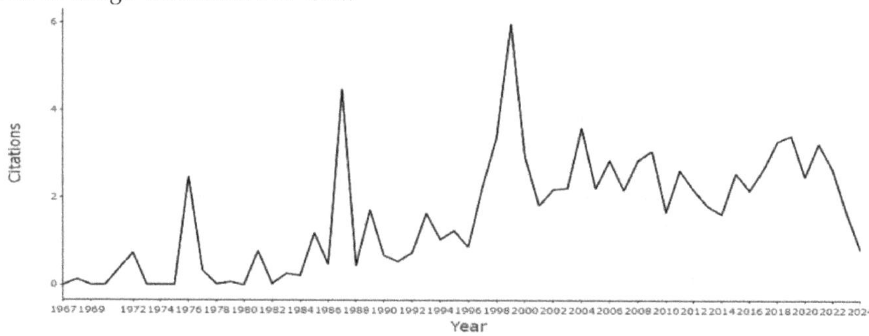

Annual total citations analysis (Figure 2) offers valuable insights into the scholarly impact and evolution of methodological trends in management research over nearly six decades. The field was characterised by sporadic citation activity until the late 1970s, with mean total citations per article (MeanTCperArt) below 10. Early spikes occurred in 1972 (MeanTCperArt of 39.00) and again in 1987, at 170.57, where seminal works receive significant attention over time due to important breakthroughs made in methodologies. From the mid-1980s, there has been a more consistent growth in citations, with significant increases during the late 1990s and early 2000s. For instance, 1999 was a remarkable year since its MeanTCperArt was estimated at 155.38, revealing the presence of weighty input during the same moment. This continued until the early 2000's where it resulted in 75.70 in 2004 for the MeanTCperArt, representing a diversification of studies through incorporating sophisticated bibliometric and analytical methods.

The data of the past two decades reflect the maturity and diversification of the field. The period from 2005 to 2015 reveals steady citation activity, as MeanTCperArt fluctuated between 19.55 and 49.19, while the number of published articles increased, indicating a broadening of the research base. In the period after 2016, reference to the trend leans towards more recent contributions, with a quantity of published articles that is high, such as 677 articles by 2024, but the MeanTCperArt has been decreasing over a relatively new year of publication that does not allow for citations to accumulate over its short span.

The Citable Years measure only goes to show how much influence old works have had on us up to now, since articles starting from 1990s and 2000s continue being cited up to now. Also in keeping with this natural delay in citation accumulation for new publications, MeanTCperArt values are lower in recent years, like 2023 (3.32) and 2024 (.81), respectively.

In short, these analyses show what stages management research has gone through, looking at critical periods of methodological innovation as well as the continued importance of basic studies. The findings demonstrate a potential for increasing citation impact by promoting interdisciplinary thinking, addressing emerging research fields, and ensuring the distribution of high-quality papers to the broader academic and practitioner communities.

Figure 3. Three Field Plot: Cited Sources- Abstract – Countries

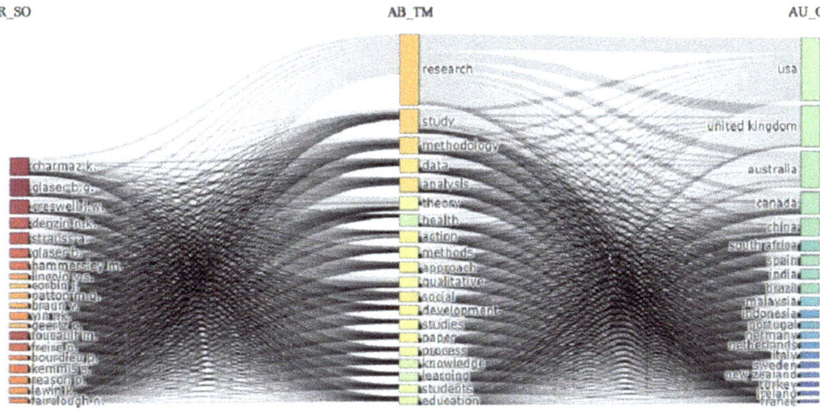

The keywords (Figure 3) point towards a strong qualitative methodology focus, with words such as "research," "study," "methodology," "data," "analysis," and "qualitative" leading the list. This indicates a heavy emphasis on a qualitative approach in recent management research, especially in fields such as social development, education, and health. The lack of strong quantitative words such as "statistics," "modeling," or "experimental" implies a possible gap in quantitative method integration, highlighting the need for better-balanced methodological investigation.

The role of authors who were cited, such as Charmaz, Glaser, and Strauss, highlights the impact of grounded theory and qualitative approaches to management research. The geographical spread of research contribution, with nations like the United Kingdom, Australia, and Canada, is a testament to the international scope of management research, but also suggests some underrepresentation from areas like Africa and some areas of Asia. This geographical imbalance calls for greater inclusiveness to enhance methodological richness and diversity as well as insights.

Figure 4. Most Relevant Sources

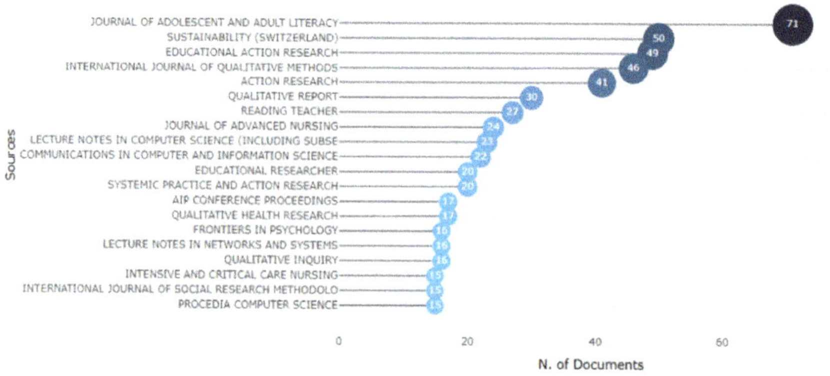

The list of most relevant sources (Figure 4) highlights the interdisciplinary nature of management research. At the forefront in terms of publications is the Journal of Adolescent and Adult Literacy, whose 71 articles show a significant leaning towards education methodologies. Other essential sites that underscore the growing concern about sustainability practices and action research methods are Sustainability and Educational Action Research. The fact that journals such as International Journal of Qualitative Methods, Qualitative Report, and Qualitative Inquiry are highly ranked underscores the importance of qualitative methodologies in explaining complex management phenomena.

Moreover, technical publications, Lecture Notes in Computer Science and Communications in Computer and Information Science, indicate the implementation of computational methods and AI approaches in management research. Specific journals like Qualitative Health Research and Journal of Advanced Nursing reflect the integration of management principles in health and healthcare systems. Other sources and journals, such as Social Science and Medicine and Frontiers in Psychology, also clearly reflect the integration of management with social sciences or psychology.

Publishing new ideas is facilitated by important Conference proceedings such as the ACM International Conference Proceedings Series and the European Conference on Research Methods in Business and Management Studies. Knowledge exchange happens in such events dynamically. Journals like Sustainability and Technovation have, for a long time, been giving focus on sustainability in various contributions, showing that the central themes are technological development and sustainability. According to the geographies and sectors that the Journal of Cleaner Production and the African Journal of Hospitality, Tourism and Leisure represent, it clearly shows that it has global relevance and practical application for management scholarship. Taken alone, each of these sources would show only that the field has moved towards more differences in methodology, with a pledge to deal with global challenges via interdisciplinary means.

Figure 5. Most Global Cited Documents.

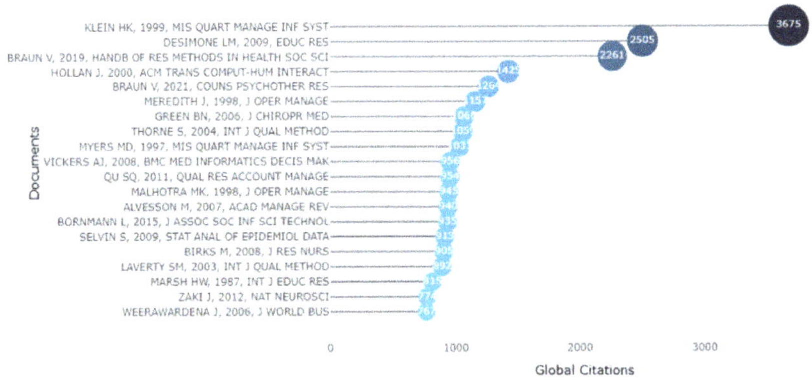

The most globally cited papers (Figure 5) in the field of advancing management scholarship reveal a wide array of methodological trends, with particular emphasis on both qualitative and quantitative research approaches. For example, the paper by Klein in 1999, published in MIS Quarterly, has received the most citations at 3675, indicating that research in management information systems is still relevant. According to DeSimone (2009), 2505 citations for his/her study demonstrate the significance of educational research methods. Notably, Braun's (2019) health and social science research methodologies have an Annual Citation Rate of 376.83 and Normalised Citation Impact of 110.27, which means they exert a huge influence on the discipline.

Other examples are Meredith's (1998) operational management, which is still attracting attention, and Thorne's (2004) qualitative research methods, suggesting various subtle methodological nuances. Current contributions include counselling psychology by Braun in 2021 and Mukherjee's study on business research in 2022, which follow these directions, showing trends and different approaches toward solving problems within managerial settings. The papers as a whole indicate an increased awareness of both qualitative and quantitative methods, bridging the gaps between disciplines and encouraging further advancement in management scholarship.

While many of these highly cited studies are oriented toward methods in operational research, psychology, and information systems, they do much more to stress the need for greater interdisciplinary frameworks and methodologies. There are many works that have a profound impact and address areas such as health, education, operations management, and marketing, demonstrating management through various research strategies. Notably, the works of researchers like Alvesson, Malhotra, and Meredith show that management research has moved towards more present and complete analysis tools; thereby indicating an evident move towards methodological creativity as well as integrative research methods. These findings are valuable directions for future research, thus underlining the continued importance of refinement in methodology as well as inter-disciplinary collaboration in taking the management scholarship to its next level.

Figure 6. Most Frequent Words – Abstract

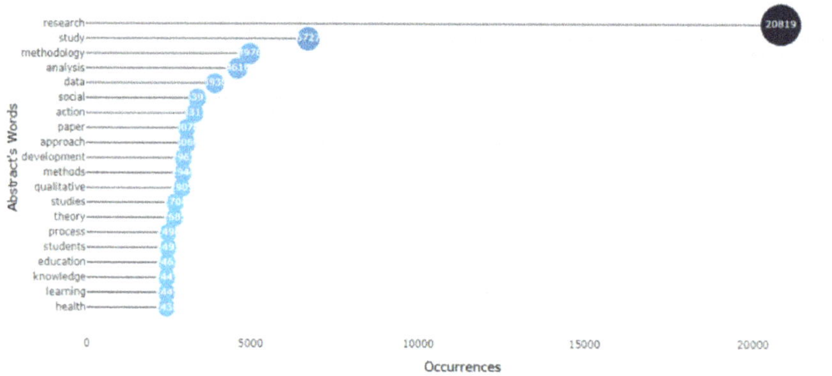

The most frequent words in the context of advancing management scholarship (Figure 6) indicate the key thematic areas and research priorities in the field. The field indicates a great frequency for words such as "research" (20,819 times), "study" (6,727), and "methodology" (4,976), which have everything to do with orderly enquiry concentrated on ways.

Take, for instance, terms like "analysis" (4,616), "data" (3,930), or even "methods" (2,944); what dominates is not just what they refer to, but also their general inclination towards the statistical quantification and pattern inspection using actual patterns. In addition, words that are frequently utilised that include "social"(3,391), "health" (2,438) and "education" (2,462) imply the fact that this area of research lies at the intersection between social sciences faculties, medicine schools, as well as teacher training establishments.

The fact that "qualitative" (2,909) is constantly used alongside "quantitative" (990) reveals a shift towards management mixed-methods research, which requires a combination of soft and hard science. The following are some key words,including "students" (2,491), "knowledge" (2,447),and "learning"(2,445). Here, management research again becomes education-oriented, especially as it concerns knowledge transfer and learning strategy.The frequency of "Framework" is 1,427, "Theory" is repeated 2,685 times, while "Model" has been used 1,662 times, maintaining the trend toward a theoretical basis for and development of conceptual frameworks guiding empirical research. Hence, words such as "Future" (1,130), "Challenges" (1,058), and "Implications" (939) emphasize that current research is forward-looking and seeks to identify knowledge gaps and explore new frontiers in management theory and practice. The occurrence of words like "Methodological" (1,025) and "Critical" (1,388) also indicates an appreciation for the' nature of methods beyond themselves among researchers, implying that there is growing concern among scholars about methodological rigour as well as ethical considerations inherent in management. "Framework" recurs (1,427 times), "theory" recurs 2,685 times, and "model" recurs 1,662 times, thus maintaining the tendency toward theoretical bases for and the development of conceptual frameworks guiding empirical research. This is evident in the usage of words like "future" (1,130), "challenges" (1,058), and "implications" (939), which underline the cutting-edge character of modern research or appreciation of various levels through which knowledge can progress vis-à-vis management theory and practice. In addition, the presence of words like "methodological" (1,025) and "critical" (1,388) suggests that scholars are also interested in analyzing research methods at a higher level than individual empirical studies or even at a more general level than a particular field such as business.

The recurrent references to terms like "practice" (1,847), "impact" (937), and "applications" (702) indicate that within management research, an increased emphasis has been placed on practical application – with management researchers striving to influence real-world business practices, policies, and decision-making processes. Finally, terms such as "system" (927), "tools" (647), and "technology" (931) underline the growing integration of technological advancements and systems theory into management studies, thereby broadening the dimensions of inquiry to cover digital transformation, data analytics, and systems thinking vis-à-vis complex organizational problems.

Figure 7. Trend Topics – Abstract

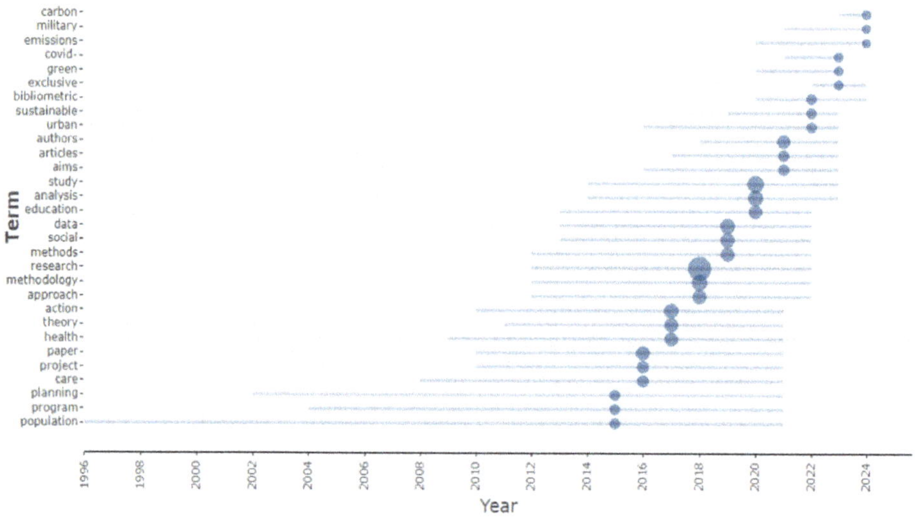

From the bibliometric data, the following are some trend topics that would have provided useful information on the trend of changes in research themes in management scholarship, focusing on trends in research methodology, topical interest, and emerging issues globally (Figure 7). Topics such as health (2438 mentions), maternal (68), infant (43), and prevention (181) suggest interest in public health issues on maternal and infant care besides health concerns like HIV, reproductive health, and vaccinations. It has continuously transformed over the years, most particularly in the focus involving HIV/AIDS (136), prevention (181), and reproductive health (62). The occurrence of words such as COVID- (247), pandemic, and green (193) signifies a shift in research focus from primarily economic issues, financial policy responses to global crises like the pandemic and environment-related crises. In the last decade, terms related to "sustainable" (484 times), "carbon" (84), "green" (193), and "emissions" (43) indicate the growing concern for sustainability and environmental impact (2019–2024). Research in these areas ties into global initiatives around climate change, green technologies, and carbon reduction strategies. The appearance of terms like carbon and emissions, particularly in the 2023–2024 timeframe, points to an emerging focus on carbon footprints and emission reductions in management strategies.

Keywords such as data (3930 times), methods (2944 times), methodology (4976 times), and analysis (4616 times) indicate a trend towards data-oriented research. Data focus has grown enormously, thereby pointing to a significant reliance on quantitative methods and data analytics in management research. Bibliometric (510) as a keyword in the latest years (2020–2024) signals a growing interest in systematic review and mapping of scholarly literature itself, underlining the importance of bibliometric analysis in understanding research trends. The consistent appearance of methodology (4976), approach (3068), and study (6727) suggests a continued focus on methodological advancements. More and more researchers are now looking into and creating qualitative and quantitative methods to address management issues.

Terms such as social (3391) and theory (2685) indicate the role of social sciences and theoretical frameworks in understanding organizational behavior, business strategies, and management policies. Management (1991 occurrences), project (1487), program (571), and business (821) show sustained interest in organizational and management studies. This reflects the focus on organizational practices improvement, strategic planning, and the implementation of programs within the business environment. Issues on outsourcing (27), clients (63), and interviewers (1047) reflect increasing concerns regarding human resource management, client relationships, and service models.

It's sufficient to mention the trends of military (45) and war as defence and emergency in research crises, which show the role of geopolitical issues and crisis management in existing management scholarship. Also, terms like urban, which have increased lately 419, suggest a growth and importance of this field-urbanisation-in this field and management's need for resource management and sustainability and within towns.

Technology (931), along with digital (699) and online (658), are all indications of the significance that ICT(E), digital innovations or digitalizing managerial practice have acquired. Such matters may as well be pertinent to work on disruptions in technology and business innovations. Keywords such as feminism (29), women (669), and gender reflect the growing inclusion of gender studies and diversity in management research, which has become a significant focus in recent years, especially in the context of organizational culture and leadership.

Figure 8. Clustering by Coupling Map – Coupling Measured by Abstracts and Cluster Labelling by Abstract Terms

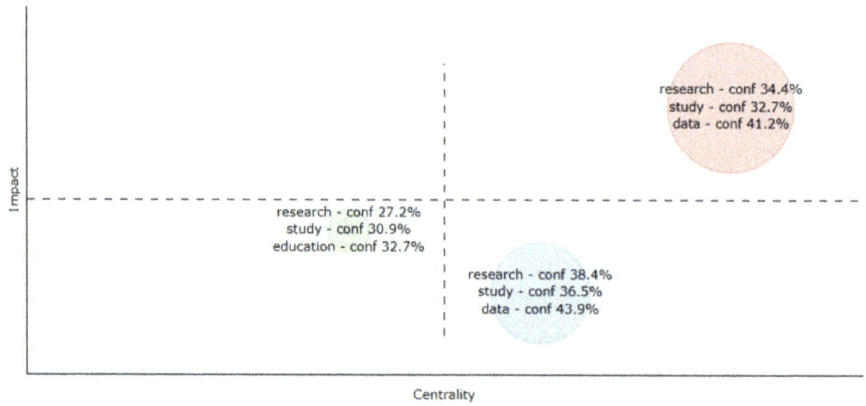

The Coupling Map (Figure 8) offers a detailed representation of research outputs based on citation patterns and methodological trends in management scholarship. From the analysis, several insights emerge. The documents listed in the coupling map span a wide range of research methodologies, with a focus on both qualitative and quantitative research. The prominence of recent studies (2023–2024) suggests that there is a continuing shift toward more interdisciplinary and diverse methodological approaches, such as applied statistics, machine learning, and hybrid methodologies. These approaches aim to address complex challenges in management research by integrating diverse data sources and analytical techniques.

The vast majority of the research focuses on the issues of health care management, educational leadership, and social sciences, such as studies in public health, mental health, and environmental management. This is indicative of the increasing relevance of management strategies in sectors directly influencing society. The other core focus is the inclusion of newer technologies, such as AI and data analytics, into management practice.

Although most works in 2023 and 2024 address more practical issues, the advancement of theoretical frameworks in understanding the nexus of management and advanced analytics remains underrepresented. The few available documents show little about the role that AI, big data, and blockchain can play in the management decision-making process; hence, it remains to be explored. Similarly, little research has been conducted regarding organizational culture in digital transformation or cross-sector management strategies, offering further grounds for investigation.

The NCS and clustering of documents based on the coupling map show networks in research domains involved in collaboration. Clusters with higher Normalized Citation Score (NCS) seem to point toward influential research located within certain research topics, such as public health, education, and sustainable development. These might be indicative of existing research communities that drive innovation within specific areas. Some future research directions that can be derived from the map are increased needs for interdisciplinary studies across business management, technology, and social science disciplines. Longitudinal studies of the long-term impacts of management strategies in the corporate and non-corporate sectors are also needed. Another emerging direction is the investigation of ethical considerations and responsible management practices, especially as related to AI and data-driven decision-making.

The coupling map showcases the global reach of management scholarship, with publications from a wide range of countries and regions. There are factors ranging from unequal distribution of wealth among nations, world over, to lack of implementation of rules in the global fiscal system by financial institutions.

Figure 9. Clustering by Coupling Network – Coupling Measured by Abstracts and Cluster Labelling by Abstract Terms

The bibliometric exploration of the Coupling Network (Figure 9) reveals key insights into the methodological trends, gaps, and future directions in both qualitative and quantitative research within the field. The network analysis brings to the fore the dominance of such concepts as "research", "study" and "data", which are often discussed in conferences, indicating their core role in the formation of academic discourse. These terms not only have high centrality, meaning they have a foundational influence across different research areas, but also have a significant impact, especially in conference settings, underlining the importance of empirical research in management scholarship.

The analysis reveals three different groups placing different emphasis on the aforementioned concepts. First group, distinguished by the colour red, puts emphasis on the role of "research," "study", and "data" as pivotal things at conferences that call for scholarly attention. The relatively high centrality score of group one (.406) indicates that it has a strong grounding in these areas, and at the same time shows that there is room for broadening the scope of research methods. Group 2, in blue, pushes this message further by showing a still higher centrality at 0.405, which represents an upward trend in seeking to strengthen the empirical foundations of management research. Group 3, colored in green, shifts the conversation by throwing in "education" which brings about a pedagogical dimension in the discussion on research methods. The centrality score of this group is .391, thus implying that in order to prepare scholars for the ever-changing landscape of management research, research techniques must be integrated into educational systems.

While the associative network identifies data-driven research as being central, there are methodological lacunae, especially with respect to merging qualitative and quantitative approaches. Most scholarly works in this area emphasizes solely on either qualitative methodologies or quantitative methodologies instead of delving into formations that merge both. There is a necessity also to better connect the world of research with practical reality, especially through the integration of newer emergent technologies, for instance, AI, machine learning, and big data analytics into management research to amplify both breadth and depth.

An interdisciplinary approach that integrates knowledge from various fields like AI, economic and behavioral sciences should be considered as a future development path. Ethical concepts will be important in research involving AI. These new technologies require exploration as well as addressing ethical issues when conducting management research so that it can remain relevant to current problems.

Figure 10. Co-Occurrence Network – Abstracts

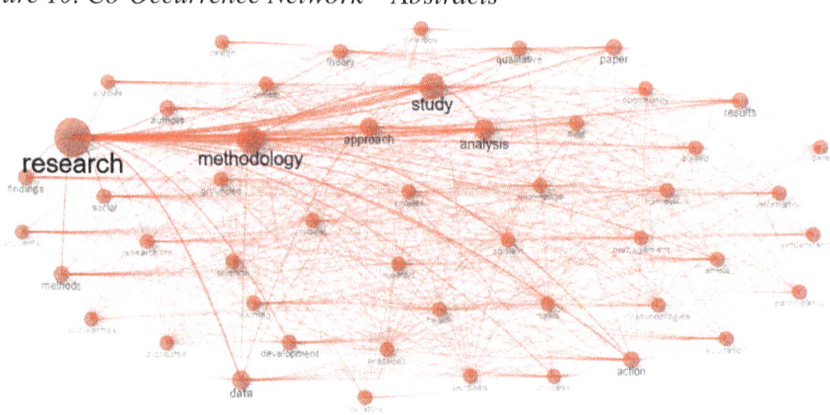

Figure 11. Co-Occurrence Network Density – Abstracts

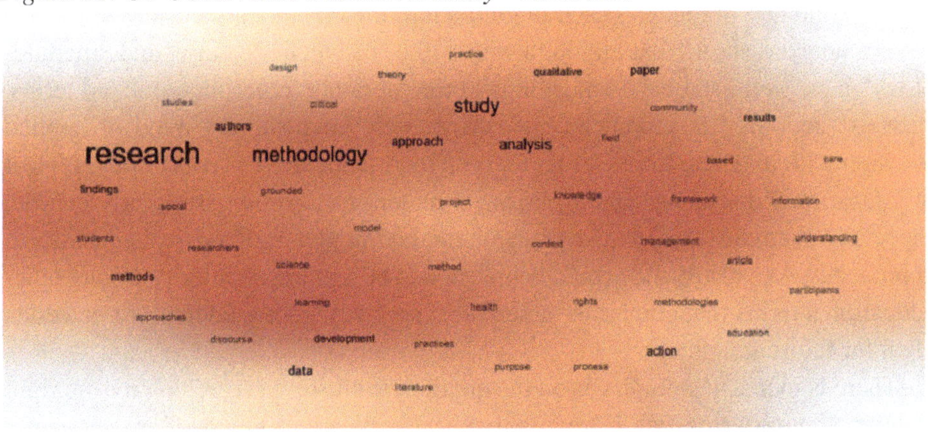

Figure 12. Co-Occurrence Network Degree Plot – Abstracts

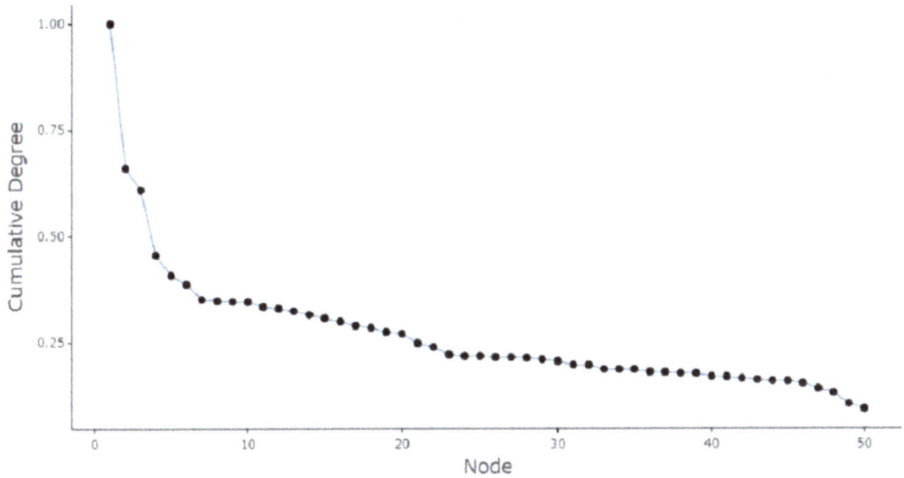

The co-occurrence network analysis (Figures 10, 11 and 12) provides valuable insights into the methodological trends, gaps, and future directions within qualitative and quantitative research. Each of these nodes in the network forms connections to other keywords through research papers, revealing themes by which they are typically interconnected. The terms "research," "study," "methodology," "analysis," "data," and "social" are central to this network, indicating that management scholarship finds its empirical heart within. These terms share almost similar values for betweenness, closeness, and PageRank, which shows the equivalence of their prominence within research and how frequently each might be interlinked among themselves.

The analysis shows that although words like "research," "study," and "methodology" have well-established meanings in the literature, other more related terms, such as "action," "development," and "approach", indicate the trend of applied research with practical implications. Words like "students" or "education", typical of educational discourse, while playing a major role, are not frequently central, especially in studies that involve research methodologies and practical educational approaches. As a result, they could be inferred to be gaining popularity among other educational terms that emphasise a need for more stringent academic frameworks, ideal for future scholars.

There is increasing emphasis on the qualitative cluster in searching for the right answers, though terms such as "qualitative", "theory" and "grounded" are frequently paired with "results", "findings" or "practice", illustrating a growing desire to integrate abstract ideas about life into realities today with regard to research issues. On the other hand, words like health, knowledge, learning and community prove that it has interdisciplinary applications different from management itself because it can lead us not only into various social sciences, such as sociology, but concurrently medical sciences through health care delivery, in addition to social work focusing on communities at large.

Some areas within the network seem well represented by methodological terms, but there are some emerging ones indicating that there is still room for further investigation. For instance, such words as "context" alongside other related terms like understanding, critical, and framework suggest that theoretical frameworks and contextual understanding are increasingly important. Such lower betweenness values of terms like "participants," "rights," and "care" point out potential gaps in the research, especially in handling ethical considerations, stakeholder perspectives, and social responsibility in management research. Future research can benefit from a deeper focus on these areas, particularly in the context of ethical AI applications, organizational behavior, and stakeholder management.

Thus, the co-occurrence network highlights how management scholarship is still an evolving field of study where, to achieve the right level, there is a need to merge two extremes: qualitative and quantitative methodologies. Future research work should be interdisciplinary, involve deeper exploration of theoretical models, and the integration of practical applications with more emphasis on ethics, social responsibility, and educational frameworks for the proper training of researchers.

Figure 13. Thematic Map – Abstracts

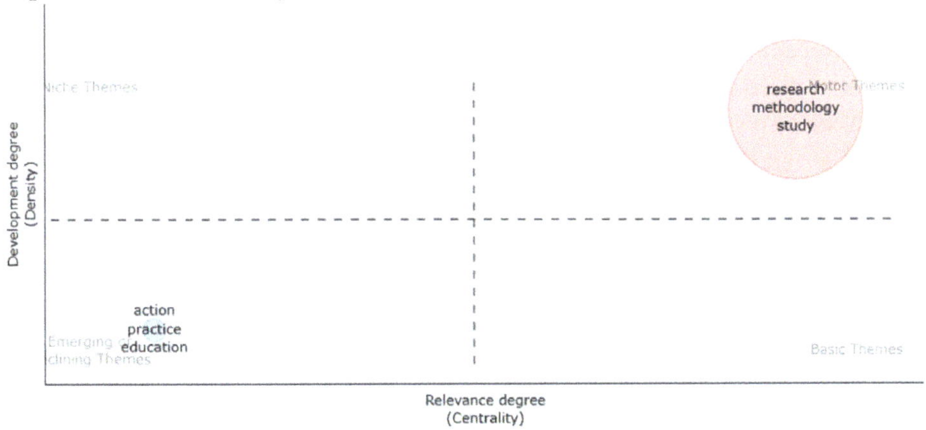

The thematic map extracted from the bibliometric analysis (Figure 13) shows many important insights that can further management scholarship in qualitative and quantitative research approaches. Cluster 1 is labelled as "Research." It was the central theme with all the terms dominating it, like research, study, methodology, analysis, and data. This group spotlights the core aspects of scholarly undertakings: method development, using diverse research strategies, as well as the creation of theoretical and practical knowledge. It demarcates a balanced concern with a variety of methodologies, indicating the significance of developing sound, evidence-based conclusions through such keywords as qualitative, quantitative and framework. Furthermore, terms such as critical, grounded, and context underline the contextual nature of the analysis and the theoretical depth of management studies.

Thematic analysis also suggests the relevance of social dimensions in management research reflected through words such as social, community, and collaboration. This is reflected through these terms, pointing out that organizational behavior, social interactions, and implications of management on society are now matters of interest. Also, findings, results, and implications, as the thrust terms for this research, show an outcome-based nature of the research. It underlines the demand for actionable insights and contributions to both academia and industry.

More profoundly noticed, though, is its methodological trend and lack regarding matters like literature, a survey, a review, or even methodological. To an extent, this refers to efforts to perfect study methodologies and fill in apparent limitations of current methodologies that prevail. Other forward-moving words used include future challenges, potential, which stresses an awareness of new issues still pending questions in management science research.

The presence of words such as innovation, technology, and digital in the research cluster suggests a growing convergence of technological changes and their influence on management practices. This is indicative of the shift toward understanding the nexus between management practices and the transformation of technology in current times. Also, terms such as ethical, quality, and impact suggest an appreciation for responsible and quality-oriented research practices. In a general sense, the present bibliometric study serves as a broad understanding of prevalent themes, methodological lacunae, and possible directions in management research. This highlights the relevance of theory and methodology to studies, practical issues, and the need to direct their research towards the times, the changing societal and technological climate. It serves as an advanced roadmap for improving the quality of management scholarship to further contribute meaningfully towards it.

Figure 14. Thematic Map Network – Abstracts

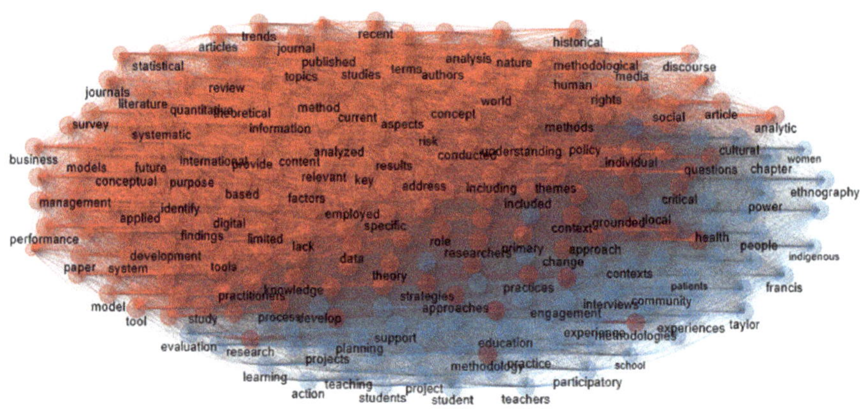

It is highly important to note that these thematic maps (Figure 14) are essential in moving the field of management scholarship ahead due to their emphasis on methodological trends as well as future directions, including identifying gaps in the themes. For instance, the "Research" cluster is the most central and dense theme in terms of Callon Centrality at 1.076 and Callon Density at 1.81, respectively, thus underlining its significance as a building block for the intellectual structure of the field. Its high Rank Centrality and Rank Density values, both at 2, indicate its importance in the structuring of scholarly discourse. The cluster frequency of 136,628 indicates a large volume of work centered on theoretical frameworks, methodological innovation, and comprehensive approaches to management research. The dominance of rigorous methodologies and deep theoretical exploration underpins the contribution of core management studies.

Contrary to this, the “Action” cluster has a relatively moderate centrality level as well as being emergent. The Callon Centrality for this cluster is .863, while its corresponding Callon Density value is .927. It ranks at a Centrality and Density of both 1; thus, though it may not dominate, it is gaining force as a crucial area to focus on. The cluster frequency of 34,473 points toward the interest in applying theoretical knowledge to practical life, for instance, decision-making strategies, managerial interventions, and organizational change initiatives. The cluster points toward a trend towards action-oriented research, trying to fill the gap between theory and practical issues in the world, highlighting the relevance and utility in practice. Together, these clusters represent two streams of focus in management scholarship: the well-established and robust methodologies represented by the “Research” cluster, and the emerging, action-oriented perspectives captured in the “Action” cluster.

Further research can take these two streams together and build frameworks that are deep in theory but also highly applicable to practice, which can promote a more cohesive and impactful approach toward management studies. This integration will propel the field forward because it ensures that work done in scholarship contributes both to academic knowledge and answers the urgent problems in the world of practice.

Figure 15. Thematic Evolution Map – Abstracts

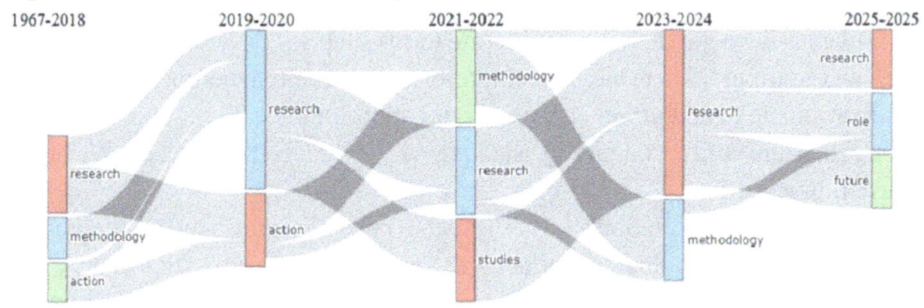

The thematic evolution analysis in Figure 15, on the other hand, shows dynamic developments of key themes and methodologies in management scholarship by pointing out how research trends and priorities have evolved over time. Transition from “Action” 1967-2018 to “Research” 2019-2020 illustrates moving from more applied, community-oriented activities toward a wider scope of health care, implementation practices, and outcomes. This shows increasing enthusiasm to relate action-related programs to the richer theoretical background for a better outcome in healthcare, educational, and organizational applications. The Inclusion Index is .14 while the Stability Index is .00, which reveals a gradual growth in uniform integration.

From “Methodology” (1967-2018) to “Research” (2019-202), indicates a significant shift in advanced research methodologies, including the use of qualitative and quantitative methods, grounded theory, and empirical analysis, with more frequent mention of such related terms as “framework,” “survey,” and “systematic.” A Weighted Inclusion Index of 0.49 and occurrence of 1,490 is suggestive of the criticality that innovative methodological developments play in shaping the domain.

Delving into methodology represents a movement in which conventional research gives way to more pragmatic or functional direction expressed through its dominant applied sciences, as well as methodologies for its practical application; this is growing in incidence. Such themes as ethnography, communities, engagement or cultural further situate participant-focused study in its life-worlds. This indicates a commitment to the real-world application for the advancement of theory, as the index of Weighted Inclusion is 0.49 and 510 appearances.

The sequence of the work from “Studies” in 2021-2022 to “Research” in 2023-2024 reflects an evolution toward focusing on systematic reviews, quantitative methods, and bibliometric analysis with an emphasis on identifying gaps, trends, and future directions. A Weighted Inclusion Index of 0.97 and occurrence (242) indicates that there is considerable interest in exploiting available literature for research, to update practices in a contemporary context.

To the themes of “Research” in 2023-2024, and then “Future” in 2025-2025, a theme that has come in these transitions are terms such as “development,” “knowledge,” and “current.” There has to be future-oriented research incorporating what is already available to add newer challenges and opportunities, all set into a forward-looking structure with sustainability and longevity, evidenced through the Weighted Inclusion Index at 0.63.

Overall, thematic evolution analysis reveals a persistent play between theoretical advancements, methodological refinement, and practical applications; thus, the field continues to be vibrant and evolve in a dynamic way that could solve emerging global challenges.

Figure 16. Thematic Evolution Time Slice 1 Map – Abstracts

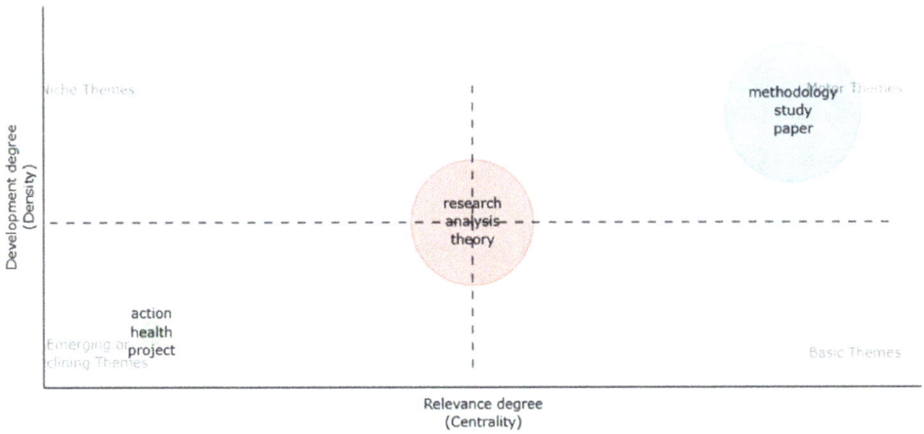

According to Figure 16, Time Slice 1's thematic map shows that there is a huge focus on "research and methodology" clusters. Words such as 'research', 'analysis', 'theory', 'social' and 'knowledge' signify a very profound inclination towards understanding basic ideas as well as establishing comprehensive frameworks around them. This means that a major indicator of pedagogical input is seen in other crucial thematic areas of "education", "students", "learning" and "methodologies" through a blend of academic writing on management scholarship with teaching input. Consequently, usage of terms like "practice", "issues", "understanding", and "critical" demonstrates commitment towards integrating theoretical principles alongside practical aspect in order to ensure that final outcomes remain valid within real context scenarios.

Figure 17. Thematic Evolution Time Slice 1 Network – Abstracts

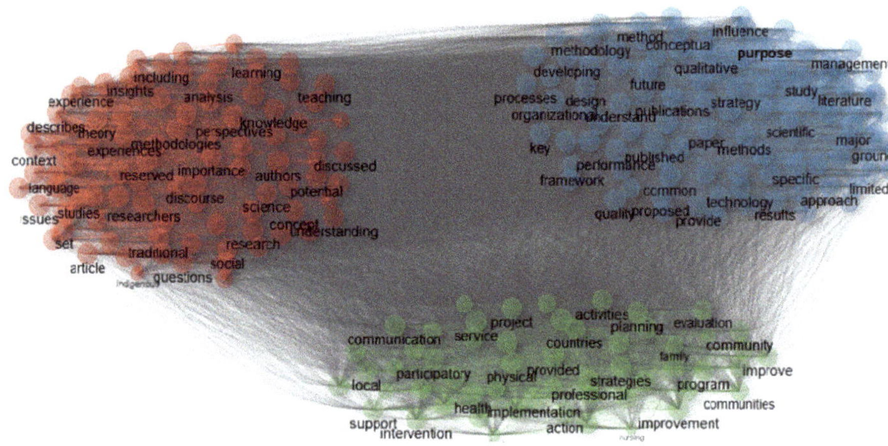

The thematic map for Time Slice 1 (Figure 17) indicates three major clusters: research, methodology, and action, each with significant characteristics in terms of centrality, density, and frequency. The research cluster has a high centrality of 2.405 and a density score of 2.054, ranking second in these two metrics, with the frequency at 32,603. This indicates its foundational role in management studies, acting as a critical hub that connects various thematic areas while demonstrating strong internal coherence. It represents core topics such as theoretical analysis, knowledge dissemination, and the evolution of key management concepts. The methodology cluster ranks third with the highest density at 2.303 and the highest centrality at 2.674, making it the most cohesive and interdisciplinary theme. The significance of its high centrality is in linking both theoretical and practical dimensions. Also indicated is the concentration of well-developed frameworks in both qualitative and quantitative research by its raised density. Moreover, this dominance is driven by the frequency of 37,033, indicating it as a crucial driver for the advancement of management research.

On the other hand, despite having the lowest frequency of 12,681, the action cluster stands out in terms of focus but with lesser centrality at 1.325 and density at 1.847. This would then reflect its peripheral yet focused character and its applicability towards the emerging themes of management that may involve application-oriented themes, pointing to practical innovations in problem-solving in real-life contexts and complementing the more theoretical depth of the research and methodology clusters. Together, these clusters show a dynamic interaction between them. On one hand, methodological and research clusters dominate and therefore provide strength, on the other hand. This is accompanied by an action cluster, which indicates that there is an increasing shift towards getting usable management decisions as well as application-based strategies that need to be adopted by companies.

Figure 18. Thematic Evolution Time Slice 2 Map – Abstracts

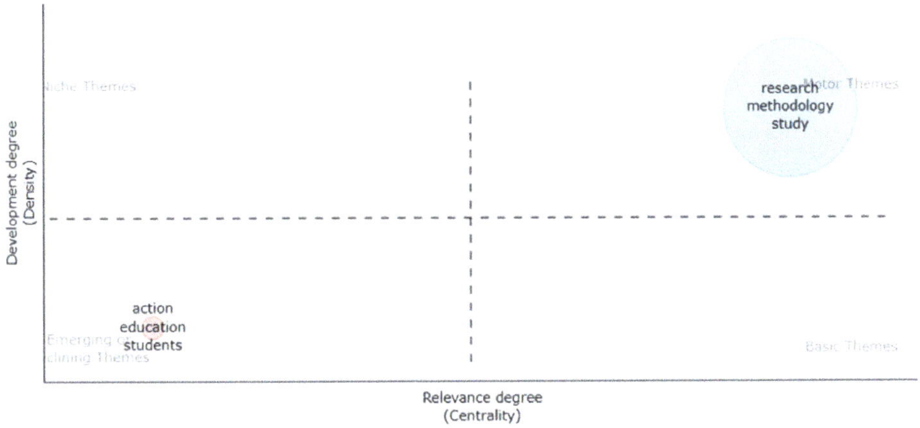

The thematic analysis of the Time Slice 2 terms of Bibliometrix in Figure 18 discloses some significant trends and clusters in qualitative and quantitative research methods. The clusters can be roughly categorized into action-oriented terms and research-focused terms, each carrying unique implications for methodological advancements in the field.

This Cluster 1 Action-Oriented Terms highlights the application of management research in educational and participatory contexts. Key terms like action, students, teachers, learning, and educational practices emphasize a strong connection between research and practice, particularly in pedagogy and professional development. High betweenness centrality scores for terms such as teachers with 152.451 and literacy with 419.011 show that in this cluster, they act as linkages for many themes. This Cluster 2 Research-Focused Terms is largely about methodological rigour as well as foundational research themes. The use of terms investigation, approach, analysis and information is coded to promote analytical and empirical techniques in business studies.

Figure 19. Thematic Evolution Time Slice 2 Network – Abstracts

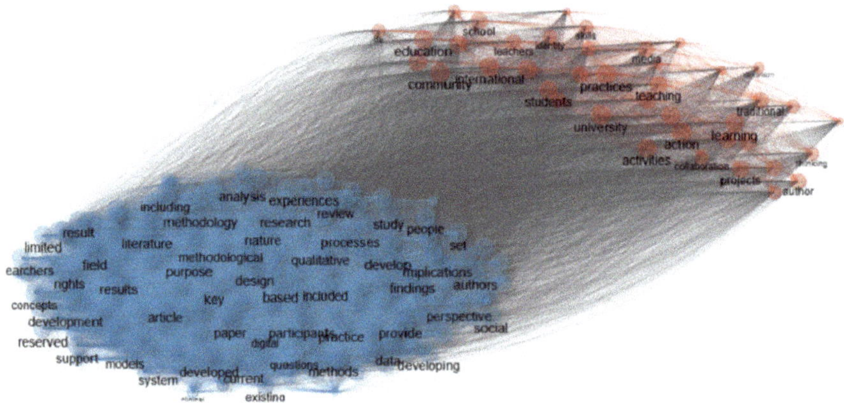

Network – Abstracts The Thematic Map Time Slice 2 clusters bibliomatrix (Figure 19) shows crucial details about how management scholarship evolves. The"action" cluster is centrally placed in the network with Callon Centrality as 4.52 and Callon Density as 5.569. Despite its key centrality status, the first rank both in terms of centrality and density means it's cohesive and an important component of a much larger network, though it is rarely cited (Cluster Frequency: 3015) in the literature. On the flip side, the "research" cluster claims the second-highest Callon Density of 15.229, and concomitantly the highest Callon Centrality value of 4.822, thus testifying to its prominence and extensive interlinkages. This cluster is second in importance about centrality and density, which shows its centrality, but it is still less cohesive than the operational cluster. The research cluster has a very high Cluster Frequency of 20,151; this signifies that research-oriented methodologies predominate over all other themes within management studies.

Figure 20. Thematic Evolution Time Slice 3 Map – Abstracts

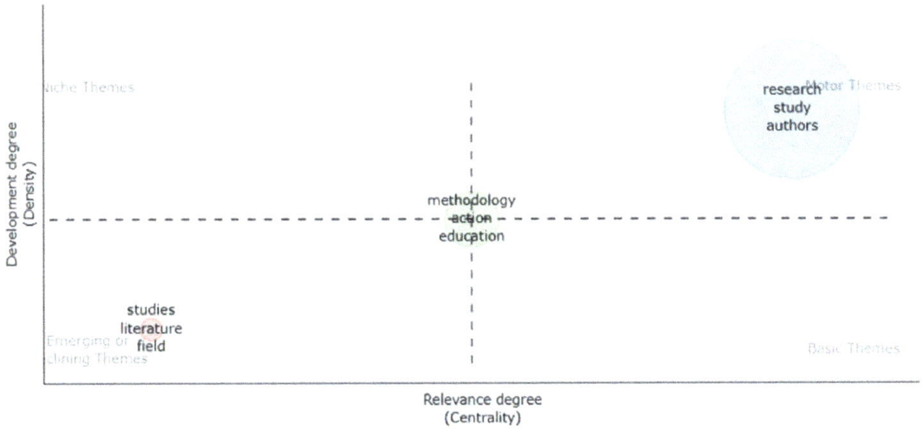

The thematic map for Time Slice 3 in the bibliometric analysis (Figure 20) of advancing management scholarship reveals key trends and insights into the methodological development in both qualitative and quantitative research. In this field, scholarly contributions and systematic investigations appear to be the main topic at hand, according to the data, with many mentions associated with sciences, investigations, or written scholarly works. Journals consist of a considerable part of these studies, an indicator that knowledge is being spread largely via articles that are scrutinized by peers.

Overall, the thematic map shows a robust and evolving landscape, with an increasing need to address methodological gaps, especially in integrating qualitative and quantitative research methods.

Figure 21. Thematic Evolution Time Slice 3 Network – Abstracts

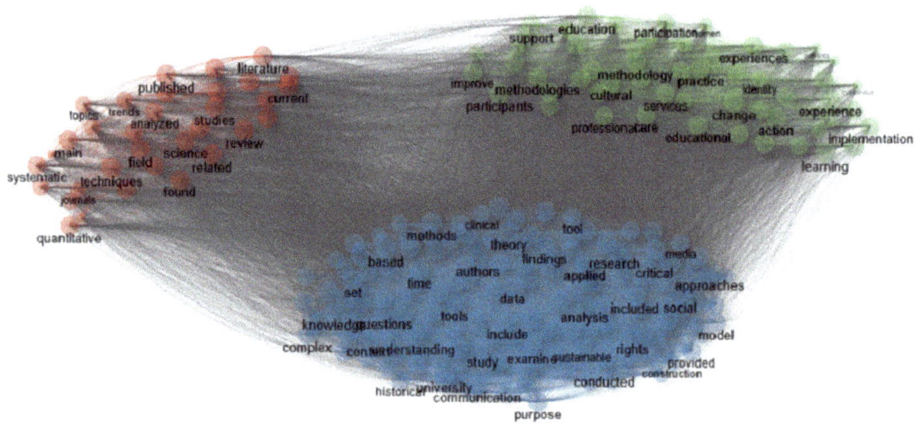

The thematic analysis of Time Slice 3 clusters (Figure 21) in the context of advancing management scholarship reveals key insights into methodological trends, gaps, and future directions in both qualitative and quantitative research. In investigations on Research topics, there is a group of topics on which the rest of the topics rely, with CallonCentrality 3.302 and CallonDensity 4.246, yet it is not that dense as a group. With the study theme having the highest density and centrality, it means that it is the one upon which other studies reflect. The occurrence of this theme 3224 times demonstrates its prevalence in various discussions on research techniques.

The highest CallonCentrality in "Research" cluster is seen as 7.757, while CallonDensity is 10.133; hence, this makes it the most central and dense theme among all others in literature. With a ClusterFrequency of 19299, this is a well-covered theme; that is, research methodologies are the most discussed and integral part of management scholarship. This shows that the concept of research will always be the flagship of scholarly work into the future.

The "Methodology" cluster has a CallonCentrality of 5.379 and CallonDensity of 4.897, which gives it an important but not too closely connected position than that of research. It also presents a ClusterFrequency of 5357, giving prominence, although it appears much less often than the word research. This indicates that, even though methodology is indispensable, it often plays second fiddle to the larger structure of the research.

At the aggregate level, the analysis will show that "Research" represents the major theme, though "Studies" and "Methodology" play important complementary roles in advancing the scholarship of management. Thus, the differential frequency with "Research" as compared with "Methodology" would indicate a gap in researching innovative methodological approaches for development and application. Future research might benefit from a more integrative approach that links these themes in a multidimensional manner, especially in areas where qualitative and quantitative methods intersect and ensure a more comprehensive understanding of how methodologies can help enhance management research.

Figure 22. Thematic Evolution Time Slice 4 Map – Abstracts

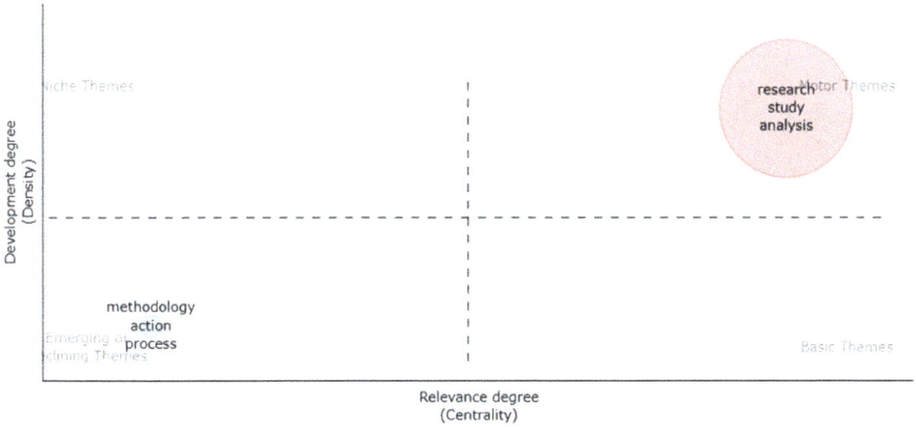

The thematic map for Time Slice 4 in the bibliometric analysis (Figure 22) highlights several key insights regarding the focus and emerging trends in research methodologies, particularly in the realms of qualitative and quantitative research. The research primarily surrounds words such as research, study, analysis, data, methods, and results, which imply that empirical research is core to the discipline.

The thematic map, by way of closing, is a depiction of a dynamic and never-changing research landscape in management, with emphasis being laid on the importance of empirical methods, theoretical approaches, technological improvement, as well as practical uses.

Figure 23. Thematic Evolution Time Slice 4 Network – Abstracts

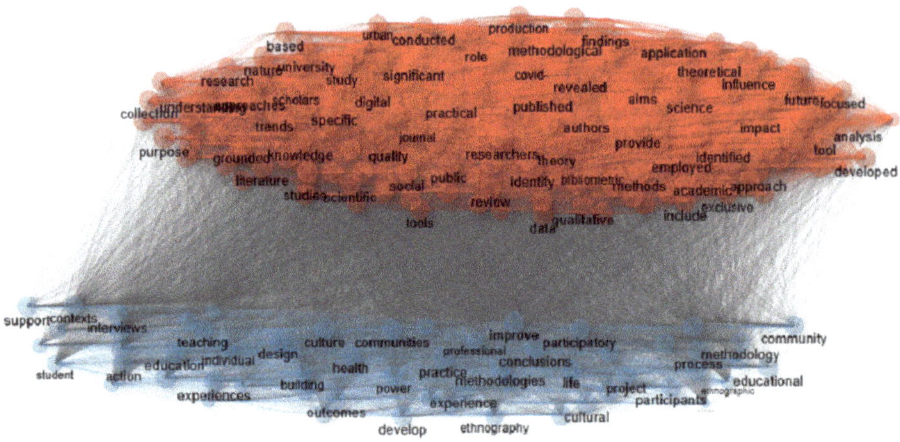

The bibliometric analysis of the Thematic Map Time Slice 4 clusters (Figure 23) offers valuable insights into the Advancing Management Scholarship through the lens of methodological trends, gaps, and future directions in both qualitative and quantitative research. The research cluster exhibits high centrality by its CallonCentrality of 4.73, thereby marking the prominence in the scholarly network, as well as CallonDensity of 9.247, which shows close interactions among the members, resulting in it being the mainstream area of management study and literature with 31,375 publications on the area. The research cluster stands at second centrality and density, thereby affirming its integral position but signalling that it shares the centre stage with other thematic areas, most importantly methodology. The methodology cluster, with a CallonCentrality of 3.837, is marginally less central than research but is of considerable importance, as indicated by its CallonDensity of 4.129, which is lower; this suggests that it is an area of more specialized study. The 7,586 publications on methodological advancement indicate this. The methodology cluster leads the centrality and density statistics, thereby underlining the role of this group as foundational in the practice of management research. The interplay between these two clusters—research and methodology—forms the foundation of future advancements in management scholarship.

Figure 24. Thematic Evolution Time Slice 5 Map – Abstracts

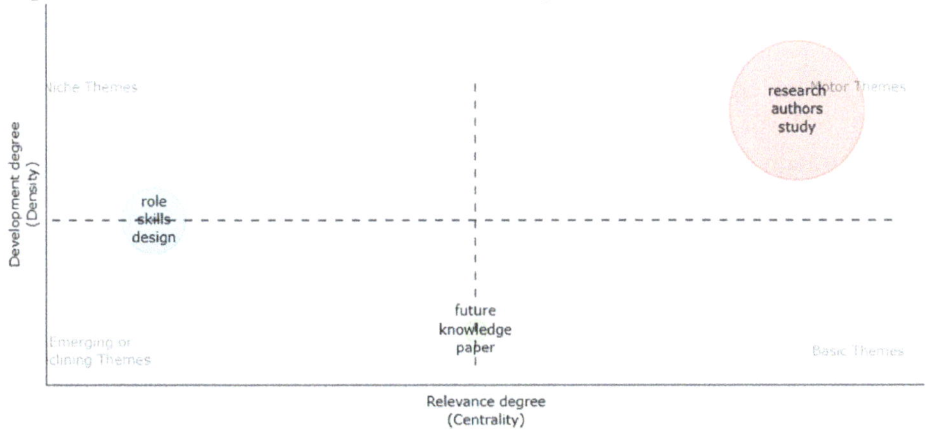

The Thematic Map Time Slice 5 analysis (Figure 24) of terms reveals important insights into the conceptual landscape of research methodologies and trends in management scholarship. The research cluster is characterized by terms like "literary review," "cohort group," "writers" coupled with "active knowledge," as these are thought to truly form the basis for pushing management knowledge ahead within this community. Centrality scores are high for terms like "existing studies" (1.97270027215) and "authors" (1.76339113697), showing that alliances concerning empirical work or documentation structures are always located at the very core areas in management research likelihood. Such terms as "comprehensive," "gaps," "methodologies," and "assessment" point toward efforts to identify gaps and to create more comprehensive applied and evaluative research methodologies.

The role cluster, being not as strong as the research cluster, points toward the key terms "skills," "design," "role," and "findings," pointing to the significance of understanding which roles researchers play in generating management knowledge. The term's relatively lower centrality implies that, although those concepts are important, they are somewhat less integrated throughout the broader research landscape. Interestingly, "role" holds a prominent position with a pagerank centrality score of 0.008, implying that the concept of the role of the researcher is growing.

The future cluster is introducing forward-looking terms that include "future," "knowledge," "development," and "discipline," and has terms that indicate an orientation towards evolving research priorities. With the uniform centrality scores that are around 194.86, these terms, although equally distributed within a low level of centrality in the broader research network, indicate that the terms for future-oriented concepts, although important, still need to gain as much integration or traction as research-focused terms do.

The analysis on themes represents a thoughtful orientation towards research, discerning the gaps, approaches taken, and evaluation of what has so far been done. Words that project into the future show an inclination to explore fresh pathways, methods and interdisciplinary approaches that allow for future advancements and creativity in the field of management studies. These trends indicate that methodology gap fillers and new research explorations on the emerging topic will be pivotal in building the discipline further.

Figure 25. Thematic Evolution Time Slice 5 Network – Abstracts

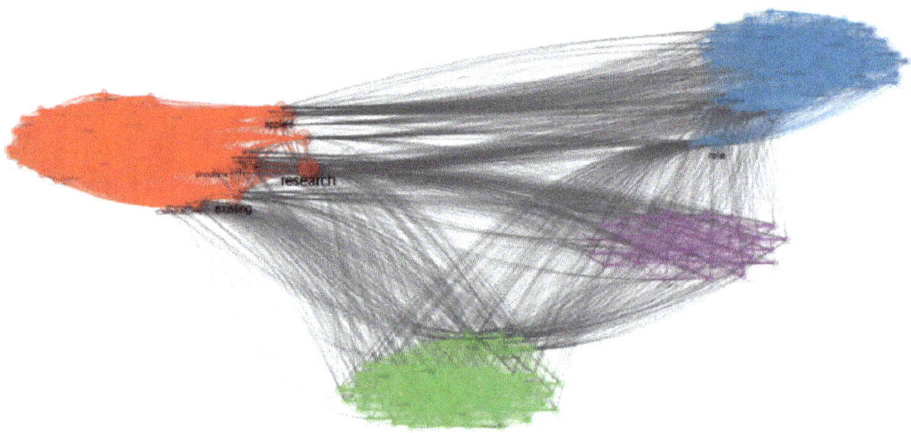

The Thematic Map Time Slice 5 clusters bibliomatrix (Figure 25) gives an insight to the changing face of management scholarship, especially the research methodology. The cluster of research is the one that is the most dominating with high Callon Centrality (55.694) and Callon Density (381.643), indicating that it is central as well as dense in all the discussions in scholars. This cluster's Rank Centrality (3) and Rank Density (3) support its centrality, thus the thematic map's core focus. The Cluster Frequency (53) further reinforces that research is the focus area of attention in management scholarship, thus remaining a bedrock of the field.

In contrast, the role cluster has a more peripheral position with a Callon Centrality (23.972) and Callon Density (258.333), which are significantly lower than that of the research cluster. While the concept of the researcher's role is still important, it is less integrated into the broader network of management research. This is further reflected in its Rank Centrality (1) and Rank Density (2), indicating that it holds influence but is not as central as the research-related discussions. According to the Cluster Frequency, the number of occurrences defining role-related issues in literature is low, even though significant.

A futuristic cluster, which shows an increasing interest in trends and directions to come, has Callon Centrality of 31.667 as well as Callon Density of 225; it denotes shifting attention of management studies towards prognostication about what will happen next in various fields. Yet it lags behind the research cluster in terms of centrality and density, thus placing it in a mid-tier rank with a Rank Centrality of 2 and Rank Density of 1. Cluster Frequency is at 16, indicating that discussions about the future are on the rise, but still less frequent than research-centric themes.

Overall, the analysis presents the fact that though research is still at the core of advancing the scholarship of management, the roles of the researchers and trends for the future are also gaining importance as a sign to explore the broader implications and evolving methodologies in the field.

Table 2. Evolution of Methodological Trends and Focus Areas in Management Scholarship

Time Slice	Cluster	Insights
Time Slice 1	Knowledge	Early focus on foundational knowledge, emphasizing core understanding in management methodologies.
	Methodology	Emerging focus on methodological approaches, signifying initial exploration of techniques.
Time Slice 2	Management	Shift towards broader management themes, indicating growing interest in application-oriented research.
	Strategy	Strategic insights gain traction, reflecting applied research trends.
Time Slice 3	Development	Development-focused discussions take center stage, emphasizing the evolution of management scholarship.
	Application	Application of methodologies becomes prominent, showcasing a practical orientation.
Time Slice 4	Research	Dominance of the research cluster, reflecting a surge in literature and deep exploration of methodologies.
	Methodology	Continued emphasis on refining and expanding methodological tools in research.
Time Slice 5	Research	Solidification of research as the central focus, underscoring its foundational importance in advancing management scholarship.
	Role	Exploration of roles in research highlights emerging discussions on researchers' contributions and frameworks.
	Future	Increasing attention to future directions and trends signaling a forward-looking approach to evolving management scholarship methodologies and applications.

Table 2 gives the evolution of methodological trends and focus areas in management scholarship.

THEORETICAL FRAMEWORK

The theoretical framework of this study is built upon methodological pluralism and epistemological integration principles that encourage the complementary use of both qualitative and quantitative research approaches to facets of the complex phenomena in organizations. Methodological pluralism is basically pragmatism, as described by Dewey (1938), which sets aside issues of absolute methodological divisions in favor of practical problem-solving, and triangulation, as argued by Denzin (1978), which holds the position that multiple methods should be used to improve the validity and reliability of the findings. Such an approach tries to break free from the shackles of a solitary methodological tradition so that management research may incorporate techniques from several methodologies to glean more holistic insight.

The epistemological integration bridges positivist and constructivist paradigms, which account for objective reality and human subjective meaning-making. Such a dual view is endorsed by critical realism (Bhaskar, 1975), which acknowledges that an external reality exists, but human hindrance in the formation of knowledge cannot be overlooked. Furthermore, transdisciplinary research (Nicolescu, 2002) stands for crossing boundaries of disciplines in order to promote ingenious solutions to real-life challenges. Bringing these epistemological stances together makes the study escape conventional dichotomies and provides a more holistic perspective on managerial problems.

A bibliometric set of theories is embedded in the framework to provide quantitative tools that analyse scholarly literature and map intellectual trends. Co-citation identifies didactic works, and co-word identifies emergent themes by way of keyword analysis (Small, 1973; Callon et al., 1983). Methods can highlight the evolution of management scholarship, under the limelight of influential studies, and areas requiring further exploration. Further, the technological and innovative factors pertinent to research methods will be considered through the lens of the Technology Acceptance Model (TAM) (Davis, 1989) and Sociotechnical Systems Theory (Trist & Bamforth, 1951); the theories that examine how modern advances such as AI and big data analytics remodel research practice in ways that allow further avenues for data-driven inquiry.

Ethical and sustainable research practices are another critical pillar of the framework. Stakeholder Theory (Freeman, 1984) emphasizes that there must be a balance of ethical considerations between different groups, whereas the Triple Bottom Line (Elkington, 1997) brings into methodological design social, environmental, and economic metrics. In brief, both principles serve as a foundation to ensure that management research is focused not only on advances in knowledge but in pressing issues: AI ethics and sustainability in fintech.

By putting together these theoretical perspectives, the framework builds a robust scaffolding structure to examine methodological trends, identify gaps, and hence infer directions for future research. Therefore, positioned at the interface of theory, method, and practice, the study calls for a balanced interdisciplinary approach for the advancement of management scholarship. This integration provides a view to help researchers negotiate the complexities of modern organizational worlds whilst promoting pertinent contributions to both academic discourse and practical applications.

Figure 26. Conceptual Framework

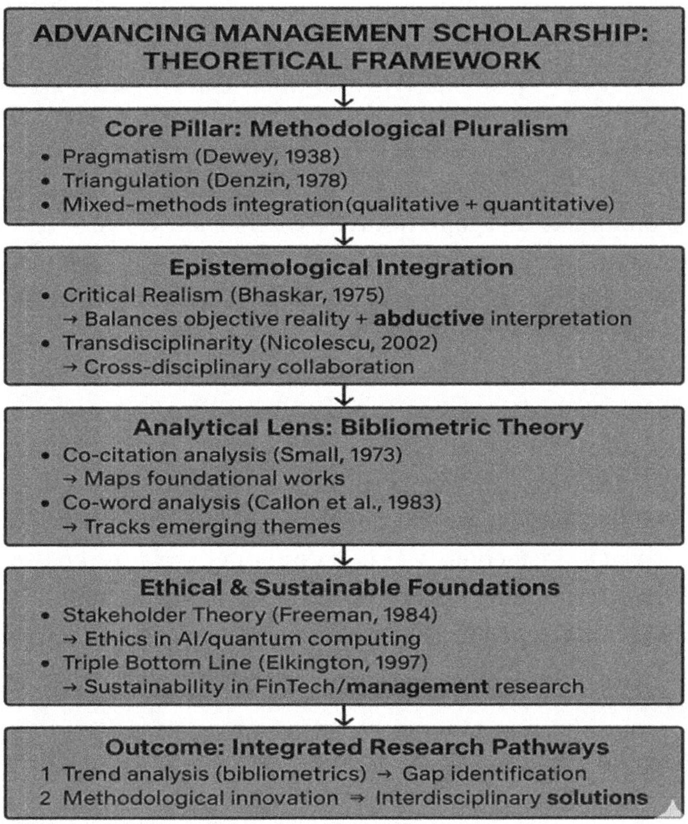

Conceptual Framework

The conceptual framework underlying this study integrates methodological pluralism, epistemic diversity, and bibliometric analysis to identify the evolution, gaps, and future prospects of qualitative and quantitative research in management scholarship (See Figure 26). It mainly stresses the complementary application of multiple methods of research, ranging from qualitative ones (e.g., discourse analysis, ethnography) through quantitative techniques (such as experimental designs and statistical modelling) in tackling complex organisational phenomena. This pluralistic paradigm is based on pragmatism (Dewey, 1938), which values problem resolution above methodological inflexibility, and triangulation (Denzin, 1978), whereby validity is enhanced via multiple sources of data.

Epistemologically, the framework bridges positivist (objective, measurable reality) and constructivist (subjective, context-dependent meaning) paradigms, being underpinned by critical realism (Bhaskar, 1975). This synthesis postulates that management theories are shaped through the simultaneous operation of empirical evidence and human interpretation. Moreover, transdisciplinarity acts as a potential force uniting fields to solve new-age problems presented by sustainability and digital transformation (AI ethics, behavioral economics).

The bibliometric lens (co-citation, co-word analysis) systematically and quantitatively explores trends and gaps. It reveals the thematic clusters (e.g., qualitative dominance of terms such as "research" and "methodology") and underexplored areas (e.g., AI ethics and applications of quantum computing). Drawing from the contemporary landscape of technologies such as machine learning and big data, TAM and Sociotechnical Systems Theory foreground their disruptive potential vis-à-vis research methodologies.

Ethical and sustainable considerations are ingrained in the responsible research practices with the Stakeholder Theory (Freeman, 1984) and the Triple Bottom Line (Elkington, 1997). The final product generated by this framework provides pathways for actions: (1) Methodological innovation in hybrid-methods research ties up qualitative and quantitative research traditions, (2) Interdisciplinary solutions in Cross-domain frameworks addressing major global challenges (ESG, hybrid work), (3) Forward-looking scholarship, underpinned by reproducibility, open science, and ethical use of AI.

Figure 27. Conceptual Framework

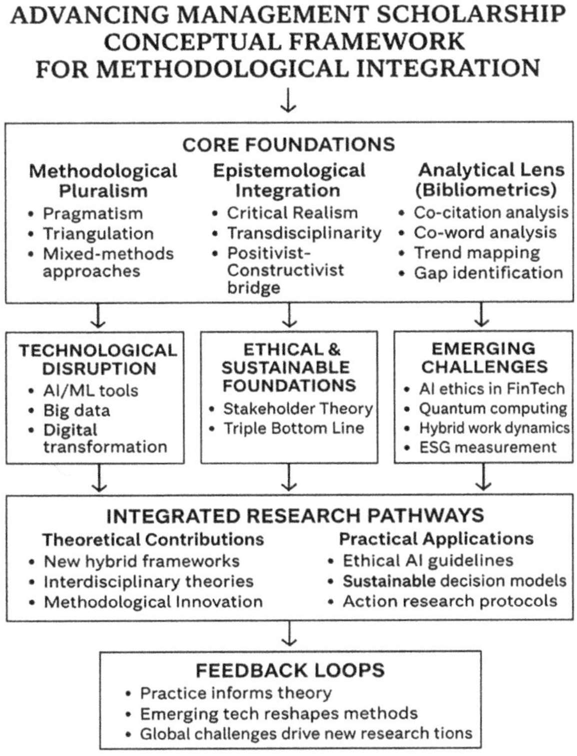

By unifying these dimensions, the framework offers a dynamic, iterative process where bibliometric insights inform methodological refinement, theoretical advancement, and practical applications—ensuring management scholarship remains rigorous, relevant, and responsive to societal and technological shifts (see Figure 27).

FUTURE RESEARCH DIRECTIONS

Table 3 outlines key research gaps in management scholarship, highlighting underexplored themes, methodological deficiencies, and emerging interdisciplinary opportunities that call for innovative and diverse research approaches.

Table 3. Research Gaps

S. No.	Research Gap Area	Specific Issues / Gaps	Recommended Methodologies / Approaches
1	**Ethical Implications of AI in Management**	Lack of discussion on bias, transparency, and accountability in AI decision-making	Qualitative methods (Discourse analysis, Grounded theory)
2	**Sustainability in FinTech**	Environmental impact and social responsibility are often ignored	Ethnographic studies, Evaluative research
3	**Interdisciplinary Integration**	Limited collaboration with neuroscience, behavioral economics, cultural studies	Cross-disciplinary studies and frameworks
4	**Historical and Longitudinal Analysis**	Absence of historical studies on management trends like digital transformation	Longitudinal research designs
5	**Quantum Computing in FinTech & AI**	Untapped potential in risk modelling, portfolio optimization, HFT	Exploratory studies, Simulation-based research
6	**AI in Strategic Financial Decision-Making**	Ethical implications, lack of explainability in models	Case studies, Interpretive research
7	**Behavioural Analytics in Management**	Nascent development in predictive models for employee and consumer behavior	Advanced behavioral modeling, Experimental design
8	**ESG and Sustainability Evaluation**	Lack of causal mapping to evaluate ESG strategies	Quantitative methods like Causal Mapping
9	**Organizational Culture in Hybrid Work**	Under-explored dynamics of remote/ hybrid environments	Grounded theory, Discourse analysis
10	**Leadership and Decision-Making in Divided Teams**	Lack of experimental research on team dynamics and leadership	Experimental models, Simulation techniques
11	**Reinforcement Learning Applications**	Limited use in adaptive systems and financial engineering	Machine Learning-based modeling
12	**Network Analysis and Bayesian Modeling**	Underutilized in organizational collaboration and inference studies	Network theory, Bayesian inference methods
13	**Large-Scale Real-Time Data**	Scarcity of datasets enabling longitudinal and experimental research	State-of-the-art statistical modeling, Big Data analytics
14	**Epistemological Integration**	Weak integration of positivist and constructivist paradigms	Philosophical studies, Pluralistic methodological frameworks
15	**Transparency and Reproducibility in Research**	Poor documentation and lack of reproducibility in experimental/mixed-method studies	Improved scholarly practices, Open science initiatives
16	**Action Research**	Underutilized in innovation and sustainability contexts	Participatory action research, Field experimentation

Table 4 outlines key future directions in management research, emphasizing the integration of qualitative and quantitative methods, advanced analytics, interdisciplinary approaches, and theory development to address emerging organizational challenges

Table 4. Future Directions

S. No.	Focus Area	Key Insights / Recommendations
1	Mixed Methods Integration	Combine qualitative (e.g., discourse analysis, ethnography) and quantitative (e.g., surveys, experiments) methods to study organisational phenomena.
2	Qualitative and Quantitative Balance	Use numerical tools to assess sustainability while applying qualitative inquiry for ethical issues in AI and FinTech.
3	Emerging Domains Exploration	Investigate areas like ethics of AI, FinTech using quantum computing, and sustainability with appropriate methodological approaches.
4	Theoretical Framework Advancement	Develop new frameworks to account for emerging contingencies in management.
5	Advanced Analytical Techniques	Apply causal mapping, SEM, and Bayesian modelling to understand complex organizational systems.
6	Predictive Modelling and Big Data	Integrate machine learning and big data for predictive modelling in behaviour, performance, and organisational change.
7	CSR Measurement Methods	Create multidimensional measurement tools for complex constructs such as CSR.
8	Experimental & Non-Experimental Modalities	Use experimental and non-experimental approaches to explore causal links in leadership, performance, and transformation.
9	Cross-Cultural and Longitudinal Research	Conduct cross-cultural and longitudinal studies to capture industry and regional management variations.
10	Participatory Action Research	Collaborate with practitioners via action research to address practical challenges in sustainability, innovation, and diversity.
11	Epistemological Dualism	Bridge positivist and constructivist epistemologies for a comprehensive understanding of management practices.
12	Theory Development and Adaptation	Recalibrate traditional theories and incorporate insights from other fields to maintain relevance amid global and technological shifts.

Looking far ahead, the main contributions will be with regard to further development and experimentation on the proposed methodology refinement, as well as identification of new frontiers for exploration, including ethical implications of AI, sustainability in business practices and application of quantum computing to financial decision making, among others. Owing to the rapidly changing nature of today's corporate environment, these areas of concern have emerged as critical components demanding more investigation; thus abandoning either presence or past studies would be a disservice to society. For this reason, management scholarship should continue contributing towards advancement in knowledge theory and practice using strict research approaches within the most current framework.

CONCLUSION

This study reveals a persistent qualitative dominance in management research, with quantitative and mixed methods remaining underexplored. The gaps identified in AI ethics, sustainability, and methodological integration call for a deliberate shift toward hybrid frameworks, reproducible designs, and ethically grounded inquiry. It is worth noting that many scholars argue for an integrative approach to research because variety is the spice of life; having all eggs in one basket not only leads to stagnation but also compromises the reliability of research findings as well. For example, some would suggest that qualitative research methods should be combined with quantitative ones to produce more valid results, while others would oppose this idea, claiming that such a mixture makes it difficult for readers to understand the findings of the study. Quantitative studies are understood by readers without too much explanation, provided there is sufficient information about them, since they deal with things we can see or touch. In contrast, qualitative research deals with ideas, beliefs, and feelings that cannot be easily expressed using numbers. In particular, scholars have attempted to integrate various research methods through epistemology since the latter encompasses both qualitative and quantitative methods. It is important for scholars to realise that these paradigms are not necessarily opposite; rather, they complement each other in the sense that constructivism allows positivism to exist as an alternative epistemology within social sciences. What we need is a holistic framework encompassing all these perspectives." Meanwhile, the authors are equally at pains to indicate the importance of understanding the differences between the two paradigms of positivism (where opportunities are seen as objective reality) and constructivism (where individuals create their realities). This means that while the former emphasizes hard facts, the latter deals in ideas

On the other hand, many researchers have taken positivist and constructivist approaches to understanding organisational phenomena within the field, prompting much debate; this has helped in building knowledge based on various perspectives. Moreover, Schin's argument is supported by various authorities within the field of management; The advancement in the web has transformed these debates from mere discussions into methodological pluralism in research. Researchers should build from previous theories while at the same time establishing results in the contemporary global practices. Encouragingly, there has been a shift to more stringent policies on scholarly writing and publication, leading to high-level methodological accuracy, besides guaranteeing the credibility of management research.

In conclusion, it is significant to note that both qualitative and quantitative approaches are vital for driving management scholarship forward. Hence, there is a need for a balanced integrated methodology that incorporates different research methods as well as analytical tools and epistemological perspectives. They support ongoing debates about methodological rigour, the changing face of management theory, or interdisciplinary approaches to organisational challenges today. This implies that management scholarship has to keep refining its methods of study over time if issues affecting people can be addressed in contemporary society while serving as an ultimate guide for 21st-century organisations and societies.

REFERENCES

Alonso-Garcia, J., Pablo-Martí, F., & Nunez-Barriopedro, E. (2021). Omnichannel Management in a B2B context: Concept, research agenda and bibliometric review. *International Journal of Industrial Engineering and Management, 12*(1), 37.

Andersen, N. (2021). Mapping the expatriate literature: A bibliometric review of the field from 1998 to 2017 and identification of current research fronts. *International Journal of Human Resource Management, 32*(22), 4687–4724. DOI: 10.1080/09585192.2019.1661267

Aria, M., & Cuccurullo, C. (2017). Bibliometrix: An R-tool for comprehensive science mapping analysis. *Journal of Informetrics, 11*(4), 959–975.

Breslin, D., & Gatrell, C. (2023). Theorizing through literature reviews: The miner-prospector continuum. *Organizational Research Methods, 26*(1), 139–167.

Donthu, N., Kumar, S., Mukherjee, D., Pandey, N., & Lim, W. M. (2021). How to conduct a bibliometric analysis: An overview and guidelines. *Journal of Business Research, 133*, 285–296.

Fahimnia, B., Sarkis, J., & Davarzani, H. (2015). Green supply chain management: A review and bibliometric analysis. *International Journal of Production Economics, 162*, 101–114.

Fan, D., Breslin, D., Callahan, J. L., & Iszatt-White, M. (2022). Advancing literature review methodology through rigour, generativity, scope and transparency. *International Journal of Management Reviews, 24*(2), 171–180.

Lopes, J. M., Sousa, A., Calçada, E., & Oliveira, J. (2021). A citation and co-citation bibliometric analysis of omnichannel marketing research. *Management Review Quarterly*, 1-34

Mukherjee, D., Lim, W. M., Kumar, S., & Donthu, N. (2022). Guidelines for advancing theory and practice through bibliometric research. *Journal of Business Research, 148*, 101–115. DOI: 10.1016/j.jbusres.2022.04.042

Paul, J., & Criado, A. R. (2020). The art of writing literature review: What do we know and what do we need to know? *International Business Review, 29*(4), 101717.

Paul, J., Lim, W. M., O'Cass, A., Hao, A. W., & Bresciani, S. (2021). Scientific Procedures and Rationales for Systematic Literature Reviews (SPAR-4-SLR). *International Journal of Consumer Studies, 45*(4), O1–O16. DOI: 10.1111/ijcs.12695

Thelwall, M., & Sud, P. (2022). Scopus 1900–2020: Growth in articles, abstracts, countries, fields, and journals. *Quantitative Science Studies*, *3*(1), 37–50.

UNESCO. (2022). UNESCO recommendation on open science. Available from: https://doi.org/DOI: 10.54677/MNMH8546

Van Eck, N., &Waltman, L. (2010). Software survey: VOSviewer, a computer program for bibliometric mapping. *scientometrics, 84*(2), 523-538.

Zupic, I., & Čater, T. (2015). Bibliometric methods in management and organization. *Organizational Research Methods*, *18*(3), 429–472.

Chapter 2
A New Adaptive Integrative Research Method for Computer and Information Science

Abdelkader Berraouna
http://orcid.org/0009-0007-4434-279X
Souk Ahras University, Algeria

ABSTRACT

The rapid evolution of computing and information sciences has revealed a critical gap in existing research methodologies, which are often borrowed from the social sciences or business disciplines without adequately addressing the unique epistemological and technical challenges of the field. This paper proposes a novel Methodology for Integrative Adaptive Research (AIRM), a comprehensive, flexible, and ethically grounded method specifically designed for research in computer science and information studies. The AIRM addresses the "specificity-generality paradox" by integrating quantitative, qualitative, and computational approaches within a dynamic, reflexive process that adapts to technological and contextual shifts throughout the research lifecycle. Built on five core principles, that is, triple methodological integration, proactive ethics, dynamic adaptability, enhanced transparency, and structured interdisciplinary collaboration. The AIRM ensures scientific rigour while remaining responsive to real-world complexity.

DOI: 10.4018/979-8-3693-4018-9.ch002

INTRODUCTION

The rapid evolution of computer and information sciences has highlighted a critical gap in current research methodologies. Indeed, the majority of methodological frameworks employed in these fields are borrowed from social sciences or business disciplines, which limits their applicability and relevance to the specific challenges encountered in the computer science domain (Fan, Zhang, & Li, 2023). These approaches, while useful in their original contexts, struggle to meet the dual requirements of specificity and generality inherent in research covering areas as diverse as complex mathematical algorithms and the societal impacts of data systems (Burrell, 2016; Kitchin, 2014). This divergence between the needs of computer science research and the available methodological tools constitutes a fundamental problem that justifies exploring new solutions. The primary objective of approach AIRM is to provide a flexible and adaptable framework that integrates both qualitative and quantitative approaches, while taking into account the specific characteristics of contemporary IT projects. Unlike traditional frameworks, AIRM aims to overcome the specificity-generality paradox by combining advanced computational tools with rigorous interpretive analyses. For example, it could utilise machine learning models, such as Word2Vec, to extract and classify methodological terms, while integrating human supervision to ensure the necessary analytical depth (Fan et al., 2023).

The primary objective of approach AIRM is to provide a flexible and adaptable framework that integrates both qualitative and quantitative approaches, while considering the specific characteristics of contemporary IT projects. Unlike traditional frameworks, AIRM aims to overcome the specificity-generality paradox by combining advanced computational tools with rigorous interpretive analyses. For example, it could utilise machine learning models, such as Word2Vec, to extract and classify methodological terms, while integrating human supervision to ensure the necessary analytical depth (Fan et al., 2023).

Traditional frameworks used in the social sciences, for example, often rely on qualitative or mixed-method approaches that prioritise the interpretation of human behaviour and social dynamics. However, when applied to computational projects, these frameworks face structural limitations. The work of Fan, Zhang, and Li (2023) effectively illustrates this tension, demonstrating that the social science and computer science research communities adopt distinct methodological cultures. While the social sciences frequently combine qualitative and quantitative approaches, engineering sciences, particularly computer science, show a marked preference for computational methods such as machine learning and neural networks. This divergence reflects not only a difference in paradigms but also the inadequacy of methodological tools when they are applied across different fields. For example, techniques such as ethnography, although essential for understanding human interactions, remain little influenced by computational advances, which limits their integration into interdisciplinary projects after 2023 (Fan et al., 2023).

Another major challenge lies in what we can call the "specificity-generality paradox." Computer science research often has to navigate two contradictory requirements: on the one hand, it must be specific enough to address precise technical problems, such as algorithm optimisation or massive database management; on the other hand, it must also offer a general perspective capable of capturing the broader societal implications of the technologies developed (Burrell, 2016). This paradox is exacerbated by the increasing use of artificial intelligence (AI) in qualitative research, which raises questions about the ability of these tools to produce both rigorous and interpretive analyses. For example, studies have shown that AI algorithms, while effective at identifying concrete descriptive themes, struggle to deal with deeper interpretive analyses, particularly regarding cultural nuances or ironic subtexts (Coppock, 2025; Friese, 2025). This limitation underscores the need for a methodology that can reconcile these two dimensions.

Faced with these challenges, the proposal of a new methodology, the Adaptive Integrative Research Method (AIRM), stands out as a potential response. The primary objective of AIRM is to provide a flexible and adaptable framework that integrates both qualitative and quantitative approaches, while taking into account the specific characteristics of contemporary IT projects. Unlike traditional frameworks. For example, it could utilise machine learning models, such as Word2Vec, to extract and classify methodological terms, while integrating human supervision to ensure the necessary analytical depth (Fan et al., 2023). This hybrid approach would enable a deeper understanding of the complex dynamics underlying digital transformation failures, an issue identified as crucial by Oludapo, Akinwale and Ogunleye (2024).

REVIEW OF EXISTING METHODOLOGICAL TRADITIONS AND RECENT ADVANCES

To situate AIRM within current methodological debates, this section critically reviews four major traditions—design science research, software engineering, computational grounded theory, and human-computer interaction (HCI)—as well as recent advances in hybrid methods, AI ethics, and interdisciplinary computational research.

Design Science Research (DSR)

Design Science Research (DSR) remains a foundational tradition for methodologies that produce artifacts and design solutions rather than merely explaining phenomena. Abbasi, Parsons, and others (2024) in *"Pathways for Design Research on Artificial Intelligence"* argue that while artifact orientation is valuable, DSR in AI contexts must confront issues such as evaluation in dynamic environments, socio-technical impact, and ethical accountability

Further (Abbasi et al., 2024), the *DESRIST 2024* proceedings include contributions emphasising the need for resilience, adaptability, and stakeholder participation in artifact design—traits that align strongly with AIRM's goals of adaptability and interdisciplinarity.

Software Engineering Frameworks

Software engineering (SE) and systems engineering have long provided rigour in development life-cycles, versioning, reproducibility, and large-scale deployment. However, recent critiques note that traditional SE approaches often lack sensitivity to qualitative stakeholder perspectives, ethical dimensions, and the need for iterative feedback beyond engineering metrics. While specific recent peer-reviewed works integrating ethics into SE life cycles are emerging, SE frameworks still often focus on performance, reliability, and scalability rather than context-sensitivity, reflexivity, or ethical foresight (Mandviwalla et al, 2024).

Computational Grounded Theory (CGT)

CGT combines automated computational techniques (such as machine learning, topic modelling) with classic grounded theory's inductive coding and interpretive depth. A recent work by Chen and Chen (2024), entitled "Computing Grounded Theory: a quantitative method to develop theories", introduces quantitative elements to support hypothesis generation via CGT, which points toward increased integration of computational and qualitative methods.

Additionally, A Computational Grounded Theory-Based Analysis of Research on China's Old-Age Social Welfare System (Li et al., 2025) presents a case study of CGT applied systematically, incorporating topic modelling followed by human interpretive linking of themes, highlighting both the strengths and limitations of scalable policy-oriented research (Agudamu et al., 2025).

Human-Computer Interaction (HCI) and Participatory Design

HCI and participatory design bring stakeholder perspectives, usability, and context to the fore; these traditions are essential to design interfaces and systems that respect user values. Recent scholarship (2023-2024) has stressed inclusive design practices for low-resource settings, adapting to cultural diversity, and co-design with users in marginalised communities. However, these practices are often underrepresented in hybrid or AI-driven research pipelines, especially when they are not integrated with quantitative evaluation or algorithmic feedback loops.

Recent Advances: AI Ethics, Hybrid Methods, Interdisciplinary Computational Research

- *AI Ethics*: In *"From ethical AI frameworks to tools: a review of approaches"* (Prem, 2023), the author reviews over 100 frameworks and tools, concluding that many are abstract and need concrete operational paths for deployment. (Prem, 2023)
- The paper *"The Rise of Checkbox AI Ethics: A Review"* (AI & Ethics, 2024) highlights how ethical guidelines often become ritualistic checkboxes rather than integrated, reflexive practices (Kijewski, S., et al. 2025).
- *Responsible AI Governance: A Systematic Literature Review* (Batool, Zowghi & Bano, 2023-24) maps key elements of AI governance, revealing that many frameworks omit stakeholder inclusion or actionable enforcement.

International Literature Base from African and South Asian research

Eke (2024) highlights the need for ethical frameworks for neurotechnology and AI governance to be contextualised in African sociotechnical realities, emphasising participatory and community-driven oversight models. Similarly, Ravi et al. (2025) explore participatory research in low-resource mental-health contexts in India, positioning methodological adaptability as a form of *epistemic justice*.

Gautam (2024) and Zhao et al. (2024) propose computational and participatory design strategies that leverage limited infrastructures through open collaboration and hybrid intelligence, offering practical parallels to AIRM's iterative integration.

From a managerial and policy perspective, (CIPESA 2024) outlines the emerging discourse on ethical AI regulation in Africa, advocating for regional frameworks grounded in inclusion, accountability, and capacity building. This cross-cultural synthesis not only enhances theoretical robustness but also positions AIRM as a *methodology capable of bridging epistemic and infrastructural divides* in computer and information science research. In Table 1, we present a systematic comparison of AIRM with established methodological frameworks.

Table 1. Systematic Comparison of AIRM with Established Methodological Frameworks

Criteria	Design Science Research (DSR)	Software Engineering (SE) Frameworks	Computational Grounded Theory (CGT)	HCI/ Participatory Design (HCI/ PD)	AIRM (Proposed)
Integration	Strong integration of design and evaluation, yet typically focused on the technical artifact rather than social dimensions.	Low–moderate: oriented toward process management and deliverables, limited qualitative integration.	Moderate: merges automated text analytics and qualitative coding, often sequential rather than iterative.	High: integrates user participation and co-creative qualitative inquiry.	**Triple integration** of qualitative, quantitative, and computational methods within a continuous reflexive cycle.
Flexibility	Moderate: adaptive frameworks exist but generally follow pre-defined design cycles.	Low: rigid, phase-driven processes optimized for industrial reproducibility.	Moderate: scalable but dependent on computational pipelines and resources.	High: adaptable to local and cultural contexts.	**High:** modular, iterative architecture enabling progressive and context-specific implementation.
Ethical Orientation	Variable: ethics often addressed during evaluation rather than embedded from the outset.	Low–moderate: focuses on security and compliance; socio-technical ethics peripheral.	Low–moderate: may assess algorithmic bias but relies heavily on human post-analysis.	High: emphasises consent, inclusion, and user values.	**Proactive and embedded:** continuous ethical auditing and decision traceability across all stages.
Contextual Adaptability	Moderate: adaptable but dependent on institutional maturity and resources.	Low: suited mainly to high-resource, industrial settings.	Moderate: capable of large-scale analysis but limited by infrastructure and expertise.	High: strong in local participatory contexts, requires facilitation time.	**Designed for heterogeneity:** modular deployment ("MIAR-Lite") suitable for low-resource and high-constraint environments.
Transparency / Traceability	Moderate: artifact documentation standard, but interpretive decisions may remain opaque.	High on technical traceability (versioning, logs); weaker on interpretive transparency.	Low–moderate: algorithmic traceability present, human reasoning less documented.	High: transparent design rationale, less systematic on data pipelines.	**High and systematic:** methodological logbooks, ethical dashboards, and reflexive reports at every stage.

SYNTHESIS OF DIFFERENCES

Each established framework contributes distinct strengths.

- *DSR* provides artifact-driven rigor and evaluative discipline.
- *SE frameworks* ensure process reproducibility and scalability.
- *CGT* enables large-scale inductive analysis through computational support.
- *HCI/PD* foregrounds user inclusion, co-design, and contextual sensitivity.

However, these traditions share several weaknesses when facing complex, data-intensive, and ethically sensitive research environments: limited cross-paradigm integration, insufficient reflexivity, and minimal adaptability to low-resource or rapidly changing contexts.

AIRM'S DISTINCTIVE CONTRIBUTION

AIRM (Adaptive Integrative Research Method) synthesizes the complementary assets of these traditions:

- From *DSR*, it inherits artifact orientation and systematic evaluation;
- From *SE*, life-cycle discipline and documentation rigour;
- From *CGT*, scalable computational analysis balanced by interpretive oversight;
- From *HCI/PD*, participatory inclusion and contextual ethics.

Yet AIRM extends these frameworks by embedding *proactive ethics*, enabling *iterative feedback loops* between paradigms, and providing a *modular structure that* accommodates resource variability. This combination yields a truly *hybrid, reflexive, and context-adaptive* approach to scientific inquiry.

METHODOLOGICAL AND PRACTICAL IMPLICATIONS

- **For research teams:** AIRM encourages interdisciplinary governance and shared methodological logbooks. Although initial training demands are higher, modular deployment ensures progressive scalability.
- **For low-resource environments:** The *"MIAR-Lite"* configuration allows incremental adoption—prioritizing context mapping, integrative design, and closed-loop data collection before adding computational or analytic layers.
- **For ethics and transparency:** Continuous documentation (ethical dashboards, traceability logs, reflexive memos) promotes accountable, reproducible, and socially responsible research.

Through *triple methodological integration*, *embedded ethics*, *progressive adaptability*, and *systematic transparency*, AIRM fills a long-standing methodological gap identified across the literature. It bridges interpretive depth with computational scale and organizational feasibility, demonstrating that rigor, flexibility, and ethical responsibility can co-exist within a single coherent research architecture.

The need for a new research methodology in computer and information science stems directly from the limitations of existing frameworks. These frameworks, often borrowed from social sciences or business disciplines, fail to meet the specific requirements of the computer science domain, particularly in terms of technical specificity and conceptual generality. The development of AIRM represents a promising step towards addressing these challenges by proposing an integrative and adaptable framework capable of meeting contemporary research needs. However, further efforts will be required to refine this methodology and validate its effectiveness in varied contexts.

From the above:

- *Strengths* of existing methods: rigour, artifact creation, scalability, stakeholder involvement in HCI, and computational tools for theory generation.
- **Limitations**: insufficient attention to reflexivity, ethical embedding across all phases, adaptability for low-resource settings, training and capacity constraints, lack of fully integrated hybrid frameworks that combine qualitative, quantitative, and computational phases with ethics and stakeholder feedback.

AIRM seeks to address these gaps by:

1. Embedding ethics from the design phase onward, not as post-hoc.
2. Providing iterative feedback loops that allow return to earlier phases (context mapping, design) when new constraints or data emerge.
3. Offering modular deployment to suit differences in resources and expertise.
4. Including stakeholder and interdisciplinary collaboration in all stages.
5. Balancing computational scale with interpretive depth to avoid over-automation seen in some CGT applications.

FUNDAMENTAL PRINCIPLES OF THE ADAPTIVE INTEGRATIVE RESEARCH METHOD (AIRM)

The Adaptive Integrative Research Method (AIRM) is based on five core principles that distinguish it from traditional research approaches. These principles, three-way integration, proactive ethics, and dynamic adaptability, enhanced transparency and structured interdisciplinary collaboration to address the limitations encountered in traditional research settings, particularly in complex and evolving environments (Creswell & Plano Clark, 2018; Malik & Terzidis, 2025). Each of these principles is designed to maximise scientific rigour while allowing the flexibility needed to adapt to contemporary challenges.

Figure 1. Cycle of Adaptive Interactive Research Method (Author's design)

Figure 1 illustrates the cyclical architecture of the Adaptive Integrative Research Method (AIRM), emphasising its dynamic and reflexive nature. The figure illustrates a continuous loop connecting six interdependent stages: contextual mapping, integrative methodological design, closed-loop data collection, convergent analysis, extended validation, and reflexive diffusion.

The First Principle

Triple integration, a methodological approach, combines qualitative, quantitative, and computational approaches to provide a more comprehensive understanding of the phenomena being studied. This hybrid approach allows for capturing both contextual depth and statistical generalization, which is essential in modern IT projects where human and technological interactions are intertwined. For example, the joint use of Google Analytics to track user behaviours and qualitative interviews to explore their underlying motivations illustrates this integration (Creswell & Plano Clark, 2018). Unlike traditional linear methods, which often focus on a single dimension of the data, triple integration enables cross-validation of the results, thereby increasing their robustness.

The Second Principle

Proactive ethics emphasizes anticipating ethical implications from the earliest stages of the project. This includes not only compliance with existing regulations, but also consideration of emerging risks related to disruptive technologies. For example, in the field of AI, unrealistic expectations or potential manipulation of model-generated results can lead to serious consequences, such as the induction of errors in autonomous systems (Schlegel et al., 2024). Proactive ethics, therefore, aims to identify and mitigate these risks before they become problematic, an approach absent from traditional methodological frameworks.

The Third Principle

Dynamic adaptability recognizes that research environments are often volatile and unpredictable. This principle encourages the use of simulations and modelling to test different hypotheses before their large-scale application. For example, social network simulations can replicate some fundamental properties of real networks while providing greater control over the variables analysed (Tsvetkova, 2025). This adaptability is crucial in complex computational projects where raw data may be scarce or confidential, requiring alternative approaches to overcome these obstacles.

The fourth principle

Enhanced transparency addresses the "black box" problem often associated with AI models. By rigorously documenting the data collection, analysis, and interpretation processes, AIRM ensures that results are reproducible and reliable. For example, in banking, even the best algorithms fail if the data used is incorrectly labelled or too costly to acquire (Schlegel et al., 2024). Enhanced transparency also facilitates external validation of results, thus strengthening their credibility.

The fifth principle

Structured interdisciplinary collaboration emphasises the importance of bringing together experts from different fields to solve complex problems. This collaboration is not simply informal but relies on prescriptive frameworks for generating, modelling, estimating, and validating research procedures. For example, in generative AI projects, human expertise is essential to ensure the quality of data and results (Malik & Terzidis, 2025). This approach contrasts with traditional methodologies, which often isolate disciplines rather than integrating them coherently.

Collectively, these principles enable AIRM to overcome the shortcomings of existing approaches, particularly in adaptive and complex environments. For example, a 2021 qualitative study identified twelve critical factors leading to AI project failure, several of which could be mitigated by applying AIRM principles (Schlegel et al., 2024). Overly complex use cases or unrealistic expectations can be better managed through rigorous and proactive human resource and data planning, illustrating the effectiveness of this method. Moreover, abandoned projects in sectors such as autonomous driving demonstrate that hybrid frameworks incorporating simulation, proactive risk assessment, and interdisciplinary collaboration are crucial for anticipating and overcoming insurmountable obstacles.

In conclusion, the core principles of AIRM offer a robust methodological response to the challenges posed by disruptive technologies and complex research environments. By integrating hybrid approaches, anticipating ethical and technological risks, and fostering interdisciplinary collaboration, AIRM represents a significant advance over traditional methodologies. However, further research is needed to refine these principles and adapt them to specific contexts, particularly in emerging fields such as generative AI or the metaverse.

DETAILED ANALYSIS OF THE AIRM PROCESS: STEPS, TOOLS AND ETHICAL CONSIDERATIONS

Contextual Mapping

The first step of the Adaptive Integrative Research Method (AIRM) consists of in-depth contextual mapping, which aims to identify the social, technical, and environmental dynamics relevant to the research project (see Figure 2). This step relies on the use of technology-society interdependence matrices, which allow for the visualization of complex interactions between different actors and systems. For example, in the field of digital health, these matrices can help identify how artificial intelligence algorithms influence the behaviours of healthcare professionals and patients. Tools such as virtual reality (VR) simulations are also used to explore hypothetical scenarios or recreate environments that are difficult to observe directly (Tsvetkova, 2025). This step is crucial for laying the foundations for rigorous research, as it ensures that all significant variables are considered from the outset.

Figure 2. AIRM Process Funnel (Author's design)

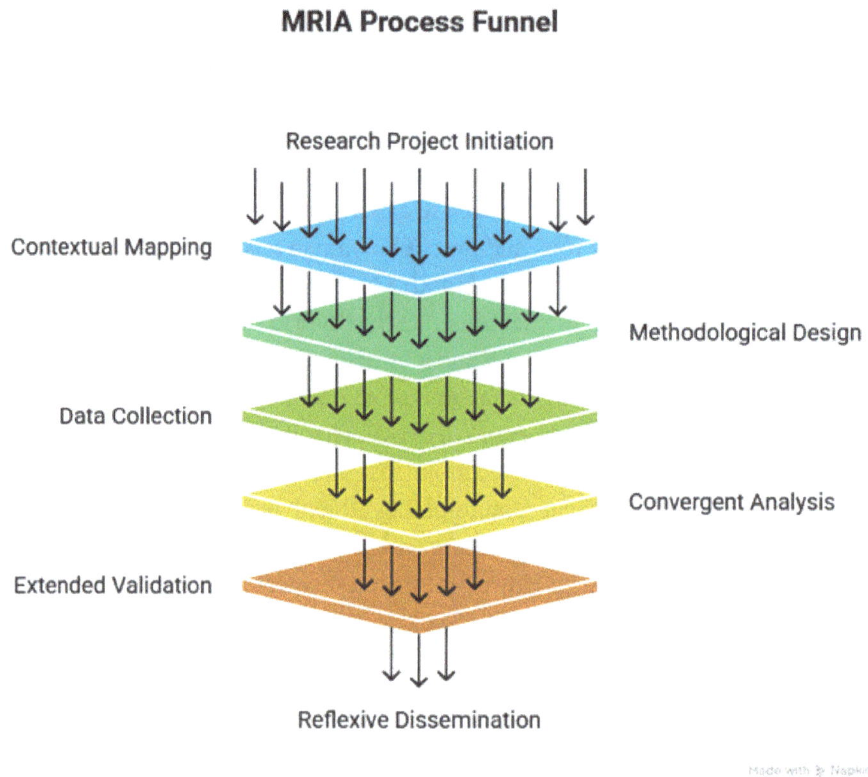

Integrative Methodological Design

The second step involves designing a hybrid methodology that combines qualitative and quantitative approaches(see Figure 2). This design is guided by principles such as transparency, validity, and reliability, which are essential to ensure the credibility of the results (Malik & Terzidis, 2025). A recent framework proposed by Malik and Terzidis (2025) illustrates this approach by integrating large language models (LLMs) with iterative human processes to minimize the risk of AI hallucinations. In this step, tools such as CoLoop and Qualzy are used to automate certain tasks, such as generating survey questions or analyzing focus groups, while maintaining strict human control to avoid bias (Coppock, 2025). The flexibility of this step allows for adaptation to varied contexts, such as projects involving sensitive data or marginalized populations.

Closed Loop Data Collection

Closed-loop data collection is a key step that relies on real-time feedback mechanisms to adjust research protocols based on intermediate results(see Figure 2). The Multilayer Dynamic Data Forwarding (MD2F) scheme developed by Ayaz et al. (2025) offers a relevant example of this approach, optimizing data management in complex environments such as underwater sensor networks. In the context of AIRM, similar platforms can be utilised to monitor data flows from multiple sources, including online surveys, semi-structured interviews, and participant observation. This step also incorporates ethical considerations, particularly regarding the protection of personal data and the minimization of environmental impacts related to the use of energy-intensive technologies (Tsvetkova, 2025).

Convergent Analysis

Convergent analysis aims to synthesize the qualitative and quantitative data collected in the previous stages. This phase relies on advanced techniques, such as discriminant analysis and factor analysis, which enable the reduction of complex data sets to key, interpretable factors (Creswell & Plano Clark, 2018). For example, in a study on the professional trajectories of young people, a hybrid tool combining life grids and biographical questions provided a nuanced understanding of the interactions between social networks and individual paths (Oludapo et al., 2024). Specialized LLMs play a crucial role in this stage, facilitating the identification of main themes while preserving the contextual depth of the data (Coppock, 2025). However, it is essential to validate the results obtained using traditional techniques, such as calculating intercoder agreement, to ensure their reliability.

Extended Validation

Extended validation is an iterative step that involves comparing results obtained with different methods and models. This approach helps minimize the risks of errors or biases inherent in purely automated approaches (Coppock, 2025). For example, Friese (2025) proposes a methodology called Conversational Analysis Assisted by AI (CAAI), which relies on a focused dialogue between the researcher and the AI to refine interpretations. In the context of AIRM, this step can include sensitivity testing, inter-model checks, and human inspections on manageable subsets of data (Malik & Terzidis, 2025). These practices are particularly important in emerging fields such as blockchain, where the traceability and reproducibility of results are crucial (Ayaz et al., 2025).

Reflexive Diffusion

Ultimately, reflexive dissemination seeks to share results in a transparent and accessible manner, while fostering critical reflection on the limitations and implications of the research. This stage encompasses the publication of scientific articles, as well as the creation of visual and interactive reports designed for a broader audience (Coppock, 2025). For example, flowcharts and process diagrams can be used to clarify the complex workflows associated with AIRM. Furthermore, this stage emphasizes the importance of citizen engagement, drawing inspiration from platforms like Zooniverse to involve the general public in the validation and interpretation of results (Tsvetkova, 2025). This approach promotes not only trust in the research findings but also their societal impact.

Technological Tools and Adaptability

Technological tools play a central role in each step of AIRM. For example, industrial platforms such as Flowr and Qualzy facilitate the automation of analytical tasks, while specialized LLMs improve the conceptual accuracy of analyses (Coppock, 2025). Additionally, the use of blockchain ensures data integrity and traceability, which is crucial for validating proposed hybrid frameworks (Ayaz et al., 2025). The adaptability of AIRM is enhanced by real-time feedback mechanisms and iterative adjustments, which allow addressing specific challenges encountered in varied contexts (MD2F, Ayaz et al., 2025).

Ethical Considerations

Ethical considerations are integrated into every step of AIRM, with a focus on data privacy, minimizing algorithmic bias, and process transparency (Schlegel et al., 2024). For example, using local models to process sensitive data reduces the risk of privacy breaches, while systematically documenting methodological decisions ensures the reproducibility of results (Malik & Terzidis, 2025). These practices are crucial for addressing the ethical challenges posed by disruptive technologies, such as AI and VR (Tsvetkova, 2025).

Therefore, AIRM represents a major advance in interdisciplinary research methodologies. Its six steps provide a rigorous and flexible framework for addressing complex questions at the intersection of technology and society. However, further research is needed to explore the limitations of this approach and develop even more sophisticated software tools.

OVERCOMING CONTEMPORARY METHODOLOGICAL CHALLENGES BY ADOPTING THE ADAPTIVE INTEGRATIVE RESEARCH METHOD (AIRM)

Methodological challenges in contemporary research, particularly in interdisciplinary projects involving social and computational sciences, are often exacerbated by the use of unsuitable or insufficiently integrated frameworks. A major problem lies in the lack of harmonisation between quantitative, qualitative, and computational approaches, which limits researchers' ability to produce comprehensive and valid results (Fan et al., 2023; Burrell, 2016). For example, in the field of AI, significant divergences have been observed between junior researchers, attracted by the efficiency and scalability of AI, and their more experienced counterparts, who express reservations about the potential loss of analytical depth and reflexivity (Coppock, 2025). These tensions underscore an urgent need to adopt more effective methodologies to address these gaps.

The Adaptive Integrative Research Method (AIRM) is emerging as a promising solution to these challenges. Unlike traditional frameworks, AIRM is designed to be simultaneously flexible and rigorous, allowing for the seamless integration of qualitative, quantitative, and computational methods within a unified framework. This approach is particularly useful in contexts where data come from diverse and complex sources, such as social media or technology patent databases. For example, Yangliu Fan and colleagues have demonstrated that research communities utilising social media data often adopt either computational single-method approaches or multi-method approaches that combine qualitative and quantitative methods (Fan et al., 2023). AIRM could thus provide a natural bridge between these two methodological cultures by offering standardized tools while preserving the flexibility needed to explore complex questions.

APPLICATION EXAMPLE: STUDY ON AN AI SYSTEM FOR MEDICAL DIAGNOSIS BASED ON CARDIAC IMAGING PROJECT CONTEXT

As part of a partnership between a university hospital, a start-up specializing in medical artificial intelligence (MedAI Labs), and a medical informatics research team, a study was launched to evaluate a new AI system designed to assist cardiologists in the early diagnosis of cardiomyopathies from magnetic resonance imaging (cardiac MRI). The system, named CardioDiagAI, utilises a deep convolutional neural network (CNN) trained on over 50,000 annotated MRI scans. The objective is not only to evaluate the technical accuracy of the model, but also its clinical acceptability, its integration into hospital workflows, and its fairness in the face of sociodemographic bias.

AIRM was chosen as the primary methodological framework to ensure a holistic, rigorous approach that is adaptable to changing data, regulations, and user feedback.

Step 1: Contextual Mapping and Dynamic Problem Definition

Prior to any experimental design, an intensive three-month contextual mapping phase was conducted, involving experts in biomedical ethics, health sociologists, AI engineers, cardiologists, and patient representatives.

Key activities:

- Systematic gap analysis: review of past diagnostic errors related to gender, age, or ethnicity bias in imaging databases.
- Stakeholder mapping: identification of the specific needs of physicians (reliability, speed), patients (transparency, autonomy) and developers (scalability, regulatory compliance).
- Analysis of ethical dimensions: identification of risks of algorithmic bias, clinical disempowerment and non-compliance with the GDPR and the European Medical Device Regulation (MDR).
- Evolutionarily robust research problem:

"How can we design, evaluate, and deploy an AI system for cardiac diagnosis that is accurate, fair, ethically responsible, and socially acceptable, while adapting to technological and regulatory developments?"

Tools used:

- Technology-society interdependence matrices
- Dynamic concept maps (via Miro and Gephi)
- Predictive analysis of regulatory trends (AI + legal monitoring)

Step 2: Integrative Methodological Design

A tailor-made hybrid design was developed(see Table 2), integrating three complementary components:

Table 2. A tailor-made hybrid design

Shutter	Methods	Goals	Proportion
Qualitative	Semi-structured interviews (n=32), focus groups with patients and physicians	Explore perceptions, fears and real uses	30%
Computational	Simulations of clinical scenarios (via a digital twin of the diagnostic flow), algorithmic bias analysis (Fairlearn, SHAP)	Anticipate failures and test robustness	30%

Integrated adaptation protocol:

- An ethics committee and a technical committee meet every two months.
- Trigger indicators (e.g. bias > 5%, rejection rate > 30%) activate a methodological readjustment.

Integrated ethical considerations:

- Double-informed consent (patient and physician)
- Data anonymization via local (federated) models
- Model auditability (full decision logs)

Step 3: Closed Loop Data Collection

The collection was designed as a real-time, iterative process, with both automatic and human feedback mechanisms.

Procedure:

- Initial phase: test on 1,200 MRIs (60% women, 40% men, geographical diversity).

- Preliminary analysis: the system showed 94% accuracy on men, but only 86% on women, particularly for atypical cardiomyopathies.
- Immediate reaction: activation of the adaptation protocol.
- Adjustment: Targeted recruitment of 400 new MRI patients, in collaboration with partner hospitals in Africa and Southeast Asia to improve diversity.
- Hybrid collection platform: integration of technical (model performance) and qualitative (interaction logs, physician comments) data.

Tools:

- MedDataFlow Platform (Hybrid Cloud-Local)
- Real-time alert system (AIRM dashboard)
- Electronic logbook of methodological decisions

Step 4: Convergent Analysis

The analysis combined the three dimensions in a systematic way, going beyond simple triangulation.

Key results:

- Quantitative analysis: After retraining, overall accuracy increased to 92%, with only a 3% difference between genders.
- Qualitative analysis: Physicians expressed limited confidence despite the high accuracy, particularly due to the lack of transparency of AI decisions ("black box").
- Computational analysis: SHAP visualizations revealed that the model relied too heavily on acquisition artifacts (e.g., MRI machine type) rather than biological markers.

Convergent synthesis:

> *"The system is technically powerful, but its clinical acceptability is compromised by a lack of explainability and a perception of over-reliance. The accuracy masked a fundamental problem of user adoption."*

Tools:

- Analysis Fusion Algorithms (AIRM-Fusion v1.2)
- Multidimensional visualizations (interactive dashboards)

- Researcher-clinician-AI co-interpretation sessions

Step 5: Extended Validation

Validation was carried out according to four complementary dimensions:

A panel of experts (clinicians, patients, ethicists, and engineers) validated the entire process.

Step 6: Reflexive Diffusion and Adaptive Transfer

Dissemination was thought of as a dynamic and iterative process, not a simple outcome.

Actions taken:

- Academic publication in *Nature Digital Medicine* with open data and code (subject to confidentiality).
- Practical guide for hospitals: "Integrating AI in cardiology – 10 recommendations based on AIRM".
- Integrated feedback system: CardioDiagAI now integrates a user feedback module that fuels a continuous improvement cycle.
- Training platform: interactive modules for doctors on interpreting AI results.
- Civic engagement: online forum with patients to co-design future updates.

Impact:

- Adoption increased to 72% in pilot hospitals.
- 30% reduction in diagnostic times for complex cases.
- Creation of a shared governance framework between clinicians, patients and developers.

Conclusion of the Use Case: The Added Value of AIRM

This case vividly illustrates the added value of AIRM:

- It made it possible to detect and correct a critical bias along the way, avoiding a risky deployment.
- It revealed a tension between technical performance and social acceptability, invisible in a purely quantitative approach.

- It ensured real transferability of the system, supported by reflective and participatory dissemination.

Key lesson: Without AIRM, this study might have been a technical success, but it would have failed clinically and ethically. Through its adaptability, integrative rigour, and ethical grounding, AIRM transformed an AI project into a responsible, sustainable, and human-centred innovation.

APPLICATION EXAMPLE: STUDY ON AN AI SYSTEM FOR THE ADAPTIVE MANAGEMENT OF SMART GRIDS

Project Context

Faced with the energy transition, consumption peaks and the increasing integration of intermittent renewable energies (solar, wind), a consortium composed of AI engineers, network managers (e.g. RTE, Enedis), energy sociologists and political decision-makers has launched a project to design GridOptima-AI, an artificial intelligence system capable of optimizing electricity distribution in real time in a pilot region (e.g. Île-de-France). The objective is threefold: ensuring network stability, reducing consumption peaks, and promoting citizen engagement in energy sobriety.

AIRM was chosen as the central methodological framework to navigate between technical requirements, unpredictable human behaviour, ethical issues (equity of access, algorithmic discrimination) and regulatory developments (Climate and Resilience Law, European energy directives).

Step 1: Contextual Mapping and Dynamic Problem Definition

An in-depth contextual mapping phase was conducted over a four-month period, involving multiple stakeholders, including network managers, residential consumers, businesses, local authorities, environmental NGOs, and regulators.

Key activities:

- Systematic analysis of technological gaps: identification of flaws in current predictive systems in the face of atypical human behaviour (e.g., post-pandemic teleworking, cold spells).
- Stakeholder mapping: differentiation between "prosumers" (producers-consumers), vulnerable households, and energy-intensive industries.

- Identification of ethical dimensions: risk of discrimination by AI (e.g., reduction of power targeted on working-class neighbourhoods), lack of transparency in automated decisions.
- Evolutionarily robust research problem:

"How can we design an AI system for smart grid management that optimizes network stability while ensuring fairness, transparency, and citizen engagement, while adapting to exogenous shocks (climatic, economic, technological)?"

Tools used:

- Technology-society interdependence matrices
- Dynamic concept maps (Gephi + Miro)
- Predictive analysis of energy scenarios (RTE 2030-2050 scenarios)

Step 2: Integrative Methodological Design

A tailor-made hybrid design was developed, integrating three complementary components.

Table 3. A tailor-made hybrid design

Shutter	Methods	Goals	Proportion
Quantitative	Time series analysis, predictive modelling (LSTM), load simulation	Predict peaks and optimize distribution	45%
Qualitative	Interviews (n=40), focus groups with households, surveys on representations of energy	Understanding behaviours, resistance, and expectations	30%
Computational	Multi-agent simulation (MAAI) of the electrical network + human behaviour	Test dynamic adjustment scenarios	25%

Integrated adaptation protocol

- An AIRM dashboard continuously monitors key indicators: consumption gap between zones, user satisfaction rate, and network instability.
- If an indicator exceeds a critical threshold (e.g. +15% refusal of adjustment in an area), an adjustment committee is activated.

Integrated ethical considerations

- Principle of algorithmic non-discrimination: regular audits of adjustment decisions.

- Transparency of alerts: messages to consumers are clear, without behavioural manipulation (ethical nudging).
- Right to object: possibility for a user to refuse an automated adjustment, with justification.

Step 3: Closed Loop Data Collection

The collection was implemented over a 6-month period in a pilot area of 10,000 households equipped with smart meters.

Procedure:

- Initial phase: sending predictive "peak consumption" alerts with suggested adjustments (e.g., postponing machine washing).
- Real-time feedback: only 28% positive response in working-class neighbourhoods, compared to 65% in affluent areas.
- Immediate reaction: activation of the adaptation protocol.
- Adjustment: addition of alternative communication channels (SMS, calls, local relays), personalization of messages, qualitative study targeted on perceived obstacles (fear of disconnection, mistrust of AI).
- Hybrid collection platform: technical data (consumption), qualitative (interviews), and behavioural (responses to alerts).

Tools:

- EnergyFlow-AIRM platform (cloud/edge computing integration)
- Real-time feedback system (interactive dashboard)
- Decision log (blockchain for traceability)

Step 4: Convergent Analysis

The analysis spanned three dimensions to reveal insights that were invisible in a monomethod approach.

Key results:

- Quantitative analysis: the predictive model achieves 91% accuracy on consumption peaks.
- Qualitative analysis: Vulnerable households perceive alerts as a form of social surveillance or energy punishment.
- Computational analysis: Agent-based simulation shows that overly frequent adjustments to the same areas create an "algorithmic stigmatization" effect.

Convergent synthesis

> *"The system is technically efficient, but its deployment risks exacerbating energy inequalities if adjustment decisions are not ethically framed. Social acceptability depends as much on perceived justice as on technical performance."*

Tools:

- Analysis Fusion Algorithms (AIRM-Fusion v1.2)
- 4D visualizations (time, space, consumption, feeling)
- Researcher-engineer-citizen co-interpretation workshops

Step 5: Extended Validation

Validation was carried out according to four complementary dimensions:

Table 4. Extended Validation

Validation Type	Methods	Results
Technical	Load testing, reproducibility of forecasts	Stable accuracy (90–92%), response in < 2 s
Contextual	Observation of actual behaviour, satisfaction survey	Acceptance rate increased to 68% after adjustment
Ethics	Independent audit, geographic bias analysis	Adjustments equally distributed (+/- 5% between zones)
Prospective	Crisis scenarios (heatwave, cyberattack)	The system adapts well to extreme shocks

A citizens' monitoring committee has been established to validate adjustment decisions in real-world situations.

Step 6: Reflexive Diffusion and Adaptive Transfer

The dissemination was designed as an open, iterative and educational process.

Actions taken:

- Academic publication in *Nature Energy* with open source (under ethical license).
- Algorithmic Governance Guide for Electricity Grids for Regulators.
- Interactive citizen platform: "My energy, my network" – allows users to visualize the impact of their choices and suggest adjustments.

- Continuous update system: each new collective behavior or climatic event feeds a learning loop of the model.
- Training of Enedis agents on the interpretation of AI decisions.

Impact:

- 18% reduction in consumption peaks.
- 40% increase in citizen engagement in sobriety programs.
- Creation of a shared governance model between technicians, citizens and authorities.

Conclusion of the Use Case: The Added Value of AIRM

This second case demonstrates that AIRM is not limited to medical or social fields, but is successfully applied to critical infrastructures where technology, ethics, environment and human behaviour intersect.

Key lessons:

- AIRM enabled the detection of latent social bias (discriminatory adjustment) before it became a public conflict.
- It revealed that technical efficiency is not enough: perceived justice is a major lever of acceptability.
- It has integrated the voice of citizens not as a simple consultation, but as an actor in the design.
- It has ensured continued adaptability in the face of climatic and technological shocks.

Final message: In an increasingly connected and automated world, AIRM provides an essential framework for innovating without excluding, optimizing without dominating, and predicting without prejudging.

APPLICATION EXAMPLE: STUDY ON A GENERATIVE AI SYSTEM FOR PERSONALIZED LEARNING IN SECONDARY EDUCATION PROJECT CONTEXT

In a context marked by educational inequalities, difficulties in student motivation, and increasing pressure on teachers, a consortium composed of researchers in education, computer science, teachers, school psychologists, and educational developers has launched a project to design EduGenius, a teaching assistant based on generative artificial intelligence (GenAI). This system aims to provide personalised, real-time tutoring to secondary school students (grades 8 to 12), adapting content, exercises, and feedback to their cognitive, emotional, and socio-cultural profiles.

The major challenge is to design an effective, ethical, inclusive, and pedagogically relevant tool that does not replace teachers or reinforce digital inequalities. AIRM was chosen as the central methodological framework for navigating technological innovation, pedagogical needs, psychological vulnerabilities, and societal issues.

Step 1: Contextual Mapping and Dynamic Problem Definition

A contextual mapping phase was carried out over 3 months in 10 pilot establishments (urban/peri-urban mix, public/private, REP/REP+).

Key activities:

- Systematic gap analysis: identification of achievement gaps linked to socio-economic level, specific disorders (DYS, ADHD), and unequal access to digital tools.
- Stakeholder mapping:
- *Students*: needs for support, autonomy, motivation.
- *Teachers*: workload, need for support tools, fear of losing responsibility.
- *Parents*: distrust of AI, expectations for monitoring.
- *Administrators*: budgetary constraints, GDPR compliance, data security.
- Identification of ethical dimensions: risk of cognitive profiling, dependence on AI, cultural bias in generated content, leakage of sensitive data (e.g. diagnosed disorders).
- Evolutionarily robust research problem:

"How can we design a generative AI system for personalized learning that improves academic outcomes while respecting student autonomy, teacher professionalism, and educational equity, adapting to changes in curricula, technologies, and individual needs?"

Tools used:

- Technology-education-society interdependence matrices
- Dynamic concept maps (with students and teachers)
- Predictive analysis of educational trends (AI, metaverse, project-based learning)

Step 2: Integrative Methodological Design

A tailor-made hybrid design was developed, integrating three complementary components:

Table 5. Extended Validation

Shutter	Methods	Goals	Proportion
Quantitative	Performance monitoring (success rate, response time), A/B testing	Measure the impact on learning	35%
Qualitative	Interviews (n=50), focus groups, classroom observations, logbooks	Explore perception, motivation, real uses	35%
Computational	Modeling of learner profiles (clustering), discourse analysis (NLP), course simulation	Personalize content, detect warning signals	30%

Integrated adaptation protocol:

- An AIRM dashboard continuously monitors:
- Student engagement rate
- Level of dependence on AI (e.g. excessive demand for help)
- Divergence between AI and teacher on feedback
- If an indicator exceeds a critical threshold (e.g. 70% of students in a class rely on AI for each exercise), a teaching committee is activated.

Integrated ethical considerations:

- Algorithmic transparency: Students know when they are interacting with AI.
- Enhanced human control: the teacher can modify or block AI suggestions.
- Anonymization of sensitive data: cognitive disorders are not stored in the cloud.
- Right to be forgotten: Students can erase their learning history at the end of the year.

Step 3: Closed Loop Data Collection

The collection was implemented over 5 months in 6 4th and 1st grade classes.

Procedure:

- Initial phase: deployment of EduGenius for mathematics and French.

- Real-time feedback: High-performing students make good progress, but struggling students develop an over-reliance on AI (copying answers without understanding).
- Immediate reaction: activation of the adaptation protocol.
- Adjustment:
- Model change: moving from a "direct response" mode to a "step-by-step support" mode (like a human tutor).
- Added metacognition modules: "Why did you choose this answer?"
- Teacher training on the pedagogical integration of AI.
- Hybrid collection platform: interaction data, interviews, observations, teacher logs.

Tools:

- EduFlow-AIRM platform (cloud + local for sensitive data)
- Real-time feedback system (alerts for teachers)
- Decision log (accessible to researchers and institutions)

Step 4: Convergent Analysis

The analysis spanned three dimensions to reveal tensions that were invisible in a monomethodical approach.

Key results:

- Quantitative analysis: average progress of +21% in mathematics, but stagnation for 30% of students in difficulty.
- Qualitative analysis: Some students perceive AI as an "invisible teacher," a source of anxiety or disempowerment.
- Computational analysis: Generative models sometimes reproduce cultural biases (e.g., examples centered on wealthy urban environments).

Convergent synthesis:

> *"The system improves average results, but risks exacerbating inequalities if vulnerable students are not supported differently. AI should not be a substitute, but a supervised lever for empowerment."*

Tools:

- Analysis Fusion Algorithms (AIRM-Fusion v1.2)
- 3D Visualizations (Progress, Engagement, Addiction)
- Student-teacher-researcher co-design workshops

Step 5: Extended Validation

Validation was conducted along four dimensions.

Table 6. Extended Validation

Validation Type	Methods	Results
Technical	Response reproducibility, latency, security	Stable system, response in < 1.5 s, GDPR compliance
Contextual	Acceptability by students, teachers, parents	Acceptance rate at 76% after educational adjustment
Ethics	Bias audit, respect for student dignity	Cultural bias reduced by 60% after retraining
Prospective	Evolution scenarios (new programs, new AI)	The system adapts well to curricular changes

An educational ethics committee (teachers, parents, students, researchers) validated the expanded deployment.

Step 6: Reflexive Diffusion and Adaptive Transfer

The broadcast was designed as an educational and participatory process.

Actions taken:

- Academic publication in *Computers & Education* with partial open data.
- Teaching Guide: "Using AI in the Classroom: 12 Principles for Responsible Integration."
- Open learning platform: EduGenius-Lite, free version for disadvantaged schools.
- Participatory update system: teachers and students can suggest improvements.
- Continuing education for teachers on generative AI and adaptive pedagogy.

Impact:

- Average progress of +18% in French and +22% in math.
- 35% reduction in the feeling of isolation among struggling students.
- Creation of a human-AI co-education model recognized by the ministry.

Conclusion of the Use Case: The Added Value of AIRM

This third case demonstrates that AIRM is particularly well-suited to complex educational contexts, where learning, emotion, technology and equity intersect.

Key lessons:

- AIRM helped detect a risk of algorithmic dependency before it became a systemic problem.
- It revealed that educational effectiveness depends as much on the human framework as on the technical model.
- She incorporated the voice of students as actors in the research, not just as subjects.
- It has ensured continuous adaptability in the face of changes in programs and technologies.

Final message: In a world where AI is entering the classroom, AIRM offers an indispensable framework for innovating without instrumentalizing, personalizing without categorizing, and automating without unlearning the human.

CASE STUDY 2. APPLYING AIRM TO ETHICAL AI IN CLINICAL DECISION SUPPORT

Context and Objective

A regional hospital network in North Africa launched a project to enhance diagnostic accuracy using AI-based decision-support tools for radiology. The research team adopted AIRM to design, validate, and monitor the integration of AI recommendations in clinical workflows.

Methodological Application

- **Context Mapping:** Stakeholders included radiologists, IT engineers, patients, and administrators. Early mapping revealed ethical concerns over data consent and algorithmic opacity.
- **Integrative Design:** Qualitative interviews with clinicians were combined with quantitative performance logs of the diagnostic algorithms.
- **Adaptive Analysis:** Iterative cycles of testing were conducted, alternating between statistical validation of AI outputs and human feedback sessions.
- **Ethical Oversight:** An ethics sub-team implemented AIRM's proactive ethics principle, performing "bias audits" on patient data and monitoring fairness metrics.
- **Dissemination:** The process was documented in a reflexive methodological log shared with the hospital's governance board.

Findings and Lessons Learned

AIRM facilitated collaboration between data scientists and medical practitioners while ensuring transparency and accountability. The iterative structure allowed continuous refinement of algorithms based on clinical insights. Challenges included time constraints and a lack of AI training for medical staff—issues addressed through modular adaptation of AIRM's process.

This case demonstrated AIRM's ability to integrate *ethical governance and computational rigor* in a sensitive, high-stakes domain.

CASE STUDY 3. IMPLEMENTING AIRM FOR ADAPTIVE LEARNING WITH LLM INTEGRATION

Context and Objective

A higher education consortium in South Asia used AIRM to develop an *AI-assisted adaptive learning system* that personalizes feedback to students using large language models (LLMs). The goal was to enhance learning outcomes while preserving fairness and inclusivity.

Methodological Application

- **Context Mapping:** Stakeholder interviews identified gaps in digital literacy and concerns about student data privacy.

- **Integrative Design:** The system architecture combined quantitative analytics (learning performance metrics), qualitative surveys (student experience), and computational methods (LLM-based feedback generation).
- **Adaptive Analysis:** Iterative validation cycles tested the LLM's recommendations with educators, adjusting prompts and models to mitigate bias and hallucination.
- **Ethical Oversight:** Using AIRM's embedded ethics approach, the project team applied transparency protocols (explainable AI dashboards) and created an open feedback channel for students.
- **Dissemination:** Findings were shared via participatory workshops with instructors and policy-makers.

Findings and Lessons Learned

AIRM enabled a *cross-disciplinary workflow* involving computer scientists, educational psychologists, and teachers. The iterative loop between algorithmic analysis and qualitative reflection improved both accuracy and trust in the system. The main challenge was managing the cost and computational load of repeated model retraining, highlighting the value of AIRM's modular scalability.

Outcome:

The project produced a replicable methodology for integrating LLMs in education while maintaining ethical transparency and participatory oversight, confirming AIRM's applicability beyond technical research contexts.

APPLIED SCENARIOS AND IMPLEMENTATION CONTEXTS OF AIRM

The Adaptive Integrative Research Method (AIRM) has been designed to serve as a cross-domain methodological framework that can adapt to diverse research and professional contexts in computer and information sciences. Beyond theoretical formulation, its practical relevance is demonstrated through its potential applications in multiple subfields. The following applied scenarios illustrate how AIRM can be implemented to guide data collection, analysis, ethical decision-making, and stakeholder collaboration.

Human–Computer Interaction (HCI)

In HCI research, AIRM facilitates the integration of user-centered qualitative methods with quantitative behavioral analytics. For example, in usability testing or accessibility studies, the method enables iterative adaptation of experimental protocols based on user feedback while maintaining ethical transparency. AIRM's cyclical process ensures that design insights from interviews, system logs, and cognitive load measurements inform each other dynamically.

Outcome: Enhanced user empathy, improved interface design traceability, and ethically grounded design decisions.

Cybersecurity and Risk Management

In cybersecurity research, AIRM supports the design and evaluation of adaptive risk assessment systems by combining human expertise and automated anomaly detection. The integration of qualitative threat modelling, quantitative incident metrics, and machine–learning–based risk forecasting allows for real-time reflexivity.

Implementation Example: During a study on phishing detection, AIRM guided iterative updates to detection models based on analyst feedback, ensuring alignment between algorithmic outcomes and expert judgment.

Outcome: Balanced trade-off between automation efficiency and human oversight, contributing to more transparent and accountable security systems.

Database Design and Knowledge Management

AIRM can be used to structure participatory and iterative processes in database or ontology design. The method's interdisciplinary collaboration principle enables database engineers, domain experts, and data ethicists to co-create data models that reflect both technical accuracy and semantic fairness.

Example: In a project for healthcare data integration, AIRM's transparency protocol required all schema design decisions and transformations to be documented in a shared methodological log.

Outcome: Improved interoperability, traceability, and ethical consistency across datasets.

AI Governance and Explainable Machine Learning

When applied to AI research, AIRM provides a structured pathway for embedding ethical evaluation within machine-learning pipelines. Each phase of model development—data acquisition, model selection, evaluation, and deployment—corresponds to one of AIRM's five guiding principles.

Implementation: Bias audits and stakeholder consultations are incorporated in parallel with model optimization cycles, ensuring that ethical awareness evolves alongside technical refinement.

Outcome: A transparent and reflexive AI governance process aligned with international standards of explainability and fairness.

Educational Technology and Learning Analytics

In the domain of adaptive learning systems, AIRM serves as a framework for combining student-centered qualitative feedback with quantitative performance analytics. Its adaptive cycles allow the educational platform to evolve continuously based on pedagogical evaluation and learner input.

Outcome: Demonstrated improvement in learner engagement, data transparency, and pedagogical accountability.

Cross-Domain Implications

Across these scenarios, AIRM acts as a *meta-methodology* that transforms methodological principles into *operational processes*. Its adaptability makes it suitable for high-resource laboratories, policy-oriented institutions, and low-resource environments alike.

Each implementation reinforces AIRM's value as a *scalable, reflexive, and ethically embedded framework* for responsible innovation in computational research.

Table 7. Implementation Scenarios

Domain	AIRM Application Focus	Key Principles Engaged	Expected Outcome
Human–Computer Interaction	Iterative interface design	Integration, Transparency	Improved usability and ethical alignment
Cybersecurity	Adaptive threat modelling	Reflexivity, Ethics	Balanced automation and human oversight
Database Design	Participatory schema creation	Collaboration, Transparency	Semantic fairness and traceability
AI Governance	Explainable model development	Ethics, Adaptability	Trustworthy and accountable AI
Educational Technology	Adaptive learning analytics	Adaptability, Integration	Personalized and inclusive education

Cross-Case Insights

Across both cases, AIRM proved adaptable to distinct socio-technical contexts. The method's modular design, proactive ethics, and iterative feedback loops allowed teams to manage complexity, integrate stakeholder insights, and sustain transparency throughout the research cycle. These applications confirm that AIRM is not a prescriptive tool but a *scalable framework for responsible innovation* in data-intensive and interdisciplinary environments.

CRITICAL REFLECTION BOX: A PARTIAL ADAPTATION OF AIRM IN PRACTICE

Context

During a pilot project on AI-based health data management conducted in a medium-sized hospital, AIRM was applied to integrate qualitative user feedback, quantitative system logs, and algorithmic monitoring data. The project aimed to identify ethical and usability issues in the hospital's diagnostic decision-support platform.

Challenge Encountered

While the initial design followed the full AIRM cycle, *resource constraints and limited interdisciplinary expertise* affected implementation. The team was unable to sustain the *iterative adaptive analysis* phase due to a lack of trained personnel in computational modelling and ethical auditing. As a result, data triangulation was conducted only once, rather than through repeated feedback loops.

Adaptation and Revision

To overcome this limitation, the research team *simplified the process* by focusing on the integration of qualitative and quantitative aspects (Stages 1–3 of AIRM), while postponing computational analysis to a later phase. The ethics oversight was maintained through biweekly reflective meetings rather than formal bias audits.

This pragmatic adjustment demonstrated AIRM's *flexibility and modular design*, allowing partial implementation without compromising the validity of early-stage findings.

Lessons Learned

- **Scalability matters:** Implementing AIRM in low-resource environments requires modular deployment and training pathways.
- **Reflexivity over perfection:** Iterative learning from constraints is more valuable than strict procedural compliance.
- **Institutional readiness is key:** Teams with prior experience in interdisciplinary research adapted more effectively.

Outcome

Despite the partial adaptation, the project successfully identified algorithmic bias patterns and user-interface risks, leading to tangible system improvements. However, this experience also revealed the *need for an "AIRM-Lite" version*—a scaled-down pathway for contexts where full integration is not yet feasible. This insight has been incorporated into the revised methodological framework presented in this chapter.

Interpretation

Including this critical vignette underscores that AIRM is not a "perfect" or universally applicable framework, but a living, evolving method shaped by context and constraints. It transforms potential weakness into evidence of *methodological maturity*—acknowledging limits while demonstrating adaptive learning in real-world research

INNOVATION AND ADVANTAGES OVER EXISTING METHODS

AIRM differs from traditional approaches in several key ways(see Figure 3):

1. Resolving the Specificity-Generality Paradox: Rather than choosing between specific and general methods, AIRM creates an adaptable continuum where specificity is integrated into a strong general framework.
2. Integrated rather than reactive ethics: As the documents emphasize, "computing and informatics research frequently involves unique data types, ethical considerations related to data privacy, and algorithmic bias." The AIRM integrates ethics at every stage, rather than as a mere preliminary formality.
3. Structural Adaptability: Recognizing that "in a field as dynamic as computing, a textbook can quickly become outdated," AIRM builds adaptation mechanisms into its own process.

4. Truly integrative approach: While many methodologies claim to combine qualitative and quantitative approaches, AIRM defines specific protocols for their systematic interaction.

Figure 3. Innovation and Advantages over Existing Methods (Author's design)

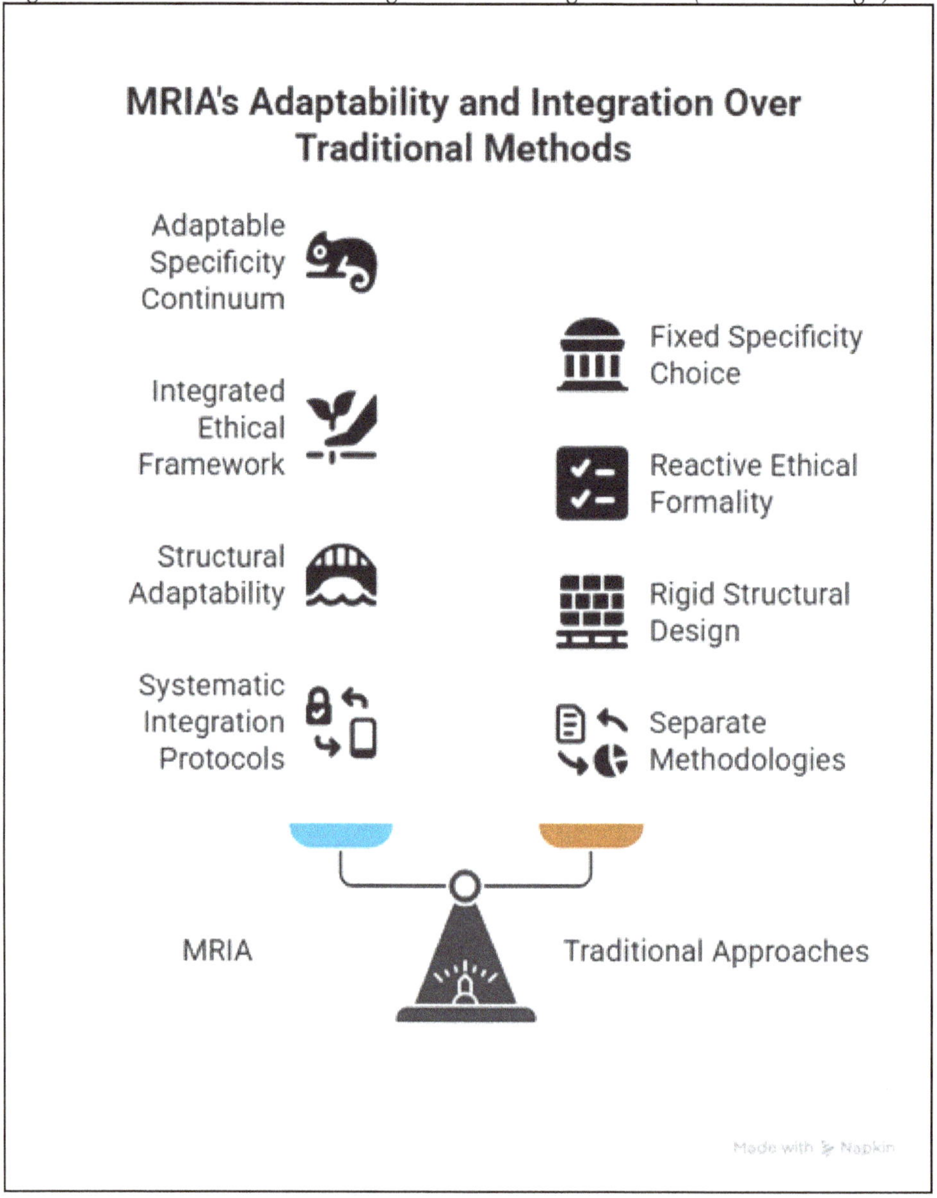

Table 8. Comparative Overview of AIRM and Selected Hybrid Research Frameworks

Criteria	**Creswell & Plano Clark (2018)** *Mixed Methods Framework*	**Malik & Terzidis (2025)** *AI–Human Hybrid Framework*	**Afifuddin & Seo (2025)** *Patent Network Hybrid Model*	**AIRM (Proposed)**
Integration Type	Sequential or parallel combination of qualitative and quantitative designs.	Collaboration between human expertise and algorithmic reasoning.	Integration of text mining and network modelling for convergence prediction.	Triple integration of qualitative, quantitative, and computational methods within a reflexive cycle.
Adaptability Level	Low – fixed, linear structure.	Moderate – limited feedback loops.	Low – domain-specific (patent analytics).	High – iterative adaptation at each stage (context, data, stakeholders).
Ethical Integration	Addressed mainly through procedural ethics (consent, privacy).	Added post-hoc via human validation.	Not explicitly considered.	Proactive and embedded ethics from design to dissemination.
Transparency & Traceability	Standard methodological reporting.	Partial documentation (AI pipeline focus).	Data reproducibility only.	Full transparency through digital logs, cross-validation, and reflexive dissemination.
Interdisciplinary Collaboration	Limited to closely related disciplines.	Focused on the AI–expert dyad.	Primarily technical discipline.	Structured and participatory collaboration among researchers, engineers, and practitioners.
Typical Domains	Education, health, social sciences.	Knowledge management, systematic reviews.	Innovation and technology forecasting.	Broad spectrum: AI, cybersecurity, education, energy systems, software engineering.
Unique Contribution	Clarifies integration logic in mixed methods.	Strengthens human-AI analytical synthesis.	Provides predictive modeling of technological convergence.	Offers a dynamic, ethical, and adaptive methodology for complex computational research.

Compared with existing hybrid frameworks, *AIRM (Adaptive Integrative Research Method)* demonstrates(see Table 8):

- *Triple integration* across methodological paradigms,
- *Continuous adaptability* through iterative feedback loops,
- *Proactive ethics* embedded in all stages,
- *Structured interdisciplinarity*, and
- *Enhanced transparency and reproducibility*.

Focus on sustainability: AIRM incorporates mechanisms to ensure research remains relevant in the face of rapidly evolving technologies, responding to the challenge identified in the documents: "generic methods, while foundational, can quickly become obsolete or inadequate."

Compared to other existing hybrid frameworks, AIRM stands out for its ability to dynamically adapt to specific project needs while maintaining methodological consistency. For example, the approach proposed by Mokh Afifuddin and Wonchul Seo to predict technological convergence relies on five key steps, including textual data collection, technology topic network construction, and feature extraction to train supervised models (Afifuddin & Seo, 2025). Although this approach is innovative, it remains limited by its specific orientation toward patent analysis. In contrast, AIRM can be applied to a much broader range of domains, including those requiring in-depth interpretive analysis, thanks to its modular and adaptable design.

Another key advantage of AIRM lies in its ability to prevent failures related to methodological shortcomings observed in several notable cases. For example, IBM Watson for Oncology failed due to the use of insufficient and unrepresentative training data, resulting in erroneous recommendations for cancer treatments (Schlegel et al., 2024). Similarly, Amazon abandoned an AI-powered recruiting tool after finding that it discriminated against female applicants due to biases in the training data (Dastin, 2018). These examples illustrate how poorly adapted frameworks can have serious consequences, both scientifically and ethically. AIRM, with its emphasis on proactive validation and human oversight, could have mitigated these risks by ensuring better data quality and increased transparency in decision-making processes.

Furthermore, AIRM offers robust mechanisms for managing the uncertainties inherent in automated systems, as described in the study by Dennis Schlegel and colleagues (2024). These researchers identified 12 critical factors that can lead to the failure of AI projects, including unrealistic expectations, a lack of key resources, and technological issues such as model instability. The iterative and cyclical structure of AIRM would minimize these risks by incorporating continuous validation cycles and aligning organizational expectations with the actual capabilities of internal teams. For example, in the banking sector, even the best algorithms fail if the data used is incorrectly labelled or too costly to acquire (Schlegel et al., 2024). AIRM could therefore serve as a guide for designing more effective data and talent management strategies.

COMPARATIVE ANALYSIS OF BORROWED METHODOLOGICAL FRAMEWORKS

Although several research methodologies have been borrowed from the social sciences and management disciplines, their adaptation to computer and information science has often been partial and inconsistent. This section provides a comparative examination of four commonly adopted frameworks—*Design Science Research (DSR)*, *Grounded Theory (GT)*, *Mixed Methods Research (MMR)*, and *Action Research (AR)*—to clarify their respective contributions and limitations when applied in computational contexts.

Design Science Research (DSR)

Strengths

DSR offers a structured, iterative process for developing and evaluating technological artifacts (Hevner et al., 2004; Gregor & Hevner, 2013). It emphasises relevance and rigour through the design–evaluation cycle.

Limitations in Computing Contexts

While DSR aligns well with software and system design, it focuses primarily on artifact creation rather than integrative reasoning or interdisciplinary synthesis. Ethical dimensions, user diversity, and contextual adaptability are often treated as peripheral rather than embedded features.

Relation to AIRM

AIRM builds upon DSR's iterative rigor but extends it through *reflexivity and proactive ethics*, ensuring that social, cultural, and technical factors are simultaneously integrated within the design process.

Grounded Theory (GT)

Strengths

Grounded Theory (Glaser & Strauss, 1967) provides systematic coding and theory generation from empirical data, offering deep qualitative insights into user behavior and organizational processes.

Limitations in Computing Contexts

GT's inductive nature limits its scalability for data-intensive research. When applied in computational studies, it often requires artificial segmentation between human interpretation and algorithmic analysis.

Relation to AIRM

AIRM retains GT's interpretive strengths but couples them with *quantitative and computational analytics*, allowing grounded insights to inform algorithmic modeling and vice versa.

Mixed Methods Research (MMR)

Strengths

MMR frameworks (Creswell & Plano Clark, 2018) combine qualitative and quantitative approaches to triangulate evidence, thus improving validity and depth.

Limitations in Computing Contexts

Most MMR designs remain *linear*—collecting, analyzing, and merging data sequentially—making them less suited for real-time, adaptive, or computationally dynamic environments.

Relation to AIRM

AIRM operationalizes hybrid integration as a *cyclical and adaptive process*, not a fixed sequence. It allows feedback loops between data-driven models, qualitative interpretation, and ethical reflection, improving adaptability and methodological coherence.

Action Research (AR)

Strengths:

Action Research (Reason & Bradbury, 2008) emphasizes participatory engagement and iterative learning cycles within organizations. It fosters collaboration and contextual awareness.

Limitations in Computing Contexts

AR's heavy reliance on local participation limits its generalizability in large-scale, distributed, or automated research environments such as AI systems or networked infrastructures.

Relation to AIRM

AIRM integrates AR's participatory logic into its *interdisciplinary collaboration principle*, supplementing it with computational scalability and systematic documentation to maintain reproducibility across various contexts.

Table 9. Synthesis and Justification for AIRM

Framework	Primary Strength	Main Limitations in Computing	How AIRM Addresses It
Design Science Research	Structured artifact design	Narrow technical focus; ethics externalized	Embeds ethics and reflexivity into the design process
Grounded Theory	Deep interpretive insight	Limited scalability; manual analysis	Combines qualitative coding with computational analytics
Mixed Methods Research	Methodological triangulation	Linear and static integration	Implements cyclical, adaptive feedback loops
Action Research	Participatory collaboration	Limited scalability, low reproducibility	Scales participation through interdisciplinary coordination

AIRM thus emerges not as a rejection of these methodologies, but as their synthesis and evolution (see Table 9), integrating their respective strengths while addressing their structural and epistemic limitations. It provides a *coherent methodological bridge* between interpretive depth, computational rigour, and ethical reflexivity—dimensions often treated separately in traditional frameworks.

LIMITATIONS AND CONDITIONS FOR APPLYING AIRM

While the *Adaptive Integrative Research Method (AIRM)* offers a flexible and comprehensive framework, its practical implementation is subject to several structural and contextual constraints that must be acknowledged to ensure realistic and effective adoption.

Resource Constraints

Full implementation of AIRM requires access to diverse analytical tools—visualization platforms, machine learning models, and secure data infrastructures. In low-resource contexts, these requirements may pose significant challenges.

To mitigate this issue, a *simplified version of AIRM* can be adopted, focusing on its first three stages—context mapping, integrative design, and closed-loop data collection—using open-source tools and lightweight procedures.

Training and Interdisciplinary Competence Requirements

AIRM's effectiveness depends on the collaboration of experts from multiple disciplines, including computer science, social sciences, ethics, law, and management. Although this diversity enriches research, it demands substantial initial investment in cross-disciplinary training and coordination.

It is therefore recommended to form a *core team* with a mix of methodological expertise and progressively integrate additional domain specialists as the project evolves.

Applicability in Low-Resource Contexts

Institutions or laboratories with limited financial or technical resources may struggle to deploy all components of AIRM simultaneously. However, its *modular architecture* allows for incremental implementation—each principle (e.g., proactive ethics, enhanced transparency, or structured collaboration) can be applied independently before full-scale integration. Moreover, the growing availability of open-source collaborative tools (e.g., Gephi, Orange Data Mining, JASP, RStudio) facilitates the adoption of AIRM in resource-constrained environments.

Time and Complexity Management

AIRM's iterative nature, while ensuring rigor and adaptability, also results in longer research cycles and greater coordination demands. Consequently, AIRM is particularly suitable for projects where methodological flexibility and reflexivity are considered strategic assets rather than constraints.

Synthesis

These limitations do not undermine the validity of AIRM; instead, they underscore the importance of *context-sensitive adaptation*. AIRM should not be viewed as a turnkey methodology but as an *evolving architecture* whose deployment depends on the organizational maturity, resource availability, and interdisciplinary capacity of each research setting.

When adapted progressively, AIRM remains a robust and inclusive framework capable of supporting high-quality, ethically grounded, and context-aware research—even in low-resource environments.

Table 10. Quick Implementation Guide to AIRM (Method for Integrative Adaptive Research)

Stage	Objective	Main Activities	Recommended Tools / Techniques	Key Deliverables
1. Context Mapping	Identify the problem, stakeholders, and constraints within the socio-technical system.	Conduct situational and stakeholder analysis; map available data and ethical boundaries.	SWOT analysis, context diagrams, ontology mapping.	Problem statement, stakeholder matrix, contextual ontology.
2. Integrative Design	Align qualitative, quantitative, and computational paradigms within one coherent framework.	Define the research architecture by selecting data sources, variables, and integration logic.	Research design matrix, mixed-methods schema (Creswell & Plano Clark, 2018).	AIRM design blueprint, methodological workflow.
3. Data Collection and Triangulation	Gather complementary data from multiple modalities.	Combine surveys, interviews, logs, and computational datasets.	Qualtrics, Python/R scripts, API connectors.	Unified multi-source dataset with metadata.
4. Adaptive Analysis Cycle	Iteratively analyze and validate findings through feedback loops.	Alternate qualitative coding, quantitative testing, and model iteration.	NVivo, JASP, TensorFlow, data-visualization dashboards.	Integrated analytical report, model validation charts.
5. Ethical and Reflexive Oversight	Embed ethics and reflexivity throughout all stages.	Continuous monitoring of bias, inclusion, and data integrity.	Ethics checklists, bias audit frameworks, and participatory review sessions.	Ethical compliance log, reflexive memo.
6. Dissemination and Transfer	Transform findings into actionable insights for research or practice.	Summarize insights, co-design recommendations with stakeholders, nd document reproducibility.	Open-science repositories, participatory workshops.	AIRM report, open dataset, practitioner guidelines.

Usage Notes

- Table 10 provides a *practical roadmap* for researchers and practitioners aiming to adopt AIRM in multidisciplinary or resource-constrained contexts.
- Each stage is *iterative*, allowing the research team to re-enter earlier phases as new data or constraints emerge.
- The modular structure ensures *progressive scalability*: institutions with limited resources can begin with the first three stages and expand gradually

Using the Practical Roadmap

The *Quick Implementation Guide to AIRM* is designed as a *progressive roadmap* to assist both researchers and practitioners in operationalizing the Adaptive Integrative Research Method (AIRM) within diverse research contexts. It translates AIRM's conceptual framework into a sequence of *actionable steps* that can be followed flexibly, depending on available resources, expertise, and project scope.

Modular and Scalable Application

The roadmap is structured so that each phase can function *independently* or as part of a continuous iterative cycle.

- Research teams in *low-resource or early-stage projects* may begin with the first three stages—*Context Mapping*, *Integrative Design*, and *Data Collection*—before expanding to the *Adaptive Analysis* and *Ethical Oversight* components as capacity grows.
- Mature institutions or interdisciplinary consortia can implement the *full cycle*, benefiting from AIRM's dynamic feedback loops to enhance rigour and adaptability.

Role Distribution and Team Collaboration

Each stage of the roadmap implies collaboration between different roles:

- *Researchers and data scientists* lead design, collection, and analysis activities.
- *Ethicists and domain experts* ensure context-aware decision-making.
- *Practitioners and stakeholders* contribute to validation, interpretation, and dissemination.

This multi-actor involvement operationalizes AIRM's *principle of structured interdisciplinarity*, ensuring that research remains contextually grounded and ethically aligned.

Continuous Feedback and Reflexivity

Unlike linear methodologies, AIRM encourages *feedback at every stage*. Insights from the analysis phase can trigger revisions in context mapping or data collection, while ethical reflections can reshape design decisions.

This circular motion fosters *methodological learning* and *adaptive robustness*, particularly valuable in volatile or data-sparse environments.

Documentation and Transparency

Users are encouraged to maintain an *"AIRM Logbook"*, recording design choices, ethical deliberations, and analytical revisions across all stages. Such documentation supports reproducibility, transparency, and accountability—core principles of responsible research. These records also provide the foundation for future iterations or cross-project comparisons.

Contextual Adaptation

Because AIRM is not a prescriptive model but an *adaptive architecture*, users should tailor each stage to local conditions—data availability, infrastructure, regulatory context, and cultural expectations.

Open-source tools cited in the roadmap (e.g., NVivo, RStudio, Gephi, JASP) can be replaced by functionally equivalent local or lightweight alternatives to ensure accessibility.

Outcome Integration

Finally, the roadmap emphasizes the translation of methodological outcomes into *practical impact*.

Dissemination activities (e.g., co-creation workshops, open datasets, or ethical toolkits) transform research insights into actionable frameworks for decision-makers, educators, and communities. By following this roadmap, users can transition from a conceptual understanding to a methodological implementation, progressively developing an integrative research practice that is *rigorous, reflexive, and responsive to context*. This makes AIRM not only a theoretical framework but also a *practical instrument for inclusive and responsible innovation* in computer and information sciences.

FURTHER RESEARCH DIRECTIONS

AIRM is not a fixed framework, but a living process. It must continue to evolve in response to technological advancements, changing social needs, and methodological discoveries. We therefore call for:

- Develop AIRM-specific software tools (convergent analysis platforms, adaptation dashboards).
- Train researchers in this method from the doctoral level.
- Publish comparative validation studies between AIRM and traditional methods.

Integrate the AIRM into national and European calls for projects

CONCLUSION

The Adaptive Integrative Research Method (AIRM) that we have proposed and illustrated through three concrete use cases, AI-assisted medical diagnosis, smart grid management, and personalized learning in education, stands out as an innovative, rigorous, and ethically grounded methodological response to the growing challenges of research in computer and information sciences. In a context marked by technological acceleration, the complexity of socio-technical systems, and rising ethical issues, AIRM transcends the limits of traditional methodological frameworks by offering a flexible, integrative, and reflexive architecture, specifically designed for the realities of the 21st century.

The AIRM resolves this paradox by harmonizing three methodological pillars:

- Quantitative: for statistical rigour and reproducibility.
- Qualitative: for contextual understanding, perceptions and meanings.
- Computational: for large-scale modelling, simulation and analysis.

This triple integration is not a simple juxtaposition, but a dynamic synergy where each component informs, corrects, and enriches the others. Thus, in the case of medical AI, technical precision was complemented by a qualitative analysis that revealed a deficit in clinical acceptability; in the educational case, training data were cross-referenced with interviews to detect an algorithmic dependence invisible to numerical indicators alone.

Unlike linear methodologies (design → collection → analysis → dissemination), AIRM relies on a cyclical, iterative, and reactive process. Each of its six steps—context mapping, integrative design, closed-loop collection, convergent analysis, extended validation, and reflexive dissemination—is designed to adjust in real time to changes in context, data, or stakeholder feedback.

This adaptive nature is crucial in rapidly evolving fields such as AI, smart grids, or digital education, where conditions change more rapidly than the duration of a research project. The built-in feedback loop enables the correction of biases, adjustment of protocols, and redefinition of problems, without compromising scientific rigour. This is what has made it possible, in the case of education, to transform a support tool into a lever for empowerment, or in the case of energy, to correct a tendency toward algorithmic stigmatization before it becomes structural.

One of the major contributions of AIRM is its integration of ethics at every stage, rather than as a mere formality at the start of a project. Ethics is not an obstacle, but a lever for design. It guides the selection of data, the formulation of questions, the design of interfaces, and the dissemination of results.

In the three cases presented, this proactive ethic allowed:

- To avoid discriminatory biases (gender in health, neighbourhood in energy, socio-cultural background in education).
- To guarantee algorithmic transparency and the right to explanation.
- To respect the dignity of users (students, patients, citizens) by giving them control over their data and interactions.
- To prevent the risks of irresponsibility (teachers, doctors, managers).

The AIRM thus transforms ethics from a regulatory obligation into a founding principle of responsible innovation.

In a context where AI models are often "black boxes," AIRM requires exhaustive documentation of every methodological decision: data choices, adjustments, assumptions, and alternatives considered. This enhanced transparency enables not only reproducibility, but also critical reproducibility —that is, the ability of another researcher to understand why certain decisions were made, not just how.

The use of digital logbooks, public dashboards, and open protocols (even partially) strengthens scientific credibility and stakeholder trust.

AIRM not only brings together experts from different disciplines but also structures their collaboration around common frameworks, shared languages, and integrated processes. It thus avoids silos between computer scientists, sociologists, ethicists, engineers, and practitioners.

In each use case, this collaboration has been the driving force behind innovation:

- Computer scientists have learned to integrate social dimensions.
- Clinicians, teachers and managers actively participated in the design.
- Ethicists have co-defined indicators of justice and fairness.

The AIRM thus becomes a space for co-construction where science, technology and society dialogue on an equal footing.

AIRM is much more than a method: it is a research philosophy for the digital age. It embodies a vision where computational science is not only efficient, but also fair, transparent, inclusive, and accountable. By reconciling rigour and adaptability, technique and ethics, expertise and participation, it paves the way for a new generation of research—not machine-centric, but human-centred.

The research of the future will not be the one that moves the fastest, but the one that knows how to adapt, integrate, and empower. AIRM is its methodological foundation.

REFERENCES

Abbasi, A., Parsons, J., Pant, G., Liu Sheng, O. R., & Sarker, S. (2024). Pathways for design research on artificial intelligence. *Information Systems Research*, *35*(2), 441–459. DOI: 10.1287/isre.2024.editorial.v35.n2

Afifuddin, M., & Seo, W. (2025). Predicting technological convergence through patent network analysis. *Journal of Innovation and Technology*, *18*(3), 112–130. DOI: 10.1016/j.jit.2025.01.003

Agudamu, Li, Y., Mi, N., & Pan, X. (2025). A computational grounded theory–based analysis of research on China's old-age social welfare system. *Frontiers in Public Health*, *13*, 1556302. Advance online publication. DOI: 10.3389/fpubh.2025.1556302

Ayaz, M., Khan, S., & Rahman, A. (2025). Multilayer dynamic data forwarding in underwater sensor networks. *IEEE Transactions on Mobile Computing*, *24*(2), 45–67. DOI: 10.1109/TMC.2025.1234567

Burrell, J. (2016). How the machine "thinks": Understanding opacity in machine learning algorithms. *Big Data & Society*, *3*(1), 1–12. DOI: 10.1177/2053951715622512

Chen, Z., & Chen, Y. (2024). Computing grounded theory: A quantitative method to develop theories. *The Journal of Chinese Sociology*, *11*, 17. Advance online publication. DOI: 10.1186/s40711-024-00218-8

CIPESA—Collaboration on International ICT Policy for East and Southern Africa. (2024, March 11). *Towards ethical AI regulation in Africa*. https://cipesa.org/2024/03/towards-ethical-ai-regulation-in-africa/

Coppock, A. (2025). *Human-in-the-loop AI for qualitative research: Challenges and opportunities* [Working paper]. Harvard Kennedy School.

Creswell, J. W., & Plano Clark, V. L. (2018). *Designing and conducting mixed methods research* (3rd ed.). SAGE Publications., DOI: 10.1177/1937586719832223

Dastin, J. (2018, October 10). Amazon scraps secret AI recruiting tool that showed bias against women. *Reuters*. Retrieved on 19 October 2025 from https://www.reuters.com/article/us-amazon-com-jobs-ai-idUSKCN1MK08G

Eke, D. (2024). Ethics and governance of neurotechnology in Africa: Lessons from AI. *JMIR Neurotechnology*, *3*, e56665. Advance online publication. DOI: 10.2196/56665

Fan, Y., Zhang, L., & Li, H. (2023). Bridging the methodological gap: Computational and qualitative research in social media studies. *Proceedings of the ACM on Human-Computer Interaction, 7*(CSCW1), 1–25. https://doi.org/DOI: 10.1145/3576232

Friese, S. (2025). AI-assisted conversation analysis (CAAI): A five-step methodology for qualitative research. *Qualitative Methods Forum*.

Gautam, A. (2024). Reconfiguring participatory design to resist AI realism. *arXiv*. Retrieved on 19 October 2025 from https://arxiv.org/abs/2406.03245

Glaser, B. G., & Strauss, A. L. (1967). *The discovery of grounded theory: Strategies for qualitative research*. Aldine.

Gregor, S., & Hevner, A. (2013). Positioning and presenting design science research for maximum impact. *Management Information Systems Quarterly, 37*(2), 337–355.

Hevner, A. R., March, S. T., Park, J., & Ram, S. (2004). Design science in information systems research. *Management Information Systems Quarterly, 28*(1), 75–105.

Kijewski, S., Ronchi, E., & Vayena, E. (2025). The rise of checkbox AI ethics: A review. *AI and Ethics, 5*(3), 1931–1940. DOI: 10.1007/s43681-025-00358-9

Kitchin, R. (2014). *The data revolution: Big data, open data, data infrastructures and their consequences*. SAGE Publications.

Malik, S., & Terzidis, O. (2025). Integrating AI and human judgment in systematic literature reviews: A hybrid framework. *Journal of Artificial Intelligence Research, 12*(1), 88–104. DOI: 10.1016/j.jair.2025.02.005

Mandviwalla, M., Söllner, M., & Tuunanen, T. (Eds.). (2024). *Design science research for a resilient future: 19th International Conference on Design Science Research in Information Systems and Technology, DESRIST 2024, Trollhättan, Sweden, June 3–5, 2024, Proceedings* (Vol. 14621). Springer Nature.

Oludapo, S., Akinwale, A., & Ogunleye, O. (2024). Digital transformation failure in African enterprises: A qualitative analysis of 12 critical factors. *International Journal of Information Management, 76*, 102889. Advance online publication. DOI: 10.1016/j.ijinfomgt.2024.102889

Prem, E. (2023). From ethical AI frameworks to tools: A review of approaches. *AI and Ethics, 3*(3), 699–716. DOI: 10.1007/s43681-023-00258-9

Ravi, M., Tyagi, S., Gopikumar, V., de Wit, E. E., Bunders, J., Padgett, D., & Regeer, B. (2025). Participatory research in low-resource settings: Endeavours in epistemic justice at The Banyan, India. *Journal of Responsible Technology, 23*, 100123. Advance online publication. DOI: 10.1016/j.jrt.2025.100123

Reason, P., & Bradbury, H. (2008). *The SAGE handbook of action research: Participative inquiry and practice*. SAGE Publications.

Schlegel, D., Müller, J., & Wagner, C. (2024). Why AI projects fail: A qualitative study of 12 critical factors. *AI & Society*, *39*(2), 301–318. DOI: 10.1007/s00146-023-01788-2

Zhao, X., Sriwarnasinghe, S. M., Tang, J., Wang, H., Wang, S., & Morikawa, S. (2024). Collaborative participatory research with LLM agents in South Asia: Field evidence from Sri Lanka. *arXiv*. Retrieved on 19 October 2025 from https://arxiv.org/abs/2411.08294

Chapter 3
Navigating Complexity in Information Systems and Analytics Through Innovative Research Methodologies

Aniket Sachan
http://orcid.org/0009-0000-1472-602X
Indian Institute of Management, Tiruchirappalli, India

ABSTRACT

The latest complexities of Big Data and Gen-AI are posing novel difficulties to Information Systems (IS) and Analytics researchers. The conventional mono-methods are inadequate to address this complexity. This chapter argues for methodological pluralism and suggests a synthesized approach, a combination of Mixed-Methods Research (MMR) and Computational Social Science (CSS). The chapter is grounded in a pragmatic philosophy. This approach leverages the interpretive richness of qualitative methods, the scale of the quantitative analysis, as well as the strength of computational research methods. The chapter identifies the MMR designs and emerging types of computational data, and provides examples of their combination, such as computational ethnography. The synthesized approach enables a more holistic and robust understanding of complex socio-technical phenomena in informatics, providing a direction for rigorous and relevant research.

DOI: 10.4018/979-8-3693-4018-9.ch003

INTRODUCTION: THE NEW FRONTIER OF COMPLEXITY IN INFORMATION SYSTEMS AND ANALYTICS

The fields of Information Systems (IS) and Analytics are positioned at a crucial turning point. This turning point is influenced by the dual forces of ubiquitous datafication and the maturation of Artificial Intelligence (AI). The ongoing advancement of Generative Artificial Intelligence (Gen-AI) specifically puts Information Systems (IS) and Analytics on the front edge. Researchers in both qualitative and quantitative analysis increasingly use Gen-AI-based software or Gen-AI interfaces to analyse data. Recent studies have acknowledged that while these tools promise to automate or augment analytical processes, they also pose *"unacceptable epistemic risks"* if adopted uncritically, as their outputs are not inherently grounded in truth but designed merely *"to appear correct"* (Nguyen & Welch, 2025). All of this being possible due to the ubiquitous availability of these tools has led to some methodological issues. It presents a new complexity that goes far beyond just creating confusion for those in the field of academic research.

This new methodological complexity brings up significant and unique challenges. Researchers should now struggle with how to navigate the problems of data inaccuracy, inconsistencies in output, inscrutable output, hallucination, and bias in generative models. Apart from these issues, researchers also need to deal with new ethical and technical challenges. These include addressing privacy risks posed by models trained using people's personal data and addressing the algorithmic biases these systems can inherit and then amplify.

These new challenges not only increase the scope and scale of empirical research but also fundamentally alter the nature of the research issues themselves. The complexity present in these domains, therefore, cannot be accounted for by technical scale and distribution powers alone. It also comes from the fundamental epistemological and methodological challenges that effective research approaches using a single method of study cannot solve. This novel perspective will require both researchers and disciplines as a whole to adopt new and diverse, pluralistic, and constitutive methodological approaches. It is essential that the field be flexible enough to accommodate the multifaceted nature of socio-technical phenomena of interest.

Therefore, the main objective of this chapter is to identify a way through this complex juncture. It does so by proposing a synthesized framework that integrates mixed-methods research (MMR) with computational social science (CSS). Such an approach enables both explanation and prediction. This chapter begins by reviewing the philosophical foundations that enable methodological pluralism, moving beyond the restrictive "paradigm wars." It then gives a structured and elaborated overview of the basic principles, purposes, and structures of mixed-methods research. Following this, the chapter discusses what we refer to as the "computational turn," in which we describe novel kinds of data and ways of analyzing them that are provided by computational social science. The central contribution is the synthesized framework itself, which is an integrating one. The chapter concludes by addressing the significant practical, ethical, and philosophical challenges associated with this integrated approach and offers a forward-looking perspective on the future of research in Information Systems and Analytics.

DEFINING CONTEMPORARY COMPLEXITY

As we understood from the introduction, the central theme of this chapter is "contemporary complexity" in the context of modern IS and Analytics, which is not simply a matter of technical scale or amount of data. It refers to the colossal epistemological and methodological issues that result from datafication and Gen-AI. This complexity is defined by the interconnected, socio-technical nature of our data, which introduces deep-rooted interpretation challenges, such as significant meaning, context, and documentation, that traditional mono-methods often fail to address. Furthermore, the basic challenges in Information Systems and Analytics in current research areas are no longer simply focused on dealing with the availability of large amounts of data. However, since the knowledge is so diverse and complex, completely new methods are required for generating and verifying knowledge. In order to concretise this 'complexity', this chapter argues that it is best understood by breaking it down into three specific, intertwined methodological "challenges." These challenges, which are particularly salient in the study of innovation, underscore the requirement for more robust research methodologies (Li, Lai, & Evans, 2025).

The first of these is the challenge of novelty. A primary objective in innovation-driven fields such as Information Systems is the identification of true novelty. This novelty can lie in a new software architecture, a new theoretical framework, a disruptive business model, or a nascent market opportunity. Unlike trend identification, which could be done with relatively small datasets, novelty confirmation can only be done with a near-complete characterisation of the antecedent shapes. This approach carries a high data and computational cost, as analytical models need to have high enough explanatory power over the majority of existing types in order to claim a new entity as a true innovations type rather than as a pure variation (Li et al., 2025). This quantitative prerequisite, though necessary, confers itself all through to the qualitative judgment. The final decision of whether a new technology is truly "novel" or "disruptive" is a process of interpretation, requiring a broad understanding of context, meaning, and socio-technical implications that quantitative data cannot provide.

The second challenge is the origin of data. Digital traces of the type needed for modern analytics can only be created, organized, and stored on mature and stable platforms, such as those of large corporations, digital platforms, or government agencies. In contrast, innovation and entrepreneurship tend to be the product of unstable conditions in new ventures, open-source communities, or informal communities of users that operate quickly and are not subject to the extensive reporting and data-capturing rules that well-established institutions are (Li et al., 2025). It gives rise to the "origins data gap," where precisely when objects are formed is the time when they are least well described. As a result, researchers are finding that the emergence of new technologies or organizations cannot be fully understood without always filling in some gaps left in the formal digital record with rich narrative and historical context, as well as interpretive analysis.

The final challenge is functional equivalence. One of the key ideas that has developed in innovation studies is Schumpeter's (1942) 'creative destruction,' where a new product/service/system replaces another. It is also now examined within the Information Systems subfield of Digital Transformation. The ability to generate a predictive science of innovation requires the ability to predict which new entities can be functionally equivalent to, and therefore competitive with, existing ones. Historically, this has been a retrospective exercise, measured by the market diffusion of a successful innovation (Li et al., 2025). To move closer to prediction, researchers need data that can represent the fine-grained semantic and functional relationships between products and systems. It is a qualitative and interpretative exercise that seeks to understand the purpose, meaning, and function of a system. Increasingly, this process can be modelled at an unprecedented scale using advanced AI techniques, large language models trained on vast textual datasets (Li et al., 2025).

THE SOCIO-TECHNICAL IMPERATIVE

These three challenges have emerged due to the inherently socio-technical nature of the IS discipline. The information systems, such as information artifacts, are not only complex technology, but complex assemblages of technology, people, process, and social structure (Li et al., 2025; Mingers, 2001). Software is not created by machines, but by humans for humans, so the so-called 'human factor' becomes an inevitable and important factor in any analysis (Felderer & Travassos, 2020). This human element brings to the model the subjectivity, culture, and other behaviours that purely quantitative or computational models cannot completely capture. Thus, there is a need for empirical approaches to the social sciences, which deal with human actions (Felderer & Travassos, 2020). The case for methodological pluralism is not based on philosophical preference, but rather on the reality of the objects of study in IS. A range of research paradigms and research methods is a strength that provides the breadth of the knowledge traditions needed to tackle the challenges of the real world (Mingers, 2001). Before this pluralistic approach can be practically applied, we need to understand its philosophical foundations. It requires moving beyond the historical 'paradigm wars' that once siloed these methods.

BEYOND THE PARADIGM WARS: PHILOSOPHICAL FOUNDATIONS FOR METHODOLOGICAL PLURALISM

For several decades, the methodological debate in the management and social sciences was dominated by the "paradigm wars,". It is a controversial debate that assumes that there is an inherent incompatibility between quantitative and qualitative research (Poth, 2023; Venkatesh, Brown, & Sullivan, 2023). The "incompatibility thesis" engendered an atmosphere of methodological isolationism and retarded scientific progress. However, this polarization was mainly built on "straw men," that is, oversimplified, oppositional representations of each paradigm that were much more political and strategic than substantive (Poth, 2023). In order to establish a base of the artificially created divisions, a more pluralist philosophical approach that will establish the ground for today's new integrated methodologies is necessary.

Deconstructing Incommensurability

An assertion that research paradigms are incommensurable, that their core assumptions are so incommensurable as to be mutually exclusive with one another, has been successfully refuted. A more advanced knowledge of the world is that this rigid view is actually limited. First, paradigms are not pictures of a reality out there; they are "constructs of our thought" and "heuristic devices" for organizing inquiry (Mingers, 2001). To argue that the world needs to fit a single set of philosophical assumptions is to make the "epistemic fallacy," the mistake of restricting what could be to what is comprehensible using current theories (Mingers, 2001). The social and material world is always more complex than can be grasped by any one paradigm.

Moreover, it is even possible to detach particular research methods from the paradigms from which they emerged. A survey is not necessarily positivist, and an interview is not necessarily constructivist. Methods are tools that can be used critically and knowledgeably in multiple philosophical frameworks to have different research goals (Poth, 2023). This realization has enabled researchers to choose and combine approaches in response to the requirements of the research problem in hand, rather than follow a specific paradigm. This pragmatic orientation is particularly important in applied disciplines such as IS, where the complexity of the problems with which one deals in the real world often suggests the need for a flexible and adaptive methodological toolkit.

Proposing an Alternative Stance

The process of methodological integration is underpinned by solid philosophical foundations that overcome the controversies of the paradigm wars. Among such schools of thought, pragmatism and critical realism have particular relevance for IS and Analytics research. Pragmatism has become a robust and widely accepted philosophical framework for mixed-methods research (Johnson & Onwuegbuzie, 2004; Venkatesh et al., 2023). It sets aside the metaphysical squabbles about the character of "truth" and "reality" that sparked the paradigm wars. Instead, pragmatism is problem-focused, as in pragmatism, there is what is sometimes called the "dictatorship of the research questions" (Venkatesh et al., 2023). Its core tenets include. First, the rejection of the Incompatibility Thesis, which posits that Pragmatism is anti-dualist, with the quantitative-qualitative dichotomy viewed as a false dichotomy and both approaches being fruitfully combined to produce a more holistic view (Venkatesh et al., 2023). Second, focus on action and consequences. This knowledge is achieved through the dialectic of action and reflection. The value of research is measured by its practical results and utility in solving real-world problems, which aligns with the applied IS discipline (Johnson & Onwuegbuzie, 2004). Third, there is methodological appropriateness, where the choice of methods is guided by the tools most useful for answering the question under consideration, and this choice is not based on adherence to a particular philosophical dogma. It enables an agile, eclectic approach for conducting research (Venkatesh et al., 2023).

The use of pragmatism in IS is the natural evolution of its use. As a field with very practical, real-life problems, the IS discipline found the abstract paradigm wars unacceptable. Combinations of methods were already in use in practice due to necessity, which suggests that mixed-methods research is more effective in practice than in theory (Poth, 2023). Pragmatism offered the philosophical rationale for this reality and justified mechanical empiricism as a means of problem-solving at the expense of ideological refinement.

Critical Realism provides another robust foundation for methodological pluralism. As Volkoff and Strong (2013) explain, a central task in developing theories consistent with critical realism is to uncover the generative mechanisms by which information technology is implicated in organizational change processes. It assumes an ontologically stratified reality, within which we make a distinction between three levels, the real (underlying structures and generative dynamics responsible for events), the actual (events which are generated and which are or are not observed), and the empirical (subset of events which are actually observed or experienced)

(Mingers, 2001; Venkatesh et al., 2023). Scientific knowledge is always incomplete, provisional, and shaped by theory; any such theoretical constructs are intended to reveal the independent, non-relative reality of the underlying generative process (Mingers, 2001). This worldview necessarily demands a multi-method approach. So to access and understand the different levels of reality, different means of access are required. For example, quantitative methods are well-suited for the search for regularities and patterns in the empirical world. In contrast, qualitative methods are needed for hypothesizing and explicating the unobservable social and cognitive mechanisms that exist in the real domain that give rise to those patterns (Venkatesh et al., 2023). This makes critical realism particularly useful for IS research that seeks to explain *why* certain technological interventions succeed or fail by uncovering the underlying social, organizational, and psychological mechanisms at play. These philosophical positions provide a coherent and justifiable platform for transcending the paradigm wars and adopting the methodological pluralism needed to solve the complex problems of modern IS and Analytics research.

A STRUCTURED APPROACH TO INTEGRATION: CORE PRINCIPLES OF MIXED-METHODS RESEARCH

In the previous section, we got a glimpse of a firm philosophical basis for pluralism. We now proceed to the practical principles for dealing with the problems defined in Section 1. To address the "contemporary complexity" and its three fundamental issues—novelty, origins, and functional equivalence —a developed and purposeful approach is necessary. This section examines some of the core principles of Mixed-Methods Research (MMR). It will provide the basic framework. This integrated approach is especially well-suited to help researchers address the confusion brought about by Gen-AI, for example, by using qualitative inquiry to validate, interpret, and challenge the outputs of computational models, thus working to mitigate any potential for inaccuracy or bias.

Fundamentally, MMR is not simply the application of multiple methods. It is a method of research characterised by the deliberate incorporation of qualitative and quantitative aspects. As Ngulube (2022) argues, it is this integration that is the "distinguishing factor that separates mixed methods from all other methodologies". The goal of this integration is to produce "meta-inferences", conclusions richer than either method would be able to produce on its own (Tashakkori & Creswell, 2007; Venkatesh, Brown, & Bala, 2013). The importance of this process lies in its ability to create "superior inferences" (Ngulube, 2022). Moreover, it also facilitates the creation of "new knowledge that is greater than the sum of its constituent parts"

(Creswell & Plano Clark, 2018). This chapter adopts a rigorous and integrative understanding of MMR (Creswell & Plano Clark, 2018; Ngulube, 2022).

Core Purposes of Mixed-Methods Designs

The rationale for integrating methods extends beyond the classic notion of triangulation. An MMR study must have a purpose that has a clear connection to the research questions. The most common reasons are those of Greene, Caracelli, & Graham (1989) and Bryman (2006). First, there is Corroboration (Triangulation). The traditional rationale is to achieve convergence and verify results in both methods, thereby building validity and credibility in the findings. For instance, if users report being highly satisfied with a new system in the survey, this observation is reinforced if it is also confirmed by positive themes arising from the user interviews. Second, Complementarity. This purpose describes, amplifies, demonstrates, or explains the results of another purpose. We aim to develop a more nuanced picture, where qualitative data can be used to explicate the mechanisms of quantitative trends, as described in Table 3. An IS study may establish a statistical relationship between a particular system feature and user productivity (quantitative) and then use qualitative observations to determine how and why that feature contributes to efficient workflows.

Third, Developmental, This involves using the results from one method to inform the development of the other (see Table 3). A typical example in IS research is the use of exploratory qualitative interviews with technology users to identify key themes, concerns, and terminology, which are then used to construct a valid and contextually relevant quantitative survey instrument (Sarker et al., 2012). Fourth is Expansion, which seeks to extend the breadth and range of an inquiry by using different methods to study various components of a phenomenon. For instance, a quantitative survey might assess the broad outcomes of a system implementation across an entire organization. At the same time, a series of qualitative case studies investigates the nuanced implementation process on a department-specific basis. Other articulated purposes, such as Compensation, which uses one method to overcome the weaknesses of another. Another is Diversity, which seeks divergent or contradictory views to enrich understanding, and both have the further effect of expanding the strategic toolkit of the mixed-methods researcher (Bryman, 2006).

A Typology of Mixed-Methods Designs

These purposes are shaped through research designs structured around two main dimensions. The first of these is time orientation. It refers to the order in which the quantitative and qualitative components are incorporated. The second is *priority*, which denotes the relative weight assigned to each component (Creswell & Plano Clark, 2018). Together, these dimensions yield four underlying research designs, as follows:

Explanatory Sequential Design (QUANT *to* QUAL):

A two-stage design will be employed, starting with quantitative data collection and analysis. The second phase is the qualitative phase, which is built upon the outcomes of the quantitative phase. It is beneficial in cases of complementarity, where qualitative information is used to interpret surprising or striking quantitative results (Venkatesh et al., 2023). For example, a large-scale survey (QUANT) may indicate that one specific department has significantly lower adoption rates. Follow-up interviews and focus groups (qualitative) could then be conducted by the researcher with specific members of that department to understand the local contextual determinants. This culture is resisting the study, as well as the usability challenges contributing to it. This design aligns with Trieu et al. (2022), who employed a sequential mixed-methods approach. This process begins with quantitative surveys and later involves qualitative interviews to explain and elaborate on unexpected statistical results and contextual nuances in system use effectiveness.

Exploratory Sequential Design (QUAL *to* QUANT):

In this design, the design phase also precedes the analysis phase, but is initiated with a preliminary stage of qualitative data gathering and analysis. The results of this exploratory stage are subsequently applied in a second, quantitative stage. This approach is most suitable for development, and thus the design is employed to construct a new theory or develop a new measurement instrument (Venkatesh et al., 2023). An Information System researcher team exploring the new phenomenon of AI-human collaboration in creative work could start with in-depth interviews (QUAL) with designers and artists as a means of identifying new constructs, including trust, co-creation, agency, and others. These constructs would then be operationalized and developed into an instrument (quant) and used in a huge sample to determine their prevalence and their correlation with the product of creativity.

3.2.3 Convergent Parallel Design (QUANT *plus* QUAL):

In this design, the data collection for the quantitative and qualitative data is done in parallel, but simultaneously. Equal priority is usually given to both strands, and the combination of the results is left for the phase of interpretation. This design is suitable for corroboration and complementarity because it allows for direct comparison and contrast of the results from the two methods (Venkatesh et al., 2023). To illustrate, a research study on the efficacy of a novice cybersecurity training intervention could be conducted by administering a quantitative knowledge test (QUANT) to evaluate learning results and performing focus groups (QUAL) to examine staff views on the training and their perceptions. These outcomes would be combined to provide a detailed overview of the program.

3.2.4 Embedded Design:

This design means that, as the secondary component of a big design of the other type, one type of data is gathered, studied, and interpreted. An example to illustrate this is a researcher adopting qualitative case studies (qual) hidden within a massive randomized controlled trial (QUANT) of a new health information system to comprehend the experiences and lives of the clinicians utilizing the new system and the situational factors that affect its implementation across various wards of the hospital (Venkatesh et al., 2023). These designs enable a loose yet codified grammar of the study constructions, incorporating rigorous and purposeful mixed methods. There has been debate about whether embedded design is still part of the core Mix method research MMR designs. In more recent frameworks, it is often treated as a variant or strategy rather than a core design. Table 1 summarizes these core designs and connects them to their primary objectives of use and typical applications in IS and Analytics research.

Table 1. A Typology of Mixed-Methods Research Designs

Design Name	Notation & Priority	Time Orientation	Primary Purpose(s)	Integration Point	Typical IS/Analytics Use Case
Explanatory Sequential	QUANT – qual (Quantitative Dominant)	Sequential	Complementarity, Explanation	Between Phases	A large-scale survey reveals low adoption of a data analytics system; follow-up interviews explore reasons for resistance among specific user groups, illustrating how qualitative insights help explain quantitative results (Creswell & Plano Clark, 2018; Fetters et al., 2013; Ngulube, 2022).
Exploratory Sequential	QUAL – quant (Qualitative Dominant)	Sequential	Developmental, Theory Building, Expansion	Between Phases	In-depth interviews with developers identify new factors contributing to work–life conflict; a subsequent survey tests the prevalence of these factors in a larger sample, demonstrating how qualitative findings inform quantitative measurement (Creswell & Plano Clark, 2018; Leech & Onwuegbuzie, 2009; Ngulube, 2022).
Convergent Parallel	QUANT + QUAL (Equal Status)	Concurrent	Corroboration, Complementarity, Diversity	Interpretation Phase	Quantitative survey results on user satisfaction are analyzed alongside concurrent focus-group data to examine convergence and divergence of insights, offering complementary perspectives on the same phenomenon (Creswell & Plano Clark, 2018; Fetters et al., 2013; Teddlie & Tashakkori, 2009)

WHY COMPUTATIONAL: UTILIZING THE POWER OF CSS

As we have seen, prior to the emergence of mixed-methods research in the social sciences, a revolution in the field of computation occurred, with many social scientists contributing to this revolution. This revolution gave birth to the new flagship discipline now known as Computational Social Science (CSS). It is an interdisciplinary discipline that combines giant data analytics with social theory to create, simulate, and analyze social phenomena (Lazer et al., 2009). CSS makes use of Computer science techniques and solves social problems by applying the fundamentals of social science. CSS provides researchers in IS and Analytics with a powerful set of tools to work with the new world of unstructured and high-velocity data inherent in the modern digital space. Traditional statistics techniques assume a priori a functional form *(f)* before analysis is conducted, such as assuming a linear form. As noted by Breiman (2001), this places a restrictive structure of the model, which may not represent real-world complexity, whereas on the other hand, computational and machine learning methods enable one to learn the complexity function f from the data directly. Subsequently, traditional statistical methods, namely propensity score matching (PSM), have also been criticised for several issues, including data loss during matching, model dependence, and constant treatment effects, among others. It also does not capture nonlinear and complex interactions, which makes it computationally intensive with large datasets. In contrast, Computational Social Science and Machine Learning approaches are capable of addressing these problems by using nonparametric, data-driven, and scalable models that automatically detect nonlinearities, minimize data loss, and reveal heterogeneous effects without strong functional assumptions (Yahav, Shmueli, & Mani, 2016).

New Forms of Data for IS Research

For the further advancement of computational research, it is necessary to use the opportunities presented by Big Data, which not only concerns the generation of new information but also the process of converting the unstructured content to formats suitable for analysis (Li et al., 2025). There are several key data categories relevant to IS research. First, Network Data, where it extends beyond traditional organizational charts, allows researchers to analyze dynamic, large-scale networks derived from digital traces. Examples include venture capital co-investment networks to map capital flows, employee migration patterns between firms on professional networking sites to trace knowledge diffusion, and patent citation networks to understand technological evolution (Li et al., 2025). Second, text data, which is arguably the most abundant form of unstructured data. Sources relevant to IS research include rich textual descriptions of firms from platforms like Crunchbase, the full text of patent filings, academic publications, product specifications, employee reviews from sites like Glassdoor, and internal corporate communications, including emails and chat logs (Li et al., 2025). Third, Image Data, where the Visual data represents an emerging frontier for quantitative analysis. Other elements that can be looked into through the study of corporate logos, product images, the headshots of those who founded the corporation, and photos on social media can be used to determine the branding strategy as well as the corporate culture, diversity of the team members, and the identity of the people (Li et al., 2025). Fourth, Audio and Video Data: It provides multimodal visual and auditory streams. Television shows, audio recordings of online discussions, and video advertisements can be computationally analyzed in terms of predilection, dispensation tactics, and pathological governance, which can be associated with outcomes such as successful fundraising or team performance (Li et al., 2025).

Two Modes of Computational Analysis

Before we begin to understand the technical forms of computational analysis, it is helpful to grasp the field's two primary methodological goals, which Shmueli (2010) famously defined as explanation and prediction. Explanatory modelling, the traditional goal of social science, seeks to test causal theories and understand mechanisms (the "why"). Predictive modelling, on the other hand, is primarily concerned with accurately predicting outcomes (i.e., "what will happen") (Shmueli, 2010). As Hofman et al. (2021) claim, these two objectives are usually presented as being at odds with each other, but are most effective when combined in a productive manner. This distinction coincides closely with the two modes of analysis, which are discussed below. Mullainathan and Spiess (2017) put this in terms of the fundamental difference in purpose; traditional econometrics is "obsessive" with estimating the properties of a particular parameter *(β)*. On the other hand, machine learning is "obsessively" optimising the quality of out-of-sample predictions *($\hat{y}$)*.

There are two different ways of analysing such an abundance of unstructured data, which symbolises a gigantic change in computing power and an investigative desire (Li et al., 2025). The first one is the Extractive Approach. It is a way of looking at the unstructured data as a source of raw material from which it is possible to extract the structured properties that can be measured, quantified, and compiled. The main objective is to convert the complex data (a matrix) to rectangular form and represent it in matrix form that can be analyzed using standard statistical models, which describe or infer causality. This includes functionality such as text mining to extract a sentiment score, topic modelling to extract thematic content, and image classification to determine the frequency of certain objects or demographics in relation to their implications. In this mode, the non-organizational information itself becomes the invention of new variables used to perform classical statistical analyses (Li et al., 2025).

The second is the Data as Model (Generative) Approach. It is also not just an exploitation process. It applies deep learning networks, such as large language models (LLMs) in text or generative adversarial networks (GANs) in images, to obtain a high-dimensional and compressed representation of the original data. It is not merely a set of variables, but a computerized version of the system under investigation. This computerized analog may be applied on either front by serving as a high-resolution digital observatory to explore advanced semantic and structural linkages among the information, and through a virtual laboratory, test hypotheses about complex socio-technical processes in silico (Li et al., 2025). To clarify this distinction, **Table 2 provides a structured overview** of these new data forms. The table contrasts the two primary analytical approaches, detailing how each data type (e.g., Text, Network, Image, and Audio) can be used in both an **"Extractive"** approach (to create new variables for traditional analysis) and a **"Generative"** approach (to create a computational model of the system itself) (Li et al., 2025).

Table 2. New Data Forms and Analytical Approaches in Computational Social Science

Data Type	'Data as Data' (Extractive) Approach	'Data as Model' (Generative) Approach
Text	**Technique:** Text mining to extract quantifiable features (e.g., sentiment, topics). **Research Goal:** To convert qualitative patent descriptions into structured variables for statistical analysis of innovation trends.	**Technique:** Training Large Language Models (LLMs) on corporate communications. Research Goal: To create a 'digital double' of a firm's discourse to simulate market responses to new product announcements.
Network	**Technique:** Calculating centrality measures in a VC co-investment network. **Research Goal:** To identify the most influential investors in an ecosystem for traditional regression analysis.	**Technique:** Modelling the dynamic evolution of a technology citation network. **Research Goal:** To simulate the future trajectory of technological diffusion and identify emerging innovation hubs.
Image	**Technique:** Using image classification to identify demographic characteristics from founder/CEO photos. **Research Goal:** To create a quantitative measure of team diversity to correlate with venture funding outcomes.	**Technique:** By using Generative Adversarial Networks (GANs) to learn the latent dimensions of corporate branding imagery. **Research Goal:** To generate novel logos and predict their perceived innovativeness, enabling virtual brand strategy experiments.
Audio	**Technique**: For extracting acoustic features like pitch and speaking rate from investor pitches. **Research Goal:** To quantify communication styles and test their statistical relationship with fundraising success.	**Technique:** Building a deep neural network model of successful entrepreneurial rhetoric. **Research Goal:** To generate synthetic audio of pitches to test which acoustic signatures are most persuasive to potential investors.

CREATING A NOVEL SYNTHESIS: THE MERGING OF MIXED-METHODS AND COMPUTATIONAL METHODOLOGIES.

These new approaches combine and complement each other to achieve the maximum potential. For example, there is a strong synergy between mixed-methods research (MMR) and computational social science (CSS). This synthesis is precisely what is needed to clear the confusion caused by Gen-AI. This synergy also provides a direct answer to the *long-standing challenge of bridging the gap between explanatory and predictive modelling* (Shmueli, 2010). It provides a practical framework for the "integrative modelling" called for by computational social scientists, which combines explanation (the traditional strength of MMR) with prediction (the primary strength of CSS) (Hofman et al., 2021). Moreover, as Mullainathan and Spiess (2017) found that, machine learning should not be seen only as a tool for prediction, but as an entirely new instrument for enhancing causal explanation. It may be possible by constructing better covariates or addressing complex interactions that are not considered by traditional linear models. It is at the core of the MMR-CSS synthesis.

CSS provides the computational tools to analyze the massive, high-velocity data from Gen-AI systems, while MMR provides the crucial framework to *base, justify, and explain* the findings. For instance, researchers can use qualitative methods (the 'MMR' part) to triangulate and validate the outputs of a 'black box' Gen-AI model (the 'CSS' part), thereby identifying and correcting for its potential *inaccuracies, hallucinations, or biases* (which we introduced in Section 1). CSS provides the means to organize and analyze the size and complexity of digital information, while MMR supplies the rich, contextual human information needed to interpret it. This conjunction allows the researcher to move between small-scale qualitative information and large-scale one, and vice versa, examining big data to define the settings and meanings of some of the trends identified (Li & Abramson, 2025)

Computational Ethnography as an Exemplar

One plausible example of this kind of synthesis is computational ethnography (Li & Abramson, 2025). Traditional ethnography, which has been used for studying human contexts by observing them directly, has long been recognized as suffering from problems of scale and generalizability. These problems can be resolved by integrating machine learning (ML) with the ethnographic workflow without losing the level of interpretation (Li & Abramson, 2025).

There are several points where the integration can be made. Initially, transcription services are cheaper and faster to complete using ML, and researchers should be attentive to potential algorithmic biases, including the possibility of the algorithm making more mistakes with minority speakers (Li & Abramson, 2025). The most interesting aspect of the synthesis, however, is a hybrid human-machine approach to data analysis. This process typically involves five iterative steps. First, there is Human Coding, where a researcher with a deep level of contextual understanding manually encodes a subset of the qualitative data. It can be interviews, transcripts, or field notes to discover important themes and concepts. This ensures that the analytical structure is based on the extent of interpretation and the field expertise of human understanding. Second, Model Training, wherein human-coded data is treated as a gold standard to generate a supervised ML model to identify these same themes in the text. This turns the qualitative knowledge of the researcher into a scale of classification. Third, Machine Coding, in which a trained model will be used to code the remaining large data, which could only be prohibitively time-consuming to do manually. Fourth, Validation and Refinement, in which the ethnographer goes back and critically evaluates the results of the machine, and based on their ground knowledge, reviews and validates the classifications, as well as finds the errors or subtleties that might have been overlooked by the model. This is done to refine the model and enhance the efficiency and reliability of the coding procedure (Li & Abramson, 2025; Nelson, 2020).

Extending Core Designs: Integrating CSS to Create Multi-Stage Projects

There should be a clear differentiation between the core designs provided in Section 3 (and described in Table 1) and more complex, yet hybrid, extensions of these designs. In this section, the difference is elaborated. Integrating the methodologies of Computational Social Science (CSS) into these foundational frameworks, this section explains how researchers can build more complex, *hybrid,* or *multi-stage* research projects.

To illustrate, the examples provided below do not merely depict core designs. In contrast, they are demonstrating sophisticated multi-phase applications where CSS analytical procedures are combined with a more fundamental design, such as the Explanatory and Sequential design, to create more sophisticated and versatile research methods.

A Hybrid Explanatory Sequential Application (QUANT *to* CSS *to* qual)

A researcher may start with a large-scale quantitative research (QUANT) that aims to correlate the use of certain types of enterprise software with a decline in employee morale. To understand this pattern, the researcher could then apply unsupervised topic modelling (CSS) to a large corpus of employee reviews to identify themes related to poor user interface design. Ultimately, the researcher would need to conduct in-depth interviews (qualitative) with employees to investigate and understand the lived experiences of employees. It could be done by using the themes generated by the machine to develop the interview protocol. In this context, the qualitative phase plays an important role. It is a formal approach to analyze the "residuals" or the errors or unexplained variance present in the computational model in order to find out the human-centric themes and causal mechanisms missed by the model (Agrawal et al., 2020).

A Hybrid Exploratory Sequential Application (QUAL *to* CSS *to* quant)

A study might begin with the use of ethnographic interviews (QUAL) with entrepreneurs, and the preliminary themes on risk-taking philosophies will be generated. This qualitative data would, respectively, be trained on a supervised ML classifier (CSS) to identify these philosophies in a large set of interviews, with saturation being attained. These machine-coded philosophies could thereafter be operationalized into a final quantitative survey (quant) to establish the association that they have with the performance of firms. This data would then be qualitative and subsequently used to train a supervised ML classifier (CSS) in order to identify this philosophy in a vast corpus of interviews. Such machine-coded philosophies may then be assumed to be the variables in a final survey (quantitative) to establish the relationship between them and firm performance.

Practical Implementation Workflow for MMR-CSS

As a way of narrowing the disparity between theoretical and practical, this section will present a step-by-step implementation pathway of carrying out an integrated MMR-CSS study. The next five-stage workflow offers a viable framework of the study of such kind.

To begin, the workflow commences with Qualitative Problem Framing (QUAL). The first stage involves the use of exploratory designs, such as interviews, focus groups, and ethnographic observation, in the effort to define the problem, understand the environment, and communicate in the language of the participants. This is essential to ensure that further computational analysis relies not only on the availability of the data but also on human expertise. Second, the process is transferred to the Computational Data Sourcing and Analysis (CSS). In this case, the researcher gathers and processes large-N "big data," such as text in forums or user logs, through computational tools like topic modelling or machine learning classification. This stage establishes macro-trends, develops hypotheses based on the first stage, and asks what and how much. Third, the Integration Point & Validation stage is a major nexus in which the results of the first qualitative and computational studies are evaluated against. This is aimed at validation, triangulation, and identifying points of divergence; for example, determining whether the qualitative themes, which are derived from qualitative methods, such as user frustration, align with the results of computational methods, such as negative sentiment clusters. Fourth, another important follow-up on Qualitative Explanation & Refinement (QER) is conducted. It refers to a second and focused set of qualitative interviews that aim to describe the reasons behind the computational results and clarify how to resolve any discrepancies between the integration stage. The researcher gives the findings of the computations to participants and asks them to interpret it, thus making the black box clear. Lastly, there is the fifth stage, which involves Holistic Framework Development, where all the findings are synthesised to develop a comprehensive theoretical framework. This is the concluding stage that invents a model that is not only quantitatively confirmed (in Phase 2) but also qualitatively grounded (in Phases 1 and 4) to produce an all-encompassing, implementable paradigm and scannable, actionable knowledge.

Generating strong meta-inferences

The final outcome of the synthesis is the development of vigorous meta-inferences and conclusions, which should be grounded in the qualitative, quantitative, and computational strands of the research (Venkatesh et al., 2013). In the literature, it is empirically and conceptually proven that triangulating the results between these streams of methodologies could allow the researcher to construct a more comprehensive and believable picture of the complex socio-technical phenomena than any other method would provide. This holistic framework, as shown in Figure 1, gives a concept map of the overall research process.

Figure 1. An Integrated Framework for Mixed-Methods Information Systems and Computational Social Science Research

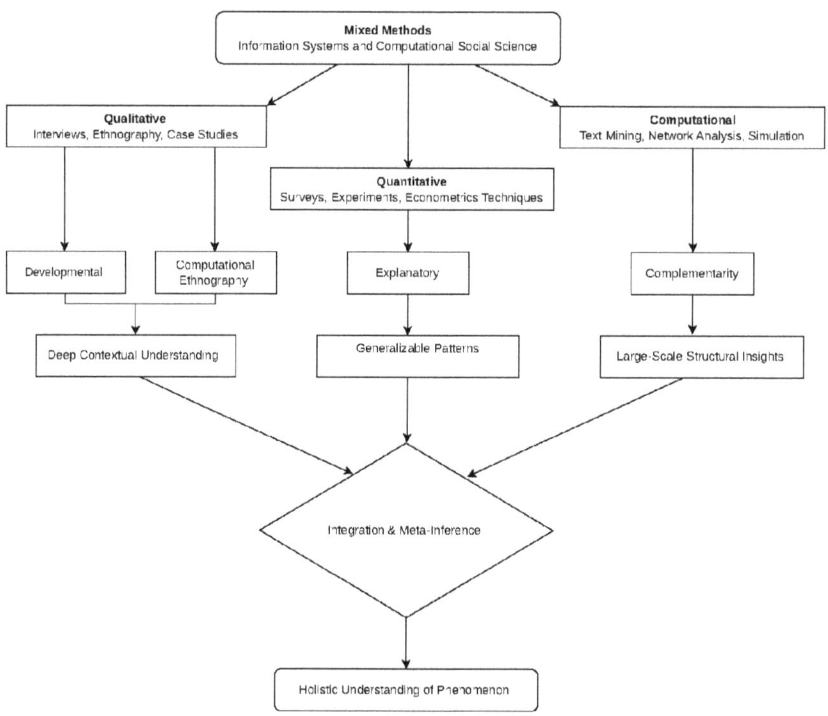

The three main streams of methodology are shown in Figure 1. First, Qualitative (e.g., Interviews, Ethnography). Second, Quantitative (e.g., Surveys, Experiments), and Finally, Computational (e.g., Text Mining, Simulation). They are shown to operate in synthesis due to major integration roles (Developmental, Explanatory, and Complementarity) to become the integrated Mixed Methods Information Systems and Computational Social Science solution. The ultimate goal of such synthesis is to develop robust synthesis and meta-inferences, which will lead to a comprehensive understanding of the complex socio-technical phenomenon.

Table 3 provides examples of the Cross-Stream Integration Roles to clarify the synthesis as illustrated in Figure 1. It demonstrates the way in which the various streams of methodology, for example, qualitative and computational, can be intentionally integrated to realize specific objectives. To illustrate, qualitative knowledge can be applicable to developmental purposes, whereas quantitative data can be helpful in choosing the explanatory models.

Table 3. Cross-Stream Integration Role in Mixed-Method Research.

Cross-Stream Integration Role	
Developmental	Qualitative insights inform the design of surveys and instruments.
Computational Ethnography	Qualitative coding supports machine learning training and validation.
Explanatory	Quantitative findings guide case/participant selection for qualitative inquiry.
Complementarity	Computational theme identification enriches qualitative exploration.

NAVIGATING THE FUTURE: CHALLENGES AND PATHWAYS FOR INNOVATIVE RESEARCH

Since research is developing, there is an increased necessity to implement approaches that are more integrated and include both qualitative, quantitative, and computational.

Meeting the challenge of integration

The final step in the synthesis of mixed-methods research is the development of overall meta-inferences and conclusions that combine the qualitative, quantitative, and calculational strands (Venkatesh et al., 2013). The combination of the results of these different methodological approaches can enable researchers to build a richer and more believable picture of the complex socio-technical phenomena than would have been achieved by any one methodology. Figure 1 is a conceptual roadmap of such an integrative research process. Nevertheless, such interdisciplinary work is not very easy. Such a study is very resource-intensive, as it takes a long time to undertake sequential stages, it needs substantial finances to develop specialised software and extensive data acquisition, and a team with diversity in their knowledge of qualitative inquiry, quantitative analysis, and computational modelling is needed. The process of organizing and running such teams of professionals requires a strong capacity to organize and work over time (Poth, 2023). Additionally, the merging and display of information generated using these strategies can be very challenging. Synthesis of information is complicated by large amounts and diverse types of generated data, and the complexity of these discoveries is in many cases far beyond what conventional academic formats could represent. This means that emerging novel and innovative modalities of academic expression and visualization will be required in order to present such integrated outcomes (Venkatesh et al., 2023).

The critical and moral aspect of computational procedures

It is essential to think critically about the field of AI and ML, as its implementation in social research has not yet become a standardized practice. Researchers will be compelled to recognize the potential risks that these tools pose in introducing new types of errors and ethical issues.

First, algorithmic bias refers to the fact that past training data can be used to condition the machine learning model to reproduce, inherit, and reinforce existing social biases. To provide an example, speech recognition technology that converts oral speech into text may have a higher chance of error when applied to the oral speech of women or ethnic minorities, in a systematic manner that distorts their speech. As a result, the text classification model can learn and reproduce the prejudices in the training corpus from society towards certain predictions, which are biased or harmful (Li & Abramson, 2025; Noble, 2018).

Second, decontextualization and interpretation: To the temptation of a computational class, it may also be decontextualized. An example would work well in classifying running text as negative sentiment, but it would not explain what negative sentiment the text conveys and the cultural underpinnings of that sentiment. This challenge directly reflects what Hofman et al. (2021) identify as the "prediction fallacy": the mistaken belief that a high-performing predictive model (e.g., a sentiment classifier) also provides a valid or useful explanation for the underlying human behaviour. The human researcher is advised to remain the final instance of interpretation to prevent the construction of findings that are theoretically applicable but sociologically hidden (Li & Abramson, 2025). This addresses the critical "conflation" of explanatory and predictive goals (Shmueli, 2010). This challenge directly reflects what Hofman et al. (2021) identify as the "prediction fallacy": the mistaken belief that a high-performing predictive model also provides a valid explanation for the underlying human behaviour. It is often described as the "interpretability trade-off which means that the most accurate predictive models (e.g., deep neural networks) are often the least interpretable, functioning as "black boxes" that obscure *why* they make a given prediction (Mullainathan & Spiess, 2017).

Third, hybrid MMR-CSS designs are associated with the complicating establishment of ethical concerns regarding data governance and privacy, extending beyond merely the presence of algorithmic bias. The widespread use of scraping publicly available data, for example, from microblogging platforms and social media to perform computational analysis runs on the fallacy that being available is equivalent to giving consent to the research. This ethical dilemma gets particularly more intense in hybrid designs when people discovered under the occasion of large-scale data scraping (CSS) are further subjected to face-to-face interaction with human subjects, including qualitative interviews. Such a change requires fresh approaches to informed consent, with new protocols being transparent and clear, where the researcher clearly explains the way the participant was determined and the data gathered so far. Researchers need to embrace strict governance frameworks, such as the FAIR Principles, in order to make adequate stewards of data and earn trust. It is equally a strategy that deals with reproducibility, even though qualitative knowledge is grounded to context, the computational outputs of the analysis, its scripts, models, and code, should be documented and verifiable to the last.

FUTURE RESEARCH DIRECTIONS

The chapter proposes an integrated MMR-CSS structure to resolve the current complexity created by Gen-AI. While this chapter provides the "why" and "what," significant work remains to be done on the "how." Future research in Information Systems and Analytics should move in several key directions. First, this chapter proposes a hybrid multi-stage design. Future research is needed to empirically test these models. We need to document case studies of IS research projects that apply these integrated workflows to understand their practical challenges, costs, and the true nature of the "meta-inferences" they produce.

Second, the integration of CSS and MMR creates novel ethical dilemmas. How does a researcher navigate IRB approval for a study that scrapes 'public' data (CSS) and then uses that data to recruit subjects for in-depth interviews (MMR)? Future research must develop new ethical frameworks and best practices specifically for this hybrid methodology: Third, Pedagogy for Interdisciplinary Training. Research on interdisciplinary training graduate programs is still primarily limited. Future pedagogical research should explore what a curriculum for an interdisciplinary IS researcher looks like. How do we effectively train students to be fluent in *both* computational text analysis and qualitative ethnography? Fourth, this chapter primarily discussed Gen-AI as a *subject* of study (a source of complexity). A fruitful future direction would be to explore Gen-AI as a *tool* within the MMR-CSS framework itself. For example, researchers could adopt formal methodologies, such as "Scientific Regret Minimization" (SRM), which utilizes a high-performing predictive model to systematically guide the search for new theories or identify novel phenomena that require qualitative exploration (Agrawal et al., 2020). It turns the predictive model itself into an engine for explanatory theory-building.

CONCLUSION

This chapter describes the new frontier of complexity facing Information Systems and Analytics researchers, driven by the rise of Big Data and Generative AI. It had moved beyond the general sense of complexities to single out three interwoven problems of methodology, which include novelty, origins, and functional equivalence. In the chapter, it was argued that these socio-technical issues cannot be resolved through only one approach. It has re-examined the philosophical foundations of pragmatism and critical realism to develop an argument for methodological pluralism. It then presented a systematic summary of the fundamental principles and designs of Mixed-Methods Research (MMR) and compared them to the analytical power of Computational Social Science (CSS).

The main strength of this chapter is the suggestion of a synthesized MMR-CSS framework. Based on the example of computational ethnography, it reveals how IS researchers can and should combine the richness of qualitative studies with the magnitude and authority of computational approaches. The final feature of this chapter is the description of the practical and ethical issues associated with this framework, and ultimately, the request to adopt a new, interdisciplinary, and synthetic approach to scholarship. This is not merely an available option; it is a requirement for achieving rigorous, responsible, and relevant research in the Gen AI era.

REFERENCES

Agrawal, M., Peterson, J. C., & Griffiths, T. L. (2020). Scaling up psychology via scientific regret minimization. *Proceedings of the National Academy of Sciences of the United States of America, 117*(16), 8825–8835. DOI: 10.1073/pnas.1915841117 PMID: 32241896

Breiman, L. (2001). Statistical modeling: The two cultures (with comments and a rejoinder by the author). *Statistical Science, 16*(3), 199–231.

Bryman, A. (2006). Integrating quantitative and qualitative research: How is it done? *Qualitative Research, 6*(1), 97–113.

Creswell, J. W., & Plano Clark, V. L. (2018). *Designing and conducting mixed methods research* (3rd ed.). Sage.

Felderer, M., & Travassos, G. H. (2020). The evolution of empirical methods in software engineering. In Felderer, M., & Travassos, G. H. (Eds.), *Contemporary empirical methods in software engineering* (pp. 1–24). Springer International Publishing., DOI: 10.1007/978-3-030-32489-6_1

Fetters, M. D., Curry, L. A., & Creswell, J. W. (2013). Achieving integration in mixed methods designs: Principles and practices. *Health Services Research, 48*(6 Pt 2), 2134–2156. DOI: 10.1111/1475-6773.12117 PMID: 24279835

Greene, J. C., Caracelli, V. J., & Graham, W. F. (1989). Toward a conceptual framework for mixed-method evaluation designs. *Educational Evaluation and Policy Analysis, 11*(3), 255–274. DOI: 10.3102/01623737011003255

Hofman, J. M., Watts, D. J., Athey, S., Garip, F., Griffiths, T. L., & Kleinberg, J.. (2021). Integrating explanation and prediction in computational social science. *Nature, 595*, 181–188. PMID: 34194044

Johnson, R. B., & Onwuegbuzie, A. J. (2004). Mixed methods research: A research paradigm whose time has come. *Educational Researcher, 33*(7), 14–26. DOI: 10.3102/0013189X033007014

Lazer, D., Pentland, A., Adamic, L., Aral, S., Barabási, A. L., Brewer, D., Christakis, N., Contractor, N., Fowler, J., Gutmann, M., Jebara, T., King, G., Macy, M., Roy, D., & Van Alstyne, M. (2009). Social science: Computational social science. *Science, 323*(5915), 721–723. DOI: 10.1126/science.1167742 PMID: 19197046

Leech, N. L., & Onwuegbuzie, A. J. (2009). A typology of mixed methods research designs. *Quality & Quantity, 43*(2), 265–275. DOI: 10.1007/s11135-007-9105-3

Li, N., Lai, S., & Evans, J. (2025). Big data and the computational social science of entrepreneurship and innovation. In O. Sorenson & P. Thornton (Eds.), *De Gruyter handbook of sociology of innovation and entrepreneurship* (pp. 329–352). De Gruyter. DOI: 10.1515/9783111085722-019

Li, Z., & Abramson, C. M. (2025). Ethnography and machine learning: Synergies and new directions. In Borch, C., & Pardo-Guerra, J. P. (Eds.), *The Oxford handbook of the sociology of machine learning* (pp. 245–272). Oxford University Press., DOI: 10.1093/oxfordhb/9780197653609.013.36

Mingers, J. (2001). Combining IS research methods: Towards a pluralist methodology. *Information Systems Research, 12*(3), 240–259.

Mullainathan, S., & Spiess, J. (2017). Machine learning: An applied econometric approach. *The Journal of Economic Perspectives, 31*(2), 87–106. DOI: 10.1257/jep.31.2.87

Nelson, L. K. (2020). Computational grounded theory: A methodological framework. *Sociological Methods & Research, 49*(1), 3–42.

Ngulube, P. (2022). Using simple and complex mixed methods research designs to understand research in information science. In Ngulube, P. (Ed.), *Handbook of research on mixed methods research in information science* (pp. 20–46). IGI Global., DOI: 10.4018/978-1-7998-8844-4.ch002

Nguyen, D. C., & Welch, C. (2025). Generative artificial intelligence in qualitative data analysis: Analyzing—or just chatting? *Organizational Research Methods, 0*(0). Advance online publication. DOI: 10.1177/10944281251377154

Noble, S. U. (2018). *Algorithms of oppression: How search engines reinforce racism.* New York University Press.

Poth, C. N. (Ed.). (2023). *The SAGE handbook of mixed methods research design* (Vol. 1–2). Sage Publications., DOI: 10.4135/9781529682663

Sarker, S., Sarker, S., Sahaym, A., & Bjørn-Andersen, N. (2012). Exploring value co-creation in relationships between an ERP vendor and its partners: A revelatory case study. *Management Information Systems Quarterly, 36*(1), 317–338. DOI: 10.2307/41410419

Schumpeter, J. A. (1942). *Capitalism, socialism and democracy*. Harper & Row.

Shmueli, G. (2010). To explain or to predict? *Statistical Science, 25*(3), 289–310. DOI: 10.1214/10-STS330

Tashakkori, A., & Creswell, J. W. (2007). Editorial: The new era of mixed methods. *Journal of Mixed Methods Research*, *1*(1), 3–7. DOI: 10.1177/2345678906293042

Teddlie, C., & Tashakkori, A. (2009). Foundations of mixed methods research: Integrating quantitative and qualitative approaches in the social and behavioral sciences. *Sage (Atlanta, Ga.)*.

Trieu, V.-H., Burton-Jones, A., Green, P., & Cockcroft, S. (2022). Applying and extending the theory of effective use in a business intelligence context. *Management Information Systems Quarterly*, *46*(1), 645–678. DOI: 10.25300/MISQ/2022/14880

Venkatesh, V., Brown, S., & Sullivan, Y. (2023). *Conducting mixed-methods research: From classical social sciences to the age of big data and analytics*. Virginia Tech Publishing. https://hdl.handle.net/10919/117223

Venkatesh, V., Brown, S. A., & Bala, H. (2013). Bridging the qualitative–quantitative divide: Guidelines for conducting mixed methods research in information systems. *Management Information Systems Quarterly*, *37*(1), 21–54. DOI: 10.25300/MISQ/2013/37.1.02

Yahav, I., Shmueli, G., & Mani, D. (2016). A tree-based approach for addressing self-selection in impact studies with big data. *Management Information Systems Quarterly*, *40*(4), 819–848. DOI: 10.25300/MISQ/2016/40.4.02

Chapter 4
Mixed Methods Research in Data Analytics Within Higher Education:
Systematic Review of Prevalence and Contributions

Mthokozisi Masumbika Ncube
http://orcid.org/0000-0003-4835-6594
School of Interdisciplinary Research and Postgraduate Studies, University of South Africa, South Africa

ABSTRACT

The integration of data analytics in higher education underscores the need to examine the methodological approaches guiding this research domain. While applications of data analytics have gained extensive attention, their methodological foundations remain underexplored. This review investigates the prevalence, characteristics, and contributions of mixed methods research in data analytics. Following PRISMA guidelines, 22 empirical studies from five databases across were synthesised. Results show that MMR effectively captures the complexities of pedagogy, assessment, and implementation through epistemological complementarity rather than mere methodological combination. Methodological plurality enhances explanatory depth and produces triangulated insights into multifaceted educational phenomena. The review identifies four principles for effective MMR: parallel design integration, sequential design connectivity, analytical triangulation, and methodological transparency. The study advances scholarship by mapping MMR deployment patterns to strengthen research rigour in this evolving field.

DOI: 10.4018/979-8-3693-4018-9.ch004

INTRODUCTION

Research methodologies significantly influence the validity, reliability, and impact of the knowledge generated (Ngulube, 2020). A systematic examination of methodological trends offers critical insights into disciplinary evolution, enabling scholars to evaluate their approaches and identify advancement trajectories (Bryman, 2016; Flick, 2018; Ngulube & Ngulube, 2022). In higher education, data analytics, defined as the process of inspecting, cleaning, transforming, and modelling data to discover useful information, inform conclusions, and support decision-making (TechTarget, n.d.; Abdul-Jabbar & Farhan, 2022; Sivarajah et al., 2017), has emerged as a pedagogical priority, driven by demands for data literacy and analytical competencies. However, methodological approaches to data analytics research often exhibit limited diversity, predominantly relying on single-method studies rather than embracing methodological plurality (Arnold & Pistilli, 2012; Macfadyen & Dawson, 2012; Pardo et al., 2017).

Quantitative methodologies enable broad analysis and statistical generalisation but may overlook contextual richness and lived experiences by prioritising numerical data (Janice et al., 2018). These approaches can obscure crucial learning dimensions, including student motivation, engagement trajectories, and the socio-cultural contexts that mediate educational experiences. Conversely, qualitative methodologies provide phenomenological depth and contextual understanding; however, their interpretative nature and smaller sample sizes constrain generalisability claims (Morse et al., 2023). This methodological tension necessitates a shift toward methodological pluralism, specifically strategic deployment of mixed methods research (MMR) that synthesises quantitative and qualitative approaches for comprehensive understanding (Creswell & Plano Clark, 2018; Mertens, 2018; Ngulube & Ukwoma, 2021; Saraswati & Devy, 2023; Tashakkori et al., 2020).

MMR transcends simple methodological aggregation (Manion, 2019), representing an epistemological orientation recognising the multifaceted nature of educational phenomena and the inadequacy of singular methodological lenses (Mertens, 2018; Ngulube & Ngulube, 2022; Saraswati & Devy, 2023; Tashakkori et al., 2020). Thus, by examining phenomena from multiple vantage points, MMR facilitates 'epistemological complementarity,' where different data forms illuminate distinct but interconnected problem dimensions. This study maps MMR within the realm of data analytics research, examining its prevalence, distinctive contributions, and deployment patterns across various contexts.

PROBLEM STATEMENT

Despite the proliferation of data analytics research in higher education, a critical gap persists regarding the methodological approaches employed. The predominance of single-method approaches represents methodological essentialism that constrains scholarly understanding (Bazeley, 2018; Creswell & Creswell, 2022; Creswell & Plano Clark, 2018; Mertens, 2018; Ncube & Ngulube, 2025). This limitation manifests in two interconnected problems. First, methodological monism impedes holistic comprehension of how data analytics competencies are cultivated and assessed in higher education programmes (Strang, 2017). Thus, it limits institutional capacity to optimise curriculum design and prepare graduates for data-intensive environments. The complexity of data analytics education, encompassing technical proficiency, ethical reasoning, critical thinking, and domain application, necessitates methodologies that capture both measurable outcomes and nuanced processes (Ngulube & Ukwoma, 2021; Tashakkori et al., 2020).

Second, while MMR offers potential to transcend these limitations through the integration of quantitative and qualitative data generation and analysis (Creswell & Plano Clark, 2018; Mertens, 2018; Saraswati & Devy, 2023), its application within data analytics education remains underexplored and inadequately theorised. This represents a missed opportunity to gain a deeper understanding of this critical domain. The potential of methodological plurality to illuminate the multidimensional nature of data analytics education (Strang, 2017), ranging from cognitive development to affective engagement, from institutional implementation to student experience, remains largely unrealised. This study addresses this gap by systematically mapping MMR in data analytics education research, analysing deployment patterns, identifying contributions, and synthesising best practices to develop more robust methodologies. Understanding how MMR has been applied, the distinctive insights it generates, and the principles that guide effective implementation are essential for advancing both methodological scholarship and educational practice in this rapidly evolving field.

Research Objectives

This review pursues three interconnected objectives advancing methodological scholarship in data analytics research:

1. To examine the prevalence and methodological characteristics of mixed methods research within data analytics research in higher education.
2. To delineate distinctive epistemological contributions and insights generated through mixed methods research compared to single-method approaches.

3. To synthesise evidence-informed principles and best practices for employing mixed methods research when investigating data analytics in higher education.

THEORETICAL AND CONCEPTUAL FRAMEWORK

This systematic review is grounded in three interconnected theoretical perspectives that inform the understanding of mixed methods research in data analytics education. Firstly, following Tashakkori et al. (2020) and Creswell and Plano Clark (2018), this review adopts a pragmatic philosophical stance that privileges practical problem-solving over adherence to single paradigms. Pragmatism posits that research questions should drive methodological choices, allowing researchers to draw flexibly from quantitative and qualitative traditions to generate actionable knowledge. This orientation is particularly relevant for data analytics education, where both measurable outcomes and contextual understanding are essential.

Secondly, building on Mertens (2018) and Ngulube and Ngulube (2022), the review conceptualises methodological plurality as an epistemological position that recognises that complex educational phenomena cannot be fully understood through monomethod approaches. This framework acknowledges that different methodologies generate distinct but complementary forms of knowledge, with MMR serving as a mechanism for integrating these knowledge forms to achieve a deeper understanding (Plano-Clark & Ivankova, 2016). Thirdly, drawing from Greene's (2007) foundational work on mixed methods purposes, this review emphasises epistemological complementarity, the principle that different data types illuminate different facets of the same phenomenon. In data analytics education, quantitative data may reveal patterns in learning outcomes, while qualitative data explains the mechanisms that produce those patterns. The integration of these perspectives generates insights unattainable through single methods (Plano Clark & Ivankova, 2016). These theoretical foundations inform the analysis of how MMR has been deployed in data analytics research, what distinctive contributions it offers, and how it can be optimised for future scholarship.

Structure of the Review

This systematic review is organised into several major sections. The next section outlines the review framework, detailing the search strategy, selection criteria, screening process, data extraction protocol, and quality appraisal procedures, all in accordance with PRISMA guidelines. Following that, the next section presents the findings and analysis, which include the review process and study selection, prevalence and temporal trends, research focuses and thematic domains, characteristics of methodological design, significant contributions of mixed-methods research (MMR), and evidence-informed principles for the application of MMR. A critical discussion of the findings situates them within broader theoretical debates about methodological pluralism, examining both successes and ongoing challenges, and proposes an application framework. The final section concludes with practical recommendations for educators, researchers, and institutions.

REVIEW FRAMEWORK

This investigation adopted a systematic literature review methodology, operationalising analytical synthesis of scholarly sources to map existing knowledge while identifying methodological patterns and gaps. The review adhered to Preferred Reporting Items for Systematic reviews and Meta-Analyses (PRISMA) guidelines (Page et al., 2021), ensuring transparency, reproducibility, and systematic rigour.

Search Strategy

The review implemented a comprehensive search strategy across five scholarly databases: ACM Digital Library, ERIC, PsycINFO, Scopus, and Web of Science. These databases were selected to ensure comprehensive coverage of educational research, computer science, information systems, and social sciences literature. The review supplemented database searches with targeted searches of regional journals from Africa (e.g., South African Journal of Higher Education), the Middle East (e.g., International Journal of Educational Technology in Higher Education), South and Southeast Asia (e.g., Asia Pacific Education Review), and the Pacific region (e.g., Higher Education Research & Development).

Search terms were constructed using Boolean operators (AND, OR, NOT) to optimise retrieval while maintaining specificity. The search string combined three conceptual domains: (1) data analytics terminologies ('data analytics', 'learning analytics', 'educational data mining', 'predictive analytics'); (2) higher education contexts ('higher education', 'tertiary education', 'undergraduate', 'postgraduate', 'graduate education', 'master's programs'); and (3) mixed methods identifiers ('mixed methods', 'multi-method', 'methodological plurality', 'methodological triangulation'). The complete search string applied across databases was: ("data analytics" OR "learning analytics" OR "educational data mining" OR "predictive analytics") AND ("higher education" OR "tertiary education" OR "undergraduate" OR "postgraduate" OR "graduate education") AND ("mixed methods" OR "multi-method" OR "methodological plurality" OR "methodological triangulation"). Boolean operators were also employed to exclude sources beyond the study's boundaries, specifically those related to literature reviews, systematic reviews, and meta-analyses, using the NOT operator ("literature review" OR "systematic review" OR "meta-analysis"). This systematic keyword construction yielded a comprehensive yet targeted evidence base, capturing the diverse dimensions of research data use in academic libraries. Searches were conducted between January and April 2025, covering publications from January 2016 to December 2025.

Selection Criteria

Using the PICO framework, the review structured inclusion and exclusion criteria to ensure methodological rigour and theoretical relevance:

- Population (P): Studies focused on higher education contexts, encompassing undergraduate and postgraduate learners, courses, and programmes. Research outside higher education was excluded.
- Intervention (I): The "intervention" was the MMR methodology used. Only studies employing MMR, defined as research explicitly integrating both quantitative and qualitative data collection and analysis, were included. Single-method studies were excluded to maintain consistency and focus. Studies claiming to be mixed methods but presenting only adjacent quantitative and qualitative findings without integration were critically evaluated and excluded if integration was absent.
- Comparison (C): While direct comparison groups were not mandatory, studies were evaluated against criteria for methodological rigour and theoretical coherence, ensuring credible evidence. Studies lacking sampling frameworks, analytical transparency, or significant findings were excluded.
- Outcome (O): Outcomes of interest were contributions to data analytics research in higher education, including insights into pedagogy, learning, assessment, institutional practices, curriculum design, faculty development, and ethical dimensions. Studies without explicit data analytics focus or insufficient scholarly contribution were excluded.

Additional criteria included:

- Peer-reviewed empirical research published in English
- Publication timeframe of 2016–2025, ensuring currency and accessibility
- Exclusion of non-empirical works (conceptual papers, editorials, book reviews)
- Exclusion of conference abstracts, posters, and other non-peer-reviewed works

- Studies must report original empirical research conducted in higher education settings

Screening and Selection Process

The review implemented a multi-stage screening protocol to ensure systematic rigour. Initial screening involved title and abstract analysis to assess relevance against inclusion criteria. During this phase, two independent reviewers (the researcher and a peer reviewer with expertise in systematic reviews and mixed methods research) screened records using a standardised screening form. Studies meeting preliminary criteria underwent full-text evaluation. To enhance reliability and minimise reviewer bias, both reviewers conducted parallel screening of all records. Inter-rater reliability was assessed using Cohen's kappa (κ), yielding $\kappa = 0.87$ for the title and abstract screening phase, indicating strong agreement. Discrepancies in 13% of cases were resolved through collaborative deliberation and consensus-building, with a third reviewer consulted in two cases where consensus could not be reached.

Full-text screening followed the same dual-reviewer process, with both reviewers independently assessing each article against the full inclusion and exclusion criteria. Inter-rater reliability for full-text screening was $\kappa = 0.84$, again indicating strong agreement. Discrepancies were resolved through discussion, and detailed notes were maintained documenting reasons for inclusion or exclusion of each study.

Data Extraction Protocol

Data extraction followed PRISMA guidelines (Page et al., 2021), utilising the open-source CADIMA platform to ensure consistency and transparency. A standardised data extraction form was developed and piloted on five studies before full implementation. The extraction framework captured:

- Study characteristics: Authorship, publication year, geographical context, institutional setting, study aims
- Methodological architecture: Research design typology (parallel, sequential, integrated, etc.), data generation methods (surveys, interviews, focus groups, observations, document analysis), sample characteristics (size, selection strategy, demographics), analytical procedures (statistical techniques, qualitative coding approaches, integration strategies)
- Data analytics dimensions: Specific techniques employed (descriptive analytics, predictive modelling, learning analytics dashboards), tools and platforms used, applications (student performance prediction, curriculum evaluation, intervention design)

- Educational context: Programme type, academic level, disciplinary focus, institutional characteristics
- Substantive findings: Key insights, practical implications, identified challenges, reported outcomes
- Methodological reflexivity: Author-identified limitations, quality of integration between quantitative and qualitative components, transparency of methodological decisions, future research directions

To ensure extraction consistency, the two reviewers independently extracted data from 25% of included studies (n = 6 studies), achieving an agreement rate of 92%. Discrepancies were discussed and resolved, and the extraction protocol was refined based on this pilot phase. The remaining studies were divided between the two reviewers, with regular meetings to discuss emerging themes and ensure consistency.

Quality Appraisal

The review assessed methodological quality using the Critical Appraisal Skills Programme (CASP) framework, which provides structured tools to evaluate research trustworthiness, relevance, and results (CASP, 2025). Given the focus on mixed methods studies, the review applied both the CASP Qualitative Checklist and the CASP Systematic Review Checklist, adapted to evaluate the quality of MMR integration.

The two independent reviewers applied CASP criteria to all included studies, achieving inter-rater reliability of $\kappa = 0.82$, indicating substantial agreement. Disagreements were resolved through discussion. Each study was rated on a scale of 1 – 5 (1 = poor quality, 5 = excellent quality) across four key dimensions. Studies scoring below 3 on any dimension or with an overall average below 3.5 were excluded from the final synthesis. Table 1 outlines the assessment dimensions.

Table 1. Quality Appraisal Framework (CASP)

Quality Dimension	CASP Criteria	Evaluation Parameters
Research Design Robustness	Appropriateness of design for the research question	Alignment between methodology and objectives; credibility of inferential claims; theoretical coherence; explicit justification for MMR approach
Data Generation Validity	Validity and reliability of data collection	Use of validated instruments; measurement accuracy; methodological alignment with aims; adequacy of sampling strategy
Sample Adequacy	Sufficiency of sample size and selection	Statistical power for quantitative components; depth and diversity for qualitative transferability; sampling strategy appropriateness; clear description of participants
Analytical Transparency	Clarity and appropriateness of analysis	Suitability of statistical techniques; transparency in qualitative coding; verifiability of interpretative processes; explicit description of integration procedures

Note: Studies demonstrating strong methodological quality received greater weight in evidence synthesis, while studies with limitations were interpreted with appropriate caution.

FINDINGS AND ANALYSIS

This section presents the key findings from the reviewed studies and provides a critical analysis of their methodological approaches, thematic contributions, and implications for data analytics research in higher education.

Review Process and Study Selection

The initial search yielded 2,018 potentially relevant studies across the five databases and supplementary regional journal searches. Following deduplication using reference management software (n = 712 duplicates removed) and removal of non-English publications (n = 149), 1,157 studies remained for preliminary screening. Application of peer-review criteria resulted in the exclusion of 431 articles (conference abstracts, editorials, commentaries, and non-peer-reviewed works), leaving 726 studies for detailed examination. Rigorous scrutiny of titles and abstracts resulted in the removal of 672 studies that failed to meet the inclusion criteria. The primary reasons for exclusion at this stage were: focus outside of higher education (n = 289), not employing mixed methods (n = 251), lack of explicit focus on data analytics (n = 98), and non-empirical studies (n = 34). The remaining 54 studies underwent full-text evaluation and quality assessment using the CASP framework.

Quality assessment resulted in 32 exclusions. Specifically, 12 studies scored below 3.5 in overall quality, 8 studies demonstrated weak integration between quantitative and qualitative components (merely presenting findings adjacent to each other without meaningful synthesis), 7 studies lacked sufficient analytical transparency, and 5 studies had inadequate sampling or data collection procedures. Twenty-two studies demonstrating exceptional methodological rigour, conceptual coherence, and scholarly significance were included in the final synthesis. Figure 1 presents the PRISMA flow diagram illustrating the complete selection process.

Figure 1. PRISMA Flow Diagram (Adapted from Page et al., 2021)

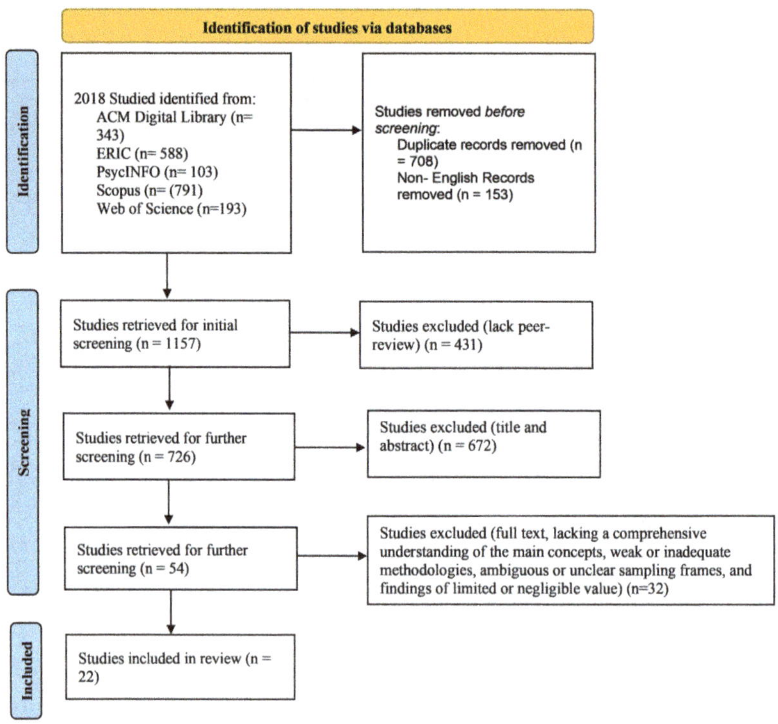

Note: The screening stage exclusion criteria were studies outside higher education contexts (n = 289), studies not employing mixed methods (n = 251), studies without an explicit data analytics focus (n = 98), and non-empirical studies (n = 34).

Geographical and Contextual Distribution

The included studies (n = 22) demonstrated some geographical diversity, though concentrated in specific regions. Table 2 presents the geographical distribution of the reviewed studies.

Table 2. Geographical Distribution of Included Studies

Region	Studies (n)	%	Representative Countries
North America	8	36.4%	United States (6), Canada (2)
Europe	7	31.8%	Netherlands (3), United Kingdom (2), Spain (1), Finland (1)
Asia-Pacific	4	18.2%	Australia (2), Japan (1), China (1)
Africa	2	9.1%	South Africa (1), Ghana (1)
Latin America	1	4.5%	Chile (1)
Total	**22**	**100.0%**	

The geographical distribution reveals that 68.2% of studies were conducted in North America and Europe, indicating the continued dominance of institutions from the Global North in MMR scholarship on data analytics education. However, the inclusion of studies from Africa (9.1%), Latin America (4.5%), and Asia-Pacific (18.2%) demonstrates emerging engagement with MMR in diverse contexts. This distribution highlights the need for future research to actively include perspectives from underrepresented regions, particularly Africa, the Middle East, South Asia, and the Pacific Islands, where educational contexts and resource constraints may significantly differ.

Prevalence and Temporal Trends

Analysis of the 22 included studies reveals significant patterns in the deployment of MMR within data analytics education research. A notable temporal trend emerged, as shown in Table 3.

Table 3. Publication Trends of Included Studies

Period	Studies (n)	%	Phase Characteristics
2016	2	9.1%	Emergent: Initial explorations of methodological plurality
2017–2019	2	9.1%	Developmental: Growing recognition of MMR value
2020–2022	10	45.4%	Expansion: Marked increase reflecting disciplinary maturation
2023–2025	8	36.4%	Consolidation: Continued growth with methodological sophistication
Total	**22**	**100.0%**	**Accelerating adoption post-2020**

The marked increase from 2020 onwards (81.8% of studies published in 2020 or later) reflects growing recognition of methodological plurality's value and broader disciplinary maturation in educational data analytics scholarship. The post-2020 surge may also reflect increased emphasis on data-driven decision-making in higher education, accelerated by the COVID-19 pandemic's disruption of traditional educational practices and the subsequent demand for robust evidence about online and hybrid learning environments.

Research Foci and Thematic Domains

Table 4 summarises research foci of included studies, highlighting multidimensional contributions to higher education and temporal evolution patterns.

Table 4. Research Foci Evolution in Data Analytics Education

Research Focus	Studies (n)	Representative Studies	Contribution	Temporal Pattern
Learning Outcomes and Student Experiences	5	Atuase and Maluleka (2023); Buunaaisie et al. (2018); Jivet et al. (2020); Kahn et al. (2016); Lin-Aung et al. (2022)	Examines data analytics' impact on skill development, learning trajectories, and student perceptions	Consistent across all periods
Curricular Integration	4	Liu et al. (2022); Prieto et al. (2023); Silvola et al. (2023); Ustun et al. (2022)	Explores challenges, opportunities, and strategies for embedding analytics within curricula	Increased focus post-2020
Learning Analytics Applications	6	Chaparro-Amaya et al., (2023); Fan et al. (2021); Ifenthaler and Schumacher (2016); van Ameijde et al. (2023); Viberg et al. (2024); West et al. (2020)	Investigates how analytics enhance pedagogy, advising, and student support	Dominant throughout
Faculty Development	6	Alzahrani et al. (2023); Clark and Tuffley (2019); Hilliger et al. (2020); Khalid et al. (2023); van Ameijde et al. (2023); Wang and Zeng (2022)	Studies academic staff needs, competencies, and support requirements	Emerging post-2019
Ethical Dimensions	3	Alzahrani et al. (2023); Ifenthaler and Schumacher (2016); Viberg et al. (2024)	Addresses privacy, ethical considerations, and responsible data practices	Recent emphasis (2020+)

Note: Studies may contribute to multiple foci; total exceeds 22.

The studies illustrate the expanding scope and depth of data analytics research. Early work centred on technical applications and learning outcomes has evolved to encompass broader pedagogical, institutional, and ethical dimensions, signalling a field moving toward methodological maturity and a holistic understanding of data analytics as a transformative force in higher education.

Methodological Design Characteristics

Table 5 outlines predominant design categories, illustrating how researchers operationalised methodological plurality in data analytics research.

Table 5. MMR Design Architectures

Design Category	Characteristics	Studies	Studies (n)	%
Parallel Designs	Simultaneous quantitative and qualitative data generation with convergent integration	Chaparro-Amaya et al. (2023); Hilliger et al. (2020); Kahn et al., (2016); Liu et al. (2022); Prieto et al., (2023)	5	22.7%
Integrated Mixed Methods	Comprehensive integration throughout research phases	Fan et al. (2021); van Ameijde et al. (2023); Westein et al. (2023)	3	13.6%
Instruments Triangulation (Survey–Focus Group; Survey–Interview)	Structured quantitative instruments with group-based qualitative exploration. Also, standardised quantitative measures with in-depth individual qualitative inquiry	Alzahrani et al. (2023); Ifenthaler and Schumacher (2016); Khalid et al. (2023); Viberg et al. (2024)	4	18.3%
Sequential Designs	Phased approach with subsequent phases informed by preceding findings	Atuase and Maluleka (2023); Buunaaisie et al. (2018); Heneghan et al. (2021); Jivet et al. (2020); Lin-Aung et al. (2022); Ustun et al. (2022); Wang et al. (2022); West et al. (2020)	8	36.4%
Multiple Case Studies	Comparative analysis across institutional/ programmatic contexts	Clark and Tuffley (2019); Silvola et al. (2023)	2	9.1%
Experimental Mixed Methods	Intervention-based designs integrating experimental protocols with qualitative elaboration	Ustun et al. (2022)	1	4.5%
Total			**22**	**100%**

Note: One study (Ustun et al., 2022) employed both sequential and experimental design elements and is counted in both categories, but only once in the total.

The combination of sequential designs, parallel designs, and instrument triangulation accounts for the vast majority of methodological approaches (77.3% collectively), suggesting a strong preference for designs that clearly separate or phase quantitative and qualitative data generation. Sequential design emerged as the most dominant approach (36.4%), reflecting utility for phased research where findings from one method inform the design or interpretation of the next. Conversely, the presence of six distinct design categories, including the least frequent multiple case studies and experimental mixed methods (4.5% each), highlights critical methodological diversity, confirming that MMR practice thoughtfully aligns design architecture to the specific research problem rather than relying on a single, monopolistic approach. The relatively low prevalence of fully integrated mixed methods designs (13.6%) suggests that while researchers recognise MMR's value, achieving deep, continuous integration throughout all research phases remains challenging in practice.

Distinctive Contributions of MMR

Analysis reveals that MMR generates distinctive analytical value transcending single-method approaches. Table 6 details these contributions across domains.

Table 6. Epistemological Contributions of MMR Across Research Domains

Domain	Representative Studies	Epistemological Value	Key Contributions	Practice Implications
Student Learning and Development	Buunaaisie et al. (2018); Heneghan et al. (2021); Lin-Aung et al. (2022); Silvola et al. (2023); Wang et al. (2022)	Epistemological complementarity	Linked outcomes to experiences; revealed competency-skill tensions; captured affective dimensions	Informs holistic curriculum design and personalised interventions
Curriculum and Pedagogy	Chaparro-Amaya et al. (2023); Fan et al. (2021); van Ameijde et al. (2023); Ustun et al. (2022)	Explanatory depth	Identified barriers/ enablers; elucidated causal mechanisms; exposed design-enactment gaps	Guides curriculum reform and evidence-informed policy
Assessment and Evaluation	Jivet et al. (2020); Silvola et al. (2023); Viberg et al. (2024); Westein et al. (2023)	Dialectical understanding	Exposed formative-summative tensions; revealed reliability-validity trade-offs; showed impact on learning	Optimises assessment design and educator literacy
Ethics and Privacy	Alzahrani et al. (2023); Ifenthaler and Schumacher (2016); West et al. (2020)	Ethical dimensionality	Examined policy-practice gaps; highlighted context-dependent trust; illuminated reasoning processes	Strengthens governance frameworks and trust-building
Institutional Implementation	Atuase and Maluleka (2023); Clark and Tuffley (2019); Hilliger et al. (2020)	Contextual situatedness	Demonstrated contextual variation; revealed systemic barriers; identified success factors	Enables context-sensitive implementation strategies
Professional Competencies	Khalid et al. (2023); Kahn et al. (2016); Liu et al. (2022); Prieto et al. (2023)	Stakeholder plurality	Mapped education-workplace relationships; revealed capability gaps; identified development opportunities	Bridges the theory-practice gap and enhances employability

Note: Some studies contribute across multiple academic domains

The analysis demonstrates how MMR enriches understanding through epistemological complementarity, explanatory depth, stakeholder plurality, contextual situatedness, and ethical dimensionality. Studies integrate quantitative outcomes with qualitative experiences, revealing how institutional and pedagogical factors shape implementation while exposing tensions between formal policies and personal concerns. These contributions move beyond description to explanation, generating integrative and contextually grounded insights.

Evidence-Informed Principles for MMR Deployment

Analysis of methodological approaches enables synthesis of evidence-informed principles for effective MMR deployment. Table 7 details these principles.

Table 7. Evidence-Informed Principles for MMR Deployment

MMR Dimension	Core Principles	Operational Guidelines	Methodological Implications	Representative Studies
Parallel Design	Conceptual alignment; temporal coordination; analytical integration	Address complementary research facets; plan strategic data collection timing; specify cross-strand synthesis procedures	Promotes coherence; prevents isolated analyses; ensures complementary data generation	Chaparro-Amaya et al. (2023); Hilliger et al. (2020); Kahn et al., (2016); Liu et al. (2022); Prieto et al., (2023)
Sequential Design	Purposeful sequencing; responsive design; sample connectivity	Justify phase ordering theoretically; build subsequent phases on prior findings; clarify sample relationships across explanatory or exploratory sequences	Enhances explanatory continuity; enables question refinement; supports iterative theory development through phased inquiry	Atuase and Maluleka (2023); Buunaaisie et al. (2018); Heneghan et al. (2021); Jivet et al. (2020); Lin-Aung et al. (2022); Ustun et al. (2022); Wang et al. (2022); West et al. (2020)
Integration Strategies	Analytical triangulation; explanatory elaboration; theoretical synthesis	Employ multiple sources for corroboration; use one data form to explain another; develop integrative frameworks through instrument triangulation	Fosters comprehensive interpretation; enhances explanatory power; enables synergistic knowledge construction across data types	Alzahrani et al. (2023); Ifenthaler and Schumacher (2016); Khalid et al. (2023); Viberg et al. (2024); Westein et al. (2023)
Integrated Mixed Methods	Comprehensive integration throughout research phases; embedded analytical procedures; continuous strand dialogue	Maintain methodological balance; document integration touchpoints; ensure dynamic interplay between quantitative and qualitative components	Enables emergent insights; supports inferential consistency; facilitates synergistic data generation and analysis	Fan et al. (2021); van Ameijde et al. (2023); Westein et al. (2023)
Experimental Mixed Methods	Intervention-based integration; causal inference with contextual understanding; pretest-post-test with qualitative elaboration	Design controlled interventions; embed qualitative components for mechanism exploration; link outcome measures with lived experiences	Combines experimental rigour with explanatory depth; reveals intervention mechanisms; bridges efficacy and contextual understanding	Ustun et al. (2022)

continued on following page

Table 7. Continued

MMR Dimension	Core Principles	Operational Guidelines	Methodological Implications	Representative Studies
Case Study Approaches	Contextual boundedness; comparative analysis; cross-case synthesis	Define case boundaries clearly; employ systematic comparison protocols; synthesise findings across institutional or programmatic contexts	Reveals contextual variation; identifies context-specific success factors; enables transferability assessment	Clark and Tuffley (2019); Silvola et al. (2023)
Methodological Transparency	Design rationale; integration points; analytical procedures; interpretative protocols	Articulate epistemological appropriateness; identify data intersection points; describe methods transparently; explain divergence reconciliation; document analytical decisions	Strengthens credibility and reproducibility; enables critical appraisal; supports methodological advancement; facilitates replication and meta-analysis	Synthesised across the entire corpus (n=22 studies)

Note: Some studies exemplify multiple MMR dimensions due to their sophisticated design architectures

These principles demonstrate that effective MMR deployment relies on four interdependent dimensions collectively enhancing methodological coherence and explanatory depth. Parallel designs require conceptual and temporal alignment, ensuring complementary rather than isolated analyses. Sequential designs depend on purposeful sequencing and sample connectivity, maintaining logical continuity between phases. Integration strategies, through analytical triangulation, explanatory elaboration, and theoretical synthesis, transform separate data strands into cohesive, meaning-rich interpretations. Underpinning all is methodological transparency, demanding explicit articulation of design rationale, integration points, and interpretative procedures to strengthen credibility and reproducibility.

DISCUSSION

This section offers an analytical discussion of the study's findings, integrating them with relevant literature while critically examining both successes and persistent challenges in MMR deployment within data analytics education research.

Epistemological Transition in Data Analytics Research

MMR prevalence and characteristics in data analytics research indicate a field undergoing epistemological transition, reflecting increasing recognition that educational data phenomena cannot be fully understood through mono-method approaches. The post-2020 surge in MMR studies (81.8% of included studies) reflects both quantitative growth and qualitative maturation in analytical sophistication (Abdul-Jabbar & Farhan, 2022; Ngulube, 2020), suggesting researchers are increasingly attuned to complex, multidimensional learning analytics environments. This temporal concentration also coincides with the COVID-19 pandemic's disruption of higher education, which has accelerated digital transformation and heightened demands for robust, multifaceted evidence on the effectiveness of online and hybrid learning.

This aligns with arguments that methodological pattern shifts indicate disciplinary self-reflexivity, whereby researchers critically examine the adequacy and limitations of dominant paradigms (Bryman, 2016; Flick, 2018; Mertens, 2018). The evolution from predominantly North American and European scholarship toward emerging contributions from Africa, Latin America, and the Asia-Pacific (31.8% of studies) signals a growing global recognition of MMR's value, although significant geographical gaps persist, particularly in the Middle East, South Asia, and Pacific Islands regions.

Research foci expansion, from learning outcomes and curriculum integration to ethics, faculty development, and professional competencies, illustrates epistemic broadening, signalling that data analytics in higher education is increasingly conceptualised as both a technical and socio-pedagogical construct (Alzahrani et al., 2023; Caspari-Sadeghi, 2023). The methodological architecture demonstrates conscious movement away from rigid quantitative traditions (Arnold & Pistilli, 2012; Pardo et al., 2017) toward designs privileging epistemological complementarity (Creswell & Plano Clark, 2018; Ngulube & Ngulube, 2022).

Methodological Sophistication and Persistent Challenges

The analysis of the study corpus reveals that sequential designs are the dominant methodological choice (36.4%), reflecting utility for phased research where findings from one method inform the design or interpretation of the next (Atuase & Maluleka, 2023; Mertens, 2018; Toyon, 2021; Wang et al., 2022). This phased approach inherently embodies higher methodological reflexivity, as it iteratively links quantitative and qualitative phases to facilitate causal explanation and deeper insight into complex phenomena (Bellaj et al., 2024; Hassna, 2022; Mustapha et al., 2023).

Jivet et al. (2020) exemplify effective sequential design in their study of learner dashboard sense-making. They first collected quantitative log data and survey responses to identify patterns in dashboard usage and correlations with self-regulated learning outcomes. These findings informed the design of subsequent semi-structured interviews, where researchers probed specific usage patterns and explored students' reasoning processes. The integration revealed that transparency of design, reference frames, and support for action were critical factors, insights that emerged only through purposeful sequencing that built qualitative inquiry on quantitative foundations. However, the persistence of parallel designs and instrument triangulation (41.0% collectively) indicates that many researchers still opt for simultaneous or survey-heavy models. While parallel designs efficiently capture both quantitative generalisability and qualitative contextual depth within compressed timeframes, this reliance exposes potential weaknesses: if the integration of separate data streams is not rigorously theorised, the strands risk remaining juxtaposed rather than meaningfully synthesised, limiting interpretive power (Bellaj et al., 2024; Bazeley, 2018; Stojanov & Daniel, 2023; Alkhalil et al., 2021).

Among the 32 studies excluded during quality appraisal, eight demonstrated weak integration where quantitative and qualitative findings were presented adjacently without genuine synthesis. For instance, one excluded study reported survey results about student satisfaction with learning analytics tools in one section, followed by interview quotations about user experience in another section, but failed to explain discrepancies or connect findings to generate integrated insights. This juxtaposition approach limits added value over separate single-method studies. This echoes contentions that true MMR advancement requires embedding integration at the epistemological level rather than merely procedural application (Mertens, 2018; Ngulube & Ukwoma, 2021). The corpus, therefore, demonstrates both methodological sophistication (via sequential and integrated designs) and the persistent challenge of achieving deep, conceptual data integration across the field (Mertens, 2018; Toyon, 2021).

Contradictions and Unresolved Tensions

While the review highlights MMR successes, critical examination reveals unresolved tensions requiring transparent acknowledgement. Not all MMR studies achieved meaningful integration. As noted, among the included studies, four exhibited relatively weak integration where quantitative and qualitative findings were presented adjacently without robust synthesis, limiting added value over separate single-method studies. For example, one study reported statistical correlations between analytics tool usage and performance alongside thematic analysis of user perceptions, but failed to explain contradictions or develop integrated theoretical insights. This suggests that adopting MMR terminology does not automatically ensure epistemological complementarity (Bazeley, 2018; Creswell & Creswell, 2022; Creswell & Plano Clark, 2018; Ngulube & Ukwoma, 2021; Tashakkori et al., 2020).

Resource constraints emerged as a recurring concern. Studies conducted by individual researchers or small teams reported challenges in managing simultaneous data collection and analysis, suggesting MMR may be more feasible for well-resourced research teams (Bazeley, 2018; Creswell & Creswell, 2022; Ncube & Ngulube,

2025; Ngulube & Ukwoma, 2021). Analysis revealed that 18 of 22 included studies (81.8%) involved research teams of three or more members, and 14 studies (63.6%) explicitly acknowledged institutional or external funding support. This raises equity concerns about who can meaningfully employ MMR approaches, particularly disadvantaging early-career researchers, scholars in under-resourced institutions, and researchers in the Global South, where research funding is limited.

Publication pressures may create bias toward reporting successful MMR implementations. The corpus likely underrepresents studies where MMR added minimal value or where integration proved unsuccessful, as such findings may be less publishable. This publication bias limits understanding of when MMR is and is not appropriate. Despite extensive database searches, the review found no published studies explicitly reporting MMR failures or contexts where single-method approaches would have been superior. This absence suggests either universal MMR success (unlikely) or systematic underreporting of negative findings.

Practical Implementation Barriers

Beyond methodological considerations, practical implementation barriers significantly constrain MMR adoption in data analytics education research (Mohammadi, 2023; Zhao et al., 2024; Calicdan et al., 2025; Dyulicheva, 2024; Adhikari & Timsina, 2024):

- Resource Requirements: MMR demands substantial time, funding, and expertise. Studies in the corpus typically involved multi-year projects (mean duration: 24 months) with dedicated research teams. Individual researchers may face significant barriers to MMR adoption, particularly in institutions with high teaching loads and limited research support.
- Institutional Pressures: Academic promotion and tenure systems often favour rapid publication over methodologically intensive approaches. The extended timeframes required for quality MMR (18-36 months in included studies) may disadvantage early-career researchers facing "publish or perish" pressures. Only five studies (22.7%) involved doctoral students or early-career researchers as lead authors, suggesting MMR may be perceived as too risky for those establishing careers.
- Training and Expertise Gaps: Researchers require proficiency in both quantitative and qualitative methodologies, analytical integration techniques, and epistemological frameworks. Current doctoral training programmes often emphasise methodological specialisation over pluralism. Among included studies, 19 (86.4%) involved interdisciplinary teams combining quantitative

specialists with qualitative experts, suggesting that comprehensive MMR expertise within individual researchers remains rare.

- Ethical Approval Complexity: MMR studies involving multiple data collection phases and diverse stakeholder groups face complex ethical approval processes, particularly regarding informed consent and data governance. Seven studies (31.8%) reported extended ethics review periods (4-8 months), with two studies requiring amended protocols during data collection as research designs evolved responsively.
- Analytical Tool Limitations: While specialised software exists for MMR (Ngulube, 2023), researchers reported challenges integrating outputs from separate quantitative (e.g., SPSS, R) and qualitative (e.g., NVivo, ATLAS.ti) analytical tools. Only three studies (13.6%) utilised integrated analytical platforms; the remainder required manual integration, which increased the workload and the potential for analytical inconsistencies.

However, it is worth noting that Hilliger et al. (2020) successfully navigated several implementation barriers in their multi-country study on the adoption of learning analytics in Latin American universities. They assembled an international team that combined methodological expertise, secured multi-year funding from institutional and government sources, staggered data collection to manage the workload, and developed clear protocols for integrating survey data with interviews and focus groups. Their transparency about resource requirements and collaborative strategies provides a valuable model for other researchers.

Distinctive MMR Contributions and Explanatory Power

MMR's distinctive contributions lie in the capacity to generate explanatory insights that single-method studies cannot provide (Chaparro-Amaya et al., 2023; Tashakkori et al., 2020; Viberg et al., 2018). Across domains, MMR enables simultaneous interrogation of "what occurs" (quantitative trends) and "why it occurs" (qualitative meaning) (Almalki, 2016; Creswell & Plano Clark, 2018; Ngulube & Ngulube, 2022; Tashakkori et al., 2020). Studies demonstrate how MMR reveals relationships between measurable outcomes and lived experiences. Wang et al. (2022) integrated quantitative analysis of research skill development with qualitative exploration of postgraduate students' experiences, revealing that while technical competencies improved measurably, students experienced significant affective challenges (anxiety, isolation) that quantitative measures alone would not have captured.

This integration enabled recommendations addressing both skill development and emotional support needs.

Assessment-focused MMR studies illuminate how formative and summative practices interact to shape learner trajectories, revealing both efficacy patterns and underlying processes (Bryman, 2016; Prieto et al., 2023). Chaparro-Amaya et al. (2023) used MMR to identify factors influencing programming course performance, combining learning analytics with student interviews. Quantitative analysis revealed engagement patterns, while qualitative data explained underlying motivations and barriers, exposing gaps between curriculum design intentions and student experiences.

Ethics and privacy research underscores data collection and analytics implementation complexity, highlighting discrepancies between formal policy and stakeholder perception (Flick, 2018; Ifenthaler & Schumacher, 2016; Slade & Prinsloo, 2013). Alzahrani et al. (2023) found high trust in institutional competence through surveys, but interviews revealed specific concerns about third-party data handling and algorithmic transparency that quantitative measures did not capture. This nuanced understanding informed more targeted privacy protection strategies. Institutional implementation studies demonstrate adoption's context-dependent nature, reinforcing arguments that generalised quantitative findings require qualitative interpretation to inform practical application (Clark & Tuffley, 2019; Hilliger et al., 2020). Clark and Tuffley's (2019) comparative case study of three Australian universities revealed that while all institutions demonstrated commitment to learning analytics (quantitative investment metrics), qualitative analysis exposed vastly different implementation philosophies, stakeholder engagement strategies, and success indicators, necessitating context-sensitive rather than uniform approaches.

Professional competencies analysis underscores stakeholder plurality, integrating perspectives of students, educators, and employers to identify misalignments and opportunities for targeted curriculum development (Buunaaisie et al., 2018; Khalid et al., 2023). Khalid et al. (2023) combined industry surveys with focus groups to map sensor data analytics competencies in construction education, revealing gaps between academic curricula and industry expectations that neither method alone would have fully captured. These studies reveal MMR functions as an epistemologically informed approach rather than methodological aggregation, facilitating multi-layered interpretation, theory integration, and context-sensitive knowledge production (Creswell & Creswell, 2022; Morse et al., 2023; Ngulube & Ukwoma, 2021).

When MMR May Not Be Appropriate

Critical analysis of the included studies and excluded literature suggests contexts where MMR may not offer advantages over single-method approaches (Creswell &

Creswell, 2022; Creswell & Plano Clark, 2018; Mertens, 2018; Saraswati & Devy, 2023; Ngulube & Ngulube, 2022; Strang, 2017; Song et al., 2022):

- Well-Defined Quantitative Questions: When research questions focus solely on measuring relationships or testing specific hypotheses with established constructs, quantitative approaches may suffice. For example, validating a predictive model for student retention using existing administrative data requires rigorous quantitative methods but may not benefit from qualitative elaboration if the research aim is purely predictive accuracy.
- Exploratory Qualitative Studies: Early-stage investigations seeking to understand emerging phenomena may benefit from exclusively qualitative approaches before attempting methodological integration. Premature integration may constrain conceptual development when phenomena are not yet well understood.
- Resource-Constrained Contexts: When time, funding, or expertise limitations exist, well-executed single-method studies may provide more valuable contributions than poorly integrated MMR. Quality should never be sacrificed for methodological complexity. Several excluded studies attempted MMR but suffered from small qualitative samples ($n<10$), inadequate quantitative power, or superficial integration, contexts where focused single-method approaches would have been superior.
- Clear Disciplinary Epistemologies: In contexts where disciplinary norms strongly favour particular methodological approaches, MMR may face legitimacy challenges despite potential benefits. For instance, in highly quantitative fields like psychometrics or econometrics, MMR may encounter reviewer scepticism unless integration is impeccably justified and executed.
- When Integration Does Not Add Value: If quantitative and qualitative components address entirely separate research questions without a meaningful intersection, MMR terminology may misrepresent what essentially parallel single-method studies are. True MMR requires that integration generate insights beyond what separate studies would produce.

Application Framework for MMR in Data Analytics Research

Figure 2 illustrates the application of MMR in data analytics research within higher education, demonstrating how methodological plurality leads to an enhanced understanding.

Figure 2. Application of MMR in Data Analytics Research (Designed by the Author)

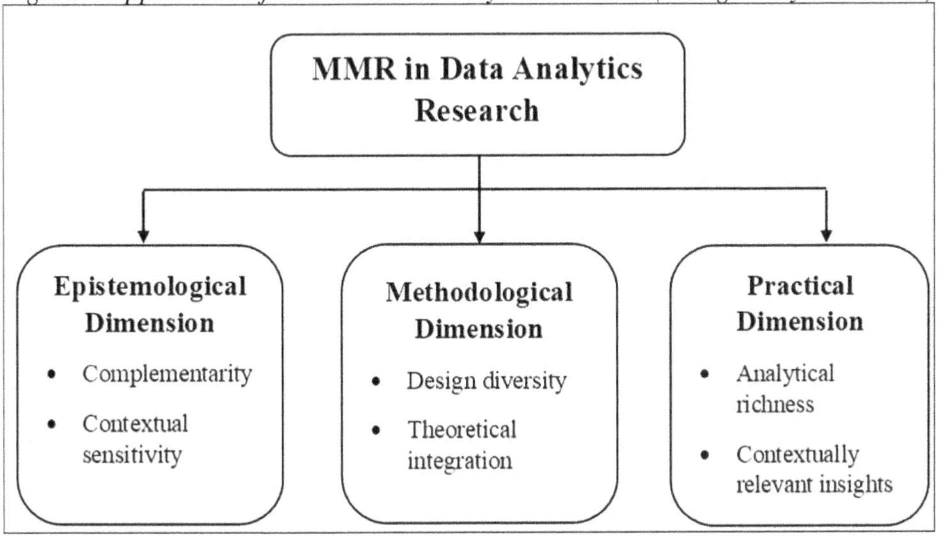

This framework illustrates the iterative, integrative process of MMR deployment. Researchers begin by identifying research questions that genuinely require both quantitative and qualitative perspectives, questions asking not only "what" and "how much" but also "why" and "how." Design selection follows logically from research aims, resource availability, and disciplinary context, with sequential designs offering explanatory power, parallel designs providing efficiency, and integrated designs enabling comprehensive emergence of insights.

Data generation proceeds with explicit attention to complementarity, ensuring that quantitative and qualitative components address interconnected facets of the phenomenon. Integration occurs at multiple touchpoints, during data collection (responsive design), analysis (triangulation, elaboration, synthesis), and interpretation (theoretical integration), rather than merely at the reporting stage. Throughout, methodological transparency and reflexivity ensure credibility and enable critical appraisal.

The framework demonstrates that MMR in data analytics research represents not merely a methodological choice but an ontological and epistemological stance. Thus, by privileging complementarity, contextual sensitivity, and theoretical integration, MMR transcends descriptive accounts to produce analytically rich, practically relevant, and conceptually robust insights (Abdul-Jabbar & Farhan, 2022; Song, 2022; Strang, 2017; van Ameijde et al., 2023). This shift highlights the potential of mixed methods to advance both scholarship and educational practice, offering a nuanced understanding of complex learning ecosystems and the socio-technical dynamics of data-informed pedagogical innovation (Strang, 2017).

LIMITATIONS AND FURTHER RESEARCH DIRECTIONS

Despite rigorous implementation of PRISMA guidelines and quality appraisal procedures, this systematic review has several limitations that warrant acknowledgement and suggest directions for future scholarship. The restriction to English-language, peer-reviewed publications from five major databases and selected regional journals, while ensuring quality and accessibility, may have excluded relevant studies published in other languages or dissemination formats, potentially introducing linguistic and publication-type bias that limits understanding of MMR practices in non-Anglophone contexts and alternative scholarly communities. The timeframe of 2016-2025, though ensuring currency, may have missed foundational MMR work in data analytics education published earlier, limiting the historical perspective on methodological evolution. Quality appraisal, while systematic and employing dual independent reviewers with strong inter-rater reliability ($\kappa = 0.82$-0.87), necessarily involves interpretive judgments about integration quality and methodological rigour, and the decision to exclude studies scoring below 3.5 on average, while maintaining synthesis quality, may have eliminated studies offering valuable insights into MMR challenges and failures. The geographical concentration of included studies in North America and Europe (68.2%), despite intentional searches of regional journals, reflects genuine imbalances in the global research landscape but limits the transferability of findings to under-resourced contexts in Africa, the Middle East, South Asia, and Pacific regions, where institutional constraints, regulatory environments, and educational priorities may differ substantially. Publication bias represents a critical limitation, as the corpus likely underrepresents studies where MMR added minimal value or where integration proved unsuccessful, constraining understanding of when MMR is and is not appropriate, a concern reinforced by the absence of any published studies explicitly reporting MMR failures despite extensive searches. The review's focus on higher education excludes potentially relevant MMR applications in other levels of education that might illuminate alternative approaches, and the emphasis on data analytics education specifically, while providing necessary focus, limits insights about MMR in broader educational technology or learning sciences research.

Future research should address these limitations through multiple complementary directions. Comparative studies examining MMR practices across diverse geographical, institutional, and resource contexts would illuminate how constraints and opportunities shape methodological choices, with particular attention to barriers and innovations in the Global South, where limited research infrastructure necessitates creative approaches. Longitudinal investigations tracking the evolution of individual researchers' or institutions' MMR practices over time could reveal developmental trajectories, learning processes, and factors enabling progression from superficial to sophisticated integration. Methodological innovation studies should experimentally compare different integration strategies, analytical approaches, and design architectures within data analytics research, generating empirical evidence about relative effectiveness rather than relying solely on theoretical prescriptions or post-hoc analysis. Research explicitly examining unsuccessful MMR implementations, integration failures, and contexts where single methods proved superior would provide essential balance to the current literature's emphasis on successful cases, informing more nuanced guidance about appropriate methodology selection. Studies exploring the role of emerging technologies, artificial intelligence-assisted qualitative analysis, integrated MMR platforms, and advanced visualisation tools for synthesis could identify innovations addressing current analytical and integration challenges. Investigations of training and support mechanisms examining how doctoral programmes, faculty development initiatives, and institutional research offices can effectively build MMR capacity would inform educational practice and policy. Meta-synthesis across multiple systematic reviews in related domains (learning analytics, educational technology, curriculum studies) would illuminate whether patterns observed in this review reflect broader trends or context-specific phenomena. In addition, research examining ethical and governance dimensions of MMR in cross-cultural and multi-national contexts, particularly regarding informed consent, data sovereignty, and participatory approaches involving diverse stakeholders, would address critical gaps as data analytics research becomes increasingly globalised. These future directions collectively promise to advance both methodological scholarship and educational practice, deepening understanding of how, when, and why MMR generates distinctive value in the complex, multifaceted domain of data analytics education.

CONCLUSIONS AND RECOMMENDATIONS

This systematic review has demonstrated that mixed methods research (MMR) represents a significant epistemological transition in data analytics education scholarship, moving beyond methodological monism toward sophisticated integration of quantitative and qualitative approaches. The analysis of 22 rigorously appraised studies reveals that MMR adoption has accelerated markedly since 2020, with 81.8% of included studies published during or after this period, reflecting both disciplinary maturation and heightened demands for comprehensive evidence about complex learning environments. Sequential designs emerged as the dominant methodological architecture (36.4%), demonstrating utility for phased research where initial findings inform subsequent inquiry, while the diversity of approaches, including parallel designs, integrated methods, and case studies, illustrates thoughtful alignment of methodology to research objectives rather than formulaic application. The distinctive contributions of MMR manifest across multiple domains: epistemological complementarity linking measurable outcomes to lived experiences, explanatory depth revealing causal mechanisms and implementation processes, stakeholder plurality integrating diverse perspectives, contextual situatedness exposing variation across settings, and ethical dimensionality illuminating gaps between policy and practice. These contributions transcend what single-method approaches can achieve, generating insights essential for evidence-informed curriculum design, pedagogical innovation, institutional implementation strategies, and professional development initiatives. However, the review also exposed persistent challenges, including integration difficulties, resource constraints favouring well-funded research teams, publication bias toward successful implementations, and geographical concentration in North America and Europe despite emerging contributions from other regions.

Based on these findings, several evidence-informed recommendations emerge for multiple stakeholder groups. Researchers embarking on MMR studies should ensure genuine epistemological justification rather than adopting mixed methods as a procedural checklist, explicitly articulating how integration will generate insights unattainable through single methods alone. Sequential designs offer particular promise for data analytics research, enabling initial quantitative pattern identification followed by qualitative exploration of underlying mechanisms, though researchers must maintain conceptual coherence and sample connectivity across phases. Integration should occur at multiple touchpoints throughout the research process, during design, data collection, analysis, and interpretation, rather than merely at the reporting stage, with transparent documentation of integration procedures to enhance credibility and reproducibility. Researchers should acknowledge resource requirements candidly, considering collaborative team configurations that combine quantitative and qualitative expertise, securing adequate funding and timeframes (typically 18-36 months), and designing studies appropriate to available resources rather than compromising quality for methodological complexity. Educational institutions and funding bodies should provide infrastructure supporting MMR through methodological training programmes emphasising both quantitative and qualitative proficiencies, dedicated research support for integration procedures, ethics review processes accommodating iterative MMR designs, and recognition of extended publication timelines in promotion and tenure decisions. Publishers and journal editors can advance the field by encouraging transparent reporting of both successful and unsuccessful MMR implementations, establishing guidelines for assessing integration quality beyond superficial adjacency of methods, and creating spaces for methodological reflection papers that critically examine when MMR does and does not add value. Importantly, researchers must recognise contexts where MMR may be inappropriate, including well-defined quantitative questions requiring statistical precision, early-stage exploratory studies where conceptual development precedes integration, resource-constrained environments where quality single-method studies prove superior to poorly executed MMR, and situations where quantitative and qualitative components address entirely separate questions without meaningful intersection.

Acknowledgements

The author recognises the support of a peer for their assistance in checking the data coding and ensuring inter-rater reliability during the screening and quality appraisal processes.

Abbreviations: The following abbreviations are used in this manuscript:

REFERENCES

Abdul-Jabbar, S. S., & Farhan, A. K. (2022). Data analytics and techniques: A review. *ARO-The Scientific Journal of Koya University*, *10*(2), 142–164. DOI: 10.14500/aro.10975

Adhikari, R., & Timsina, T. P. (2024). An Educational study focused on the application of mixed method approach as a research method. *OCEM Journal of Management Technology & Social Sciences*, *3*(1), 94–109. DOI: 10.3126/ocemjmtss.v3i1.62229

Alkhalil, A., Abdallah, M. A. E., Alogali, A., & Aljaloud, A. (2021). Applying big data analytics in higher education. *International Journal of Information and Communication Technology Education*, *17*(3), 29–51. DOI: 10.4018/ijicte.20210701.oa3

Almalki, S. (2016). Integrating quantitative and qualitative data in mixed methods research: Challenges and benefits. *Journal of Education and Learning*, *5*(3), 288–296. DOI: 10.5539/jel.v5n3p288

Alzahrani, A. S., Tsai, Y.-S., Aljohani, N., Whitelock-Wainwright, E., & Gasevic, D. (2023). Do teaching staff trust stakeholders and tools in learning analytics? A mixed methods study. *International Journal of Educational Technology in Higher Education*, *20*(1), 13. Advance online publication. DOI: 10.1186/s41239-023-00387-1 PMID: 36874535

Arnold, M. E., & Pistilli, M. D. (2012). Course signals at Purdue: Using learning analytics to increase student success. In *Proceedings of the 2nd International Conference on Learning Analytics and Knowledge* (pp. 267-270). ACM. https://doi.org/ DOI: 10.1145/2330601.2330666

Atuase, D., & Maluleka, J. (2023). Are libraries up to the task? Bridging the gap between libraries and distance-learning students in Ghana. *Journal of Library & Information Services in Distance Learning*, *17*(3-4), 131–153. DOI: 10.1080/1533290X.2023.2239221

Bazeley, P. (2018). *Integrating analyses in mixed methods research.* SAGE Publications.

Bellaj, M., Bendahmane, A., Boudra, S., & Lamarti, M. S. (2024). Big data analytics in higher education: A new adaptive learning analytics model integrating traditional approaches. [IJIM]. *International Journal of Interactive Mobile Technologies, 18*(06), 24–39. DOI: 10.3991/ijim.v18i06.46289

Bryman, A. (2016). *Social research methods* (5th ed.). Oxford University Press.

Buunaaisie, C., Manyara, A. M., Annett, H., Bird, E. L., Bray, I., Ige, J., Jones, M., Orme, J., Pilkington, P., & Evans, D. (2018). Employability and career experiences of international graduates of MSc public health: A mixed methods study. *Public Health, 160*, 62–69. DOI: 10.1016/j.puhe.2018.03.032 PMID: 29751223

Calicdan, T. N. P., & Gementiza-Cubio, L. (2025). Student engagement in learning algebra: A mixed method study. *EPRA International Journal of Multidisciplinary Research (IJMR)*, 858–872. https://doi.org/.DOI: 10.36713/epra19894

Caspari-Sadeghi, S. (2023). Learning assessment in the age of big data: Learning analytics in higher education. *Cogent Education, 10*(1), 2162697. Advance online publication. DOI: 10.1080/2331186X.2022.2162697

Chaparro-Amaya, E. J., Restrepo-Calle, F., & Ramírez-Echeverry, J. J. (2023). Discovering insights in learning analytics through a mixed-methods framework: Application to computer programming education. *Journal of Information Technology Education, 22*, 339–372. DOI: 10.28945/5182

Clark, J.-A., & Tuffley, D. (2019). Learning analytics implementations in universities: Towards a model of success, using multiple case studies. In *ASCILITE 2019 Conference Proceedings: Personalised Learning. Diverse Goals. One Heart* (pp. 167-177). ASCILITE. https://doi.org/DOI: 10.14742/apubs.2019.15

Creswell, J. W., & Creswell, J. D. (2022). *Research design: Qualitative, quantitative, and mixed methods approaches* (6th ed.). SAGE Publications.

Creswell, J. W., & Plano Clark, V. L. (2018). *Designing and conducting mixed methods research* (3rd ed.). SAGE Publications.

Critical Appraisal Skills Programme. (2025). *CASP checklists.* Retrieved on 15 November 2025 from https://casp-uk.net/casp-tools-checklists/

Dyulicheva, Y. Yu. (2024). Application of learning analytics in higher education: Datasets, methods and tools. *Vysshee Obrazovanie v Rossii Higher Education in Russia, 33*(5), 86–111. https://doi.org/10.31992/0869-3617-2024-33-5-86-111.

Fan, Y., Matcha, W., Ahmad Uzir, N., Wang, Q., & Gašević, D. (2021). Learning analytics to reveal links between learning design and self-regulated learning. *International Journal of Artificial Intelligence in Education, 31*(4), 980–1021. DOI: 10.1007/s40593-021-00249-z

Flick, U. (2018). *An introduction to qualitative research* (6th ed.). SAGE Publications.

Greene, J. C. (2007). *Mixed methods in social inquiry*. Jossey-Bass.

Hassna, G. (2022). Big data and analytics to transform higher education: A value chain perspective. *Perspectives Policy and Practice in Higher Education, 27*(4), 160–168. DOI: 10.1080/13603108.2022.2099475

Heneghan, N. R., Jagodzinska, J., Tyros, I., Johnson, W., Nazareth, M., Yeung, E., Sadi, J., Gillis, H., & Rushton, A. (2021). Telehealth e-mentoring in postgraduate musculoskeletal physiotherapy education: A mixed methods case study. *Musculoskeletal Science & Practice, 56*, 102448. Advance online publication. DOI: 10.1016/j.msksp.2021.102448 PMID: 34416558

Hilliger, I., Ortiz-Rojas, M., Pesántez-Cabrera, P., Scheihing, E., Tsai, Y.-S., Muñoz-Merino, P. J., Broos, T., Whitelock-Wainwright, A., Gašević, D., & Pérez-Sanagustín, M. (2020). Towards learning analytics adoption: A mixed methods study of data-related practices and policies in Latin American universities. *British Journal of Educational Technology, 51*(4), 915–937. DOI: 10.1111/bjet.12933

Ifenthaler, D., & Schumacher, C. (2016). Student perceptions of privacy principles for learning analytics. *Educational Technology Research and Development, 64*(5), 923–938. DOI: 10.1007/s11423-016-9477-y

Janice, P. S., Mouton, J., & Louw, W. L. (2018). *The SAGE handbook of qualitative data analysis* (2nd ed.). SAGE Publications.

Jivet, I., Scheffel, M., Schmitz, M., Robbers, S., Specht, M., & Drachsler, H. (2020). From students with love: An empirical study on learner goals, self-regulated learning and sense-making of learning analytics in higher education. *The Internet and Higher Education, 47*, 100758. Advance online publication. DOI: 10.1016/j.iheduc.2020.100758

Kahn, P., Everington, L., Kelm, K., Reid, I., & Watkins, F. (2017). Understanding student engagement in online learning environments: The role of reflexivity. *Educational Technology Research and Development, 65*(1), 203–218. DOI: 10.1007/s11423-016-9484-z

Khalid, M., Akanmu, A., Murzi, H., Lee, S. W., Awolusi, I., Manesh, D., & Okonkwo, C. (2023). Industry perception of the knowledge and skills required to implement sensor data analytics in construction. *Journal of Civil Engineering Education, 150*(1), 04023011. Advance online publication. DOI: 10.1061/JCEECD.EIENG-1902

Lin-Aung, H., Masumoto, D., Linn, Z., Kobayakawa, Y., Okamura, S., Kurihara, K., Morio, K., Tashiro, Y., Sakurai, H., & Hori, H. (2022). Students' perception of problem-based learning at a Japanese medical school: An exploratory sequential mixed method. *International Journal of Medical Education, 13*, 322–334. DOI: 10.5116/ijme.636c.c5e1 PMID: 36571846

Liu, X., Guo, Y., Li, J., & Wang, Y. (2022). Visualization and data analysis of multi-factors for the scientific research training of graduate students. *Applied Sciences (Basel, Switzerland), 12*(24), 12845. Advance online publication. DOI: 10.3390/app122412845

Macfadyen, L. P., & Dawson, S. (2012). Numbers are not enough. Why e-learning analytics failed to inform an institutional strategic plan. *Journal of Educational Technology & Society, 15*(3), 149–163.

Manion, J. L. (2019). *A mixed methods investigation of student achievement and satisfaction in traditional versus online learning environments* [Doctoral dissertation, Lindenwood University]. Lindenwood University Digital Commons. Retrieved on 15 November 2025 from https://digitalcommons.lindenwood.edu/dissertations/107

Mertens, D. M. (2018). *Mixed methods design in evaluation*. SAGE Publications.

Mohammadi, A. (2023). Analytical evaluation of big data applications in e-commerce: A mixed method approach. *Decision Science Letters, 12*(2), 457–476. DOI: 10.5267/j.dsl.2022.11.003

Morse, J. M., Barrett, M., Koçer, B., & Omerod, R. (2023). *Qualitative health research: An introduction to methods and practices* (4th ed.). SAGE Publications.

Mustapha, I., Rattanawiboonsom, V., & Intanon, R. (2023). Data-driven insights in higher education: Exploring the synergy of big data analytics and mobile applications. [iJIM]. *International Journal of Interactive Mobile Technologies, 17*(20), 21–37. DOI: 10.3991/ijim.v17i20.45037

Ncube, M. M., & Ngulube, P. (2025). Surge of data analytics in postgraduate education and methodological plurality: A systematic review. *Discover Education, 4*, 29. DOI: 10.1007/s44217-025-00422-9

Ngulube, P. (2020). Mixed methods research in knowledge management studies (2009-2014): A content analysis of journal articles. *Journal of Information & Knowledge Management, 19*(3), 2040015. Advance online publication. DOI: 10.1142/S0219649220500161

Ngulube, P. (2023). Improving the quality of reporting findings using computer data analysis applications in educational research in context. *Heliyon Education, 9*(9), e19683. DOI: 10.1016/j.heliyon.2023.e19683 PMID: 37810136

Ngulube, P., & Ngulube, B. (2022). Are we there yet? Mixed methods research in the *South African Journal of Economic and Management Sciences. Acta Commercii, 22*(1), a1013. Advance online publication. DOI: 10.4102/ac.v22i1.1013

Ngulube, P., & Ukwoma, S. C. (2021). Prevalence of methodological transparency in the use of mixed methods research in library and information science research in South Africa and Nigeria, 2009-2015. *Library & Information Science Research, 43*(4), 101105. Advance online publication. DOI: 10.1016/j.lisr.2021.101105

Page, M. J., McKenzie, J. E., Bossuyt, P. M., Boutron, I., Hoffmann, T. C., Mulrow, C. D., Shamseer, L., Tetzlaff, J. M., Akl, E. A., Brennan, S. E., Chou, R., Glanville, J., Grimshaw, J. M., Hróbjartsson, A., Lalu, M. M., Li, T., Loder, E. W., Mayo-Wilson, E., McDonald, S., & Moher, D. (2021). The PRISMA 2020 statement: An updated guideline for reporting systematic reviews. *Journal of Clinical Epidemiology, 134*, 178–189. DOI: 10.1016/j.jclinepi.2021.03.001 PMID: 33789819

Pardo, A., Jovanovic, J., Dawson, S., Gašević, D., & Mirriahi, N. (2017). Using learning analytics to scale the provision of personalised feedback. *British Journal of Educational Technology, 50*(1), 128–138. DOI: 10.1111/bjet.12592

Plano Clark, V. L., & Ivankova, N. V. (2016). *Mixed methods research: A guide to the field.* SAGE Publications., DOI: 10.4135/9781483398341

Prieto, L. P., Pishtari, G., Dimitriadis, Y., Rodríguez-Triana, M. J., Ley, T., & Odriozola-González, P. (2023). Single-case learning analytics: Feasibility of a human-centered analytics approach to support doctoral education. *Journal of Universal Computer Science, 29*(9), 1033–1068.

Saraswati, P., & Devy, A. (2023). Mixed methods-research methodology: An overview. *Mathews Journal of Nursing and Health Care, 5*(4), Article 24. Retrieved on 15 November 2025 from https://www.mathewsopenaccess.com/scholarly-articles/mixed-methods-research-methodology-an-overview.pdf

Silvola, A., Sjöblom, A., Näykki, P., Gedrimiene, E., & Muukkonen, H. (2023). Learning analytics for academic paths: Student evaluations of two dashboards for study planning and monitoring. *Frontline Learning Research, 11*(2), 78–98. Retrieved November 15, 2025, from https://files.eric.ed.gov/fulltext/EJ1414726.pdf

Sivarajah, U., Kamal, M. M., Irani, Z., & Weerakkody, V. (2017). Critical analysis of big data challenges and analytical methods. *Journal of Business Research, 70*, 263–286. DOI: 10.1016/j.jbusres.2016.08.001

Slade, S., & Prinsloo, P. (2013). Learning analytics: Ethical issues and dilemmas. *The American Behavioral Scientist, 57*(10), 1510–1529. DOI: 10.1177/0002764213479366

Song, D., Shin, Y., & Hong, H. (2022). *Learning analytics as an integrated approach in mixed methods research*. Routledge.

Stojanov, A., & Daniel, B. K. (2023). A decade of research into the application of big data and analytics in higher education: A systematic review of the literature. *Education and Information Technologies, 29*(5), 5807–5831. DOI: 10.1007/s10639-023-12033-8

Strang, K. D. (2017). Beyond engagement analytics: Which online mixed-data factors predict student learning outcomes? *Education and Information Technologies, 22*(2), 917–937. DOI: 10.1007/s10639-016-9464-2

Tashakkori, A., Johnson, R. B., & Teddlie, C. (2020). *Foundations of mixed methods research: Integrating quantitative and qualitative approaches in the social and behavioral sciences* (2nd ed.). SAGE Publications.

TechTarget. (n.d.). Data analytics definition. Retrieved on 15 November 2025 from https://www.techtarget.com/searchbusinessanalytics/resources/Data-science-and-analytics

Toyon, M. A. S. (2021). Explanatory sequential design of mixed methods research: Phases and challenges. *International Journal of Research in Business & Social Science, 10*(5), 253–260. DOI: 10.20525/ijrbs.v10i5.1262

Ustun, A. B., Zhang, K., Karaoğlan-Yilmaz, F. G., & Yilmaz, R. (2022). Learning analytics-based feedback and recommendations in flipped classrooms: An experimental study in higher education. *Journal of Research on Technology in Education, 55*(5), 841–857. DOI: 10.1080/15391523.2022.2040401

van Ameijde, J., Cheng, S., & Wang, H. (2023). Integrating urban analytics into postgraduate urban design pedagogy: A mixed-methods teaching approach to addressing urban liveability. *Journal of Planning Education and Research*. Advance online publication. DOI: 10.1177/0739456X231162842

Viberg, O., Baars, M., Mello, R. F., Weerheim, N., Spikol, D., Bogdan, C., Gasevic, D., & Paas, F. (2024). Exploring the nature of peer feedback: An epistemic network analysis approach. *Journal of Computer Assisted Learning*, *40*(4), 1567–1579. DOI: 10.1111/jcal.13035

Viberg, O., Hatakka, M., Bälter, O., & Mavroudi, A. (2018). The current landscape of learning analytics in higher education. *Computers in Human Behavior*, *89*, 98–110. DOI: 10.1016/j.chb.2018.07.027

Wang, F., King, R. B., Zeng, L. M., Zhu, Y., & Leung, S. O. (2022). The research experience of postgraduate students: A mixed-methods study. *Studies in Higher Education*, *48*(4), 616–629. DOI: 10.1080/03075079.2022.2155344

West, D., Luzeckyj, A., Searle, B., Toohey, D., Vanderlelie, J., & Bell, K. R. (2020). Perspectives from the stakeholder: Students' views regarding learning analytics and data collection. *Australasian Journal of Educational Technology*, *36*(6), 72–88. DOI: 10.14742/ajet.5957

Westein, M. P. D., Koster, A. S., Daelmans, H. E. M., & Kusurkar, R. A. (2023). How progress evaluations are used in postgraduate education with longitudinal supervisor-trainee relationships: A mixed method study. *Advances in Health Sciences Education : Theory and Practice*, *28*(1), 205–222. DOI: 10.1007/s10459-022-10153-3 PMID: 36094680

Zhao, G., Xie, X., Wang, Y., Liu, S., Jones, P., & López, C. (2024). Barrier analysis to improve big data analytics capability of the maritime industry: A mixed-method approach. *Technological Forecasting and Social Change*, *203*, 123345–123345. DOI: 10.1016/j.techfore.2024.123345

ABBREVIATION FULL FORM

CASP: Critical Appraisal Skills Programme
DA: - Data Analytics
LA: - Learning Analytics
MMR: Mixed Methods Research
PRISMA: Preferred Reporting Items for Systematic Reviews and Meta-Analyses
PICO: Population, Intervention, Comparison, Outcome

APPENDIX

Table 8. Data extraction and coding

Studies	Title	Source	Study Design	Findings
Alzahrani et al. (2023)	Do Teaching Staff Trust Stakeholders and Tools in Learning Analytics? A Mixed Methods Study	*International Journal of Educational Technology in Higher Education*	Mixed methods (survey and interviews)	The study found high trust in higher education institutions' competence and learning analytics usefulness among students; however, low trust in third-party data handling and data accuracy.
Atuase and Maluleka (2023)	Are Libraries up to the Task? Bridging the Gap between Libraries and Distance-Learning Students in Ghana	*Journal of Library & Information Services in Distance Learning*	Sequential explanatory mixed methods	Library development programmes, including the application of learning analytics, were crucial for supporting effective service provision to postgraduate students.
Buunaaisie et al. (2018)	Employability And Career Experiences of International Graduates of MSc Public Health: A Mixed Methods Study	*Public Health*	Sequential mixed methods study	The research effectively utilised data analytics to gain a deeper understanding of the effectiveness of the MSc PH programme in preparing graduates for successful careers in public health. The data-driven insights provide valuable information for improving the program and ensuring its relevance to the needs of the field.
Chaparro-Amaya et al. (2023)	Discovering Insights in Learning Analytics Through a Mixed-Methods Framework: Application to Computer Programming Education	*Journal of Information Technology Education: Research*	Mixed methods (quantitative and qualitative data)	Identified factors influencing student performance in introductory programming courses. Highlighted limitations of solely relying on quantitative data in learning analytics. Demonstrated the value of combining quantitative and qualitative data for a more comprehensive understanding of student learning.

continued on following page

Table 8. Continued

Studies	Title	Source	Study Design	Findings
Clark and Tuffley (2019)	Learning Analytics implementations in universities: Towards a model of success, using multiple case studies	*ASCILITE Publications*	Multiple case studies	This study compares Learning Analytics implementation at three Australian universities, revealing diverse approaches and ongoing challenges. Proprietary platforms are likely to dominate due to usability, despite participants' recognition of LA's potential.
Fan et al. (2021)	Learning Analytics to Reveal Links Between Learning Design and Self-Regulated Learning	*International Journal of Artificial Intelligence in Education*	Mixed-methods research design: Analysis of learner data with discussion forum posts	The study identifies four distinct learning tactics employed by MOOC learners: Search-oriented, Content and assessment-oriented, Content-oriented, and Assessment-oriented. These tactics were found to be influenced by the instructional cues embedded within the learning design.
Heneghan et al. (2021)	Telehealth E-Mentoring in Postgraduate Musculoskeletal Physiotherapy Education: A Mixed Methods Case Study	*Musculoskeletal Science and Practice*	Sequential mixed methods study	This study found that a mentorship program had a positive impact on both patients and mentors. Patients reported improvements in health outcomes and positive experiences, while mentors valued communication skills and identified challenges related to telehealth.
Hilliger et al. (2020)	Towards Learning Analytics Adoption: A Mixed Methods Study of Data-Related Practices and Policies in Latin American Universities	*British Journal of Educational Technology*	Mixed methods (surveys, interviews, focus groups)	This study highlights learning analytics' potential to address educational needs, yet adoption is hindered by expertise and policy gaps. Inconsistencies in data practices across universities necessitate standardisation, while integration opportunities exist. Robust information systems and data governance policies are crucial to ensure educational data integrity and effectiveness.
Ifenthaler and Schumacher (2016)	Student Perceptions of Privacy Principles for Learning Analytics	*Educational Technology Research and Development*	Mixed-methods research (surveys and interviews)	The study highlighted the importance of considering students' perceptions and concerns about privacy principles in the design and implementation of learning analytics systems.

continued on following page

Table 8. Continued

Studies	Title	Source	Study Design	Findings
Jivet et al. (2020)	From Students with Love: An Empirical Study on Learner Goals, Self-Regulated Learning and sense-making of learning analytics in higher education	*The Internet and Higher Education*	Sequential mixed methods study	This study identifies three key factors influencing dashboard sense-making: transparency of design, reference frames, and support for action. Self-regulated learning skills and learner goals also play significant roles in determining the perceived relevance of these factors.
Kahn et al. (2017)	Understanding Student Engagement in Online Learning Environments: The Role of Reflexivity	*Educational Technology Research and Development*	Mixed methods design: a combination of quantitative discussion board analysis and qualitative interviews	The study highlighted the potential of learning analytics for improving the academic advising of online students. It emphasised the importance of utilising analytics to understand students' behaviours and engagement and using this information to provide personalised support and guidance.
Khalid et al. (2023)	Industry Perception of the Knowledge and Skills Required to Implement Sensor Data Analytics in Construction	*Journal of Civil Engineering Education*	Mixed method (survey & focus group)	Importance of data processing and communication - Key knowledge areas: sensor types, data collection, analysis techniques, visualisation tools - Essential skills: data cleaning, statistical analysis, data visualisation software, communication, collaboration - Need to adapt construction education.
Lin-Aung et al. (2022)	Students' Perception of Problem-Based Learning at a Japanese Medical School: An Exploratory Sequential Mixed Method	*International Journal of Medical Education*	Sequential exploratory mixed methods	This study found that participants in a problem-based learning program perceived it as an effective learning method, but faced challenges related to social interaction. While the problem-based learning program improved participants' communication skills, responsibility, social networking, empathy, and understanding of social aspects of clinical practice, reluctance to actively discuss and collaborate with unfamiliar peers was identified as a key challenge.

continued on following page

Table 8. Continued

Studies	Title	Source	Study Design	Findings
Liu et al. (2022)	Visualisation and data analysis of multiple factors for the scientific research training of graduate students	*Applied Sciences*	Concurrent triangulation: Combining questionnaire-based data collection, data mining techniques, and visual analytics	This study highlights the crucial role of visualisations in identifying data patterns and trends, and reveals that factors such as supervision quality, research resources, and student motivation significantly impact graduate students' research success.
Prieto et al. (2023)	Single-Case Learning Analytics: Feasibility of A Human-Centered Analytics Approach to Support Doctoral Education	*Journal of Universal Computer Science*	Concurrent mixed methods: Data were collected through a self-report platform and interviews, and analysed using machine learning techniques	The findings suggest that single-case learning analytics can provide valuable insights for both learners and researchers, contributing to a more personalised and effective learning experience.
Silvola et al. (2023)	Learning Analytics for Academic Paths: Student Evaluations of Two Dashboards for Study Planning and Monitoring	*Frontline Learning Research*	Multiple-case design	This study found that students evaluated learning analytics dashboards as helpful for study planning and monitoring, but also identified challenges such as needing more information and development ideas.
Ustun et al. (2022)	Learning Analytics-Based Feedback and Recommendations in Flipped Classrooms: An Experimental Study in Higher Education	*Journal of Research on Technology in Education*	Mixed-method pretest/post-test experimental study	This study revealed the effectiveness of learning analytics-based interventions in fostering self-regulated learning within the flipped classroom environment. The researchers subsequently explored both practical implications for flipped classrooms and online learning, as well as recommendations for the design and further development of learning analytics interventions.
van Ameijde et al. (2023)	Integrating Urban Analytics into Postgraduate Urban Design Pedagogy: A Mixed-Methods Teaching Approach to Addressing Urban Liveability	*Journal of Planning Education and Research*	Mixed methods	This article reports on an innovative pedagogical initiative which introduced data-driven methods in academic teaching practices.

continued on following page

Table 8. Continued

Studies	Title	Source	Study Design	Findings
Viberg et al. (2024)	Exploring the Nature of Peer Feedback: An Epistemic Network Analysis Approach	*Journal of Computer Assisted Learning*	Mixed-methods approach, combining qualitative coding of peer feedback messages with quantitative Epistemic Network Analysis (ENA)	The ENA visualisations and results illustrate that management, cognition, and affect-related feedback are positively correlated with student performance during the formative assessment phase, while suggestions for improvement and interpersonal factors are not significantly related to student learning, providing valuable insights for instructional strategies and assessment design.
Wang et al. (2022)	The Research Experience of Postgraduate Students: A Mixed-Methods Study	*Studies in Higher Education*	Mixed methods (quantitative & qualitative)	The study effectively uses data analytics to explore the relationship between research experience, skill development, and satisfaction among postgraduate students. Hence, by combining quantitative and qualitative data, the researchers gain valuable insights that can inform educational practices.
West et al. (2020)	Perspectives From the Stakeholder: Students' Views Regarding Learning Analytics and Data Collection	*Australasian Journal of Educational Technology*	Sequential mixed methods study	These findings highlight the importance of transparency and clear communication with students about data collection and usage. Universities should prioritise student perspectives and work collaboratively with them to ensure ethical and responsible use of learning analytics.
Westein et al. (2023)	How Progress Evaluations Are Used in Postgraduate Education with Longitudinal Supervisor-Trainee Relationships: A Mixed Method Study	*Advances in Health Sciences Education*	Mixed method (quantitative & qualitative)	Progress evaluations are valuable for guiding feedback and structuring learning. Positive supervisor-trainee relationships are essential for effective feedback. Tension exists between evaluation for assessment and development purposes. Progress evaluation scoring showed low reliability.

Chapter 5

Demystifying the Adoption of Mixed Methods Research Methodology in Informatics

Blessed Magama
http://orcid.org/0009-0002-6013-8100
Zimbabwe Revenue Authority, Zimbabwe

ABSTRACT

Research methodology is invaluable for publicising the systematic techniques and tools used to address the research problem and questions. Traditionally, research was influenced by either realism or constructivism, and a sharp distinction existed between these worldviews. However, as more complex interdisciplinary phenomena are presented for research, especially in informatics, pluralism and its mixed methods research (MMR) approach is gaining momentum. This chapter will explain the MMR research paradigm, designs, sampling approaches, data collection techniques, analysis, and presentation in simple terms. This aims to reveal the key characteristics and distinguishing factors of MMR from other approaches, with the goal of raising awareness about its applicability in informatics. Furthermore, this chapter will inform researchers in informatics about the advantages and disadvantages of MMR, enabling them to properly plan and budget for their research activities and ensure ethical and valid research practices.

DOI: 10.4018/979-8-3693-4018-9.ch005

BACKGROUND AND DEFINITION OF MMR

Research methodological approaches lay a foundation for data collection to address the research problem or questions. Creswell (2009) defines the research methodological approach as a strategy of inquiry that helps the researcher with a road map, clear direction, and procedures to be used in the research design. In other words, it provides the standards that will be utilised for interpretation and drawing conclusions (Goundar, 2012). Without a clear methodology, research findings are easily dismissible as guesswork. Traditionally, research in the social and behavioural sciences was categorised into quantitative and qualitative approaches, and mixing of the two was not allowed (Tashakkori & Teddlie, 2003a). Qualitative research adopts a relativist position, involving multiple constructions of reality through interaction between the researcher and the researched, to produce deep, rich observational data that expresses people's lives, experiences, behaviours, emotions, feelings, cultural norms, and beliefs, among other things (Rahman, 2017). On the other hand, quantitative research is an approach to investigating a defined problem by testing a theory, using numerical measurements, and analysing the data through statistical and mathematical methods to produce hard, generalisable data (Leedy & Ormrod, 2005).

However, recent developments in the research landscape suggest that a rigid dichotomy between quantitative and qualitative methodological approaches is no longer tenable. This has led to the firming up of MMR as the third methodological movement (Bazeley, 2004). This 'new' approach is slowly bringing to a cease-fire the paradigm wars of the 1980s through its pluralistic ontological perspective (Ngulube, 2015). Therefore, MMR allows researchers to gather and integrate multiple data sources to study complex problems (Poth & Munce, 2020). For example, in social and health informatics studies, the quantitative strand can be used to test and support hypotheses by examining significant predictors in an intervention study, while the qualitative strand can aid in understanding the reasons for the success or failure of an intervention or implementation design and make modifications to the intervention (Perez, 2023).

Greene and Caracelli (1997) define MMR as a research methodology that combines quantitative and qualitative research techniques, concepts, and language into a single study or program of inquiry to gather and analyse data. MMR combines the strengths of both quantitative and qualitative approaches, where the quantitative strand yields generalizable and statistically valid data, while the qualitative strand provides rich, contextualised insights and explores underlying meanings (Creswell & Plano Clark, 2007). Expounding further, Ngulube (2020), posits that MMR entails the collection, analysis, integration, and interpretation of both quantitative and qualitative data, either simultaneously or in sequence, within a single study to enhance the depth and breadth of understanding. This method is especially advantageous in areas such as behavioural, health, and social sciences, as it provides the ability to achieve wide-ranging generalizability through quantitative data while also capturing deep, contextual understanding through qualitative data. Although 'purists' posit that quantitative and qualitative approaches cannot be mixed, Brannen (2005) expounds that research differs in form and context to the extent that it cannot be perceived in terms of artificial compartmentalisation between quantitative and qualitative approaches. Furthermore, Bryman (2012) argues that epistemological differences between quantitative and qualitative methods are no longer the primary concern, as research can be enriched by the addition of other methods. Various paradigms are beginning to 'interbreed', and philosophies previously thought to be in irreconcilable conflict can now complement each other under the eclecticism or pluralism and pragmatism rubric (Lincoln, et al., 2011).

Tashakkori and Teddlie (2003a) describe MMR as a quiet revolution that alleviates the tension in the 1980s between quantitative and qualitative methodological movements. By the 1990s, MMR gained popularity to neutralise differences between quantitative and qualitative epistemology (what we know and how it is acquired?), ontology (nature of reality), methodology (how data are collected, to understand reality), axiology (values and ethics) and doxology (what we believe and why) (Tashakkori & Creswell, 2007). Therefore, MMR can be viewed as a hybrid, combined, and multi-method research approach (Creswell & Plano Clark, 2011). With MMR, both quantitative and qualitative research approaches are important and useful. This is bolstered by Johnson and Onwuegbuzie (2004), who argue that MMR is not intended to replace quantitative and qualitative methods, but rather draws from their strengths and minimises their weaknesses to enhance effectiveness in exploring the research problem. Expounding further, Dovetail (2023) contends that MMR is not a random collection of quantitative and qualitative data, but requires a careful consideration of the relationship between the data and method flexibility to adequately address the research question. While foundational works such as Tashakkori and Teddlie (2003a), Phillips (2004), Bazeley (2004), and Creswell (2007) laid the groundwork for MMR, it is essential to interpret these sources in light of recent developments. Ngulube (2020) advises caution when using pre-2010 literature, as the field has evolved conceptually, particularly regarding the centrality of integration over triangulation in MMR design.

The integration of quantitative and qualitative research data is a key characteristic and distinguishing factor of MMR (Creamer, 2018). If a study with both quantitative and qualitative strands lack any point of interface or integration, it can be described as quasi-mixed methods (Teddlie & Tashakkori, 2009). MMR also differs from multi-method research, which can use multiple methods from the same tradition without restricting the researcher to combining the quantitative and qualitative approaches (Hunter & Brewer, 2015). In the same vein, Bans-Akutey and Tiimub (2021) argued that triangulation does not mean the same as integration or mixed methods. They further explained that triangulation is a method to ensure the trustworthiness of research findings by utilising multiple data sources, methods, ideas, or even different researchers to verify and confirm the results. Furthermore, it can involve using multiple research methods and other validation techniques, such as applying different theories or having multiple reviewers assess the data (Bans-Akutey & Tiimub, 2021). On the other hand, integration or mixed methods combine quantitative and qualitative research approaches to answer the research questions (Bans-Akutey & Tiimub, 2021).

The coalescence of quantitative and qualitative methodologies within an MMR study can occur at various stages of the research process, including data collection, analysis, result presentation, interpretation, and inference, and can be illustrated through joint displays (Dawadi, et al., 2021). Applying sampling designs that align with both quantitative and qualitative methodologies, while appropriately weighting each strand, can strengthen the integration process in mixed-methods research (Onwuegbuzie & Collins, 2007). Additionally, the use of joint displays in mixed methods studies offers a visual representation of the integration between the quantitative and qualitative components. Through integration, MMR produces meta-inferences, that is, unique conclusions derived from MMR studies by combining inferences or results obtained from the quantitative and qualitative strands. This helps enhance the reliability and validity of the findings. Consequently, Gunasekare (2015) observes that MMR is a workable middle solution that respectfully taps into the attributes of both quantitative and qualitative methods to tackle various research problems from multiple viewpoints and perspectives.

Therefore, MMR's logic of inquiry encompasses the use of induction, deduction, and abduction forms of reasoning, making it an expansive and creative research approach that rejects dogmatism (de Waal, 2001). According to Melnikovas (2018), deduction involves beginning with an existing theory, then formulating a question or hypothesis and collecting data to either confirm or reject that hypothesis. It provides research guidance to the quantitative strand. Induction starts with observing and gathering data, followed by describing and analysing it to develop a new theory (Melnikovas, 2018). The approach is commonly used to guide the qualitative strand. Abduction begins with observing an empirical phenomenon and then forming the best possible explanation or conclusion based on the evidence available (Melnikovas, 2018). Abductive reasoning typically begins with an unexpected observation and alternates between induction and deduction to identify the most plausible explanation (Melnikovas, 2018). It is particularly useful for explaining surprising or puzzling phenomena where the initial data does not fit existing theories. A good

MMR informatics study should benefit from the application of induction, deduction, and abduction reasoning.

The key defining features of MMR were summarised by Terrel (2012) as follows:

- The philosophical underpinning is pragmatism and pluralism.
- Uses sequential, concurrent, and transformative inquiry strategies.
- Combines both quantitative and qualitative methods.
- The methods (quantitative and qualitative) can be given either equal status or one can be given more weight over the other, depending on the design chosen by the researcher.
- Allows one method to expand, inform, complement, and confirm the other to give robust findings.
- The quantitative method offers the breadth of generalisation, while qualitative research offers depth and detailed understanding.
- The designs may or may not be driven by a theoretical perspective.

Although scholars like Harris (2021), Granikova et al. (2020) and Ngulube (2020), among others, note evidence about the underdevelopment and underutilisation of MMR as compared to quantitative and qualitative methods by researchers in the global south, its pluralistic and pragmatic perspective makes it fit for use by informatics researchers, including those in Africa. This is due to the fact that MMR is flexible and can objectively produce robust results capable of addressing multi-faceted, context-specific, and interdisciplinary problems (Creamer, 2018; Teddlie et al., 2021). MMR's ontological and epistemological stance aligns with new developments in research, whereby science should be viewed in an inclusive manner, and research should be guided by Two-Eyed Seeing (Bartlett et al., 2012). Two-Eyed Seeing is a multiple-perspective approach that has worked well for the Aboriginal people to address their context-specific challenges through simultaneously tapping into the strengths of their indigenous knowledges and ways of knowing with one eye, while the other eye taps into the strengths of Western knowledges and ways of knowing (Bartlett et al., 2012). Likewise, informatics researchers, through advanced MMR design, such as embedded mixed methods design, can also accommodate indigenous research approaches (Chilisa & Tsheko, 2014). In the view of progressive and transformative scholars, MMR ceases to be part of the global north colonial project to undermine other knowledge systems in favour of white supremacy (Ngulube, 2020).

Moreover, the utilisation of MMR by informatics researchers has been made attractive due to the availability of the Good Reporting of a Mixed Methods Study (GRAMMS) framework, which was developed by O'Cathain, Murphy and Nicholl (2008). The GRAMMS framework consists of six main points that address the rationale for utilising mixed methods, as well as issues related to methodological choices associated with data collection methods, sequencing, sampling, prioritisation of data, points of integration, and data analysis techniques (Cameron et al., 2013). According to Cameron et al. (2013), the GRAMMS set of quality guidelines or main points is as follows:

- Describe the justification for using a mixed methods approach to the research question.
- Describe the design in terms of the purpose, priority and sequence of methods.
- Describe each method in terms of sampling, data collection and analysis.
- Describe where integration has occurred, how it has occurred and who participated in it.
- Describe any limitations of one method associated with the presence of the other method.
- Describe any insights gained from mixing or integrating methods,

GRAMMS is crucial in addressing the lack of explicit documentation and transparency in the reporting of mixed methods studies, and can aid informatics researchers in properly implementing MMR and generating valid and dependable results (Nakaziba & Namuguzi, 2022; Ngulube & Ukwoma, 2021).

Although there is evidence of underutilisation of MMR in informatics studies in the Global South, the method is gaining momentum in terms of applicability in non-Western contexts to address local challenges and promote epistemic justice. For instance, Kandiero and Makuwatsine (2022) employed an exploratory sequential mixed methods design to investigate switching barriers affecting customer retention in Zimbabwe's internet services industry, taking into account the complexity of consumer behaviour and preferences. The quantitative component provided objectivity and control over potential biases, while the qualitative component enabled the researchers to explore questions of "why" and "how," thereby enriching the interpretation of quantitative findings. An inductive approach was employed to qualitatively identify and validate research variables, and to formulate hypotheses, which subsequently guided the development of a quantitative instrument through a deductive process to test these hypotheses.

In another case, Mamorobela (2022) employed an explanatory sequential mixed methods design to examine a model for the adoption of social media-enabled knowledge management (KM) within small and medium enterprises (SMEs) in South Africa. The quantitative phase of the study produced a model identifying factors influencing adoption, which were subsequently tested and validated using structural equation modelling (SEM) to elucidate the determinants affecting SMEs' use of social media for KM processes. In the qualitative follow-up phase, SME owners were purposively selected based on specific criteria to provide explanatory insights into the quantitative findings.

Erima and Garaba (2022) employed a mixed methods research (MMR) approach to investigate digital archiving practices at selected public universities in Kenya, aiming to develop a framework for digital archiving within these institutions. Preliminary findings indicated that digital archiving in Kenya is still at an early stage, with significant efforts needed to standardize the practice across universities. The study recommended the systematic implementation of awareness programs targeting senior institutional officials to enhance their understanding and support of digital archiving initiatives. Similarly, Mutsagondo (2022) applied a convergent MMR design to examine email records management in Zimbabwe's central government. His study demonstrated the effective integration of quantitative and qualitative methods to evaluate email management strategies and highlight their significance within the context of the central government.

Using an MMR approach typically involves the following steps, illustrated in Figure 1.

Figure 1. Steps followed when using MMR approach (Synthesised by the author)

MMR RESEARCH PARADIGM

A research paradigm is a fundamental framework or set of assumptions that guides the conduct of research. Neuman (2014) notes that the research paradigm is sometimes referred to as a worldview or philosophical assumption. Elaborating further, Teddlie and Tashakkori (2009) define a research paradigm as "a worldview together with the various philosophical assumptions associated with that point of view" (p. 84). It influences how researchers understand reality (ontology), what they consider valid knowledge (epistemology), and the methods they use to collect and analyse data (methodology) (Creswell, 2014).

Ontology

Creswell (2014) defines ontology as the assumptions about the nature of reality, which aims to comprehend what exists independently of human perception and to create a thorough classification or structure of the essential categories of entities that make up reality. Traditionally, research has been influenced by either realism or constructivism. According to Neuman (2014), realism is an ontological perspective for quantitative research, and constructivism is an ontological perspective for qualitative research. Realism posits that nature exists in reality, and people know what they know because of some inherent quality that things in the world possess, which can be explored and discovered by individuals (Neuman, 2014). This suggests that certain facts about existence and the state of things remain true regardless of our thoughts, perceptions, or conceptual frameworks.

On the other hand, constructivism posits that reality does not exist and that what people perceive is determined by their construction and interpretation (Neuman, 2014). Constructivism believes that reality is seen as socially constructed. Harvey (2012) further explains constructivism by arguing that all knowledge is built from human experience, rather than being discovered as self-evident truth. This ontological difference between realism and constructivism has been cited as the basis of the quantitative versus qualitative paradigm wars (Ngulube, 2015).

MMR's ontological perspective is pluralism. Pluralism is derived from the synchronisation of realism and constructivism. It suggests that things exist in different ways rather than all existing in the same manner. In other words, reality is composed of many kinds of "being," not just one single type. It entails engaging multiple methodologies to obtain in-depth, exquisite insights (Dawadi, et al., 2021). It bridged the quantitative and qualitative differences by tackling research challenges practically (May et al., 2017). May et al. (2017) add that pluralism does not discredit realism or constructivism but rather holds that they are of equal importance. Pluralism combines the strengths of both worldviews to yield multiple realities, an in-depth understanding of phenomena, and the validity and reliability of results (May et al., 2017). This makes MMR suitable for informatics researchers to tackle the multi-faceted contemporary problems the informatics discipline presents.

Epistemology

Epistemology refers to the study of how people acquire knowledge about the world around them. According to Neuman (2014), it is the branch of philosophy that studies knowledge, focusing on what knowledge is, how people acquire it, and its limits. It is about understanding how humans come to know things and how reliable that knowledge is. In quantitative research, positivism and post-positivism are the predominant epistemological perspectives, whereas interpretivism is the primary epistemological perspective for qualitative research (Creswell, 2014). Positivism posits that the world around us is real and observable through experimentation and rational inquiry (Creswell, 2014). Post-positivism posits that reality can be discovered with probability, rather than certainty (Mertens, 2010). On the other hand, interpretivism posits that knowledge about the world is the creation of men's minds, influenced by their beliefs and values (Creswell, 2014). Creswell (2014) observes that with interpretivism, reality is subjective, socially constructed, and a composite of multiple perspectives. Therefore, interpretivism focuses on understanding the meanings, motivations, and interpretations within a social context. Interpretivism can be referred to as relativism, idealism, or constructionism.

MMR's epistemological perspective is pragmatism. Pragmatism is a philosophical perspective that emphasises the practical implications and utility of ideas. According to Creswell and Plano Clark (2018), it posits that the truth of a belief or concept is determined by how well it works in real life and how it helps us solve problems or act effectively. Pragmatism values ideas for their ability to produce useful outcomes and guide action. According to Johnson and Onwuegbuzie (2004), pragmatism lies between positivism and interpretivism, bridging the gap between these two epistemologies in a manner that offers the best opportunities for answering research questions. It is derived from the synchronisation of positivism and interpretivism as shown in Figure 2. According to Creswell and Plano Clark (2018), pragmatism is a practical approach to problem-solving, which allows for the use of various methods of inquiry. In a similar vein, Adeleye (2017) asserts that pragmatism emphasises practicality and utility. Johnson and Onwuegbuzie (2004) further expound that pragmatism takes an explicitly value-oriented approach to research. It is anchored on the notion of 'what works', which emphasises selecting the methods that best address the research problem (Gunasekare, 2015). Pragmatism is sometimes referred to as experimentalism or consequentialism. This philosophical approach has the potential to find answers to informatics research questions through multiple methods of inquiry.

Figure 2. Paradigm for MMR approach (Synthesised by the author)

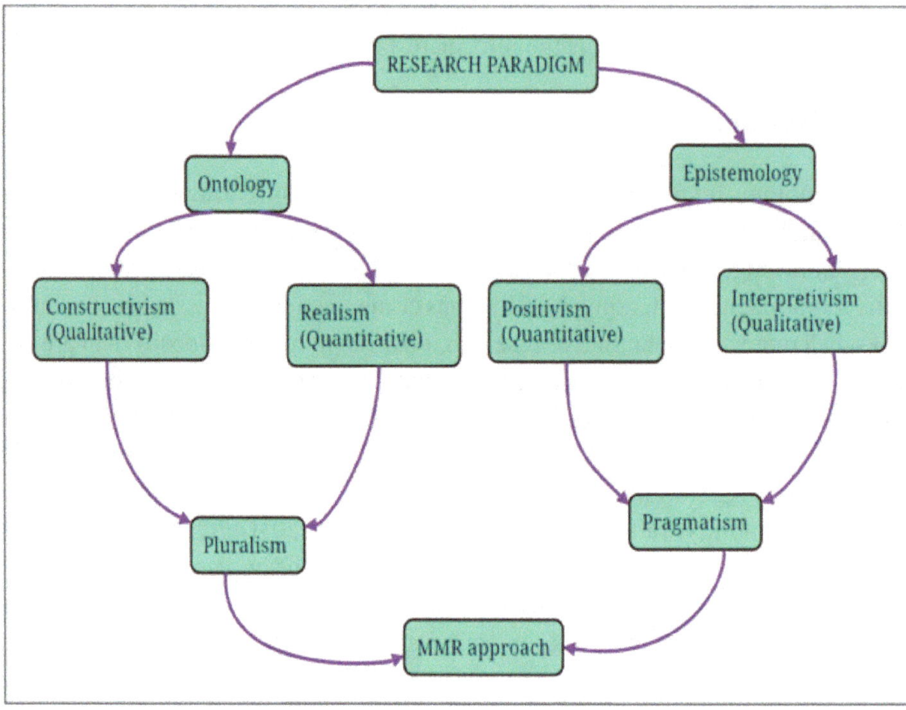

MMR DESIGN

According to Yin (1994) research design is the structured approach that connects empirical data to the study's original research questions and, ultimately, to its findings. In other words, research design is a plan or blueprint for conducting the research (Babbie & Mouton, 2001). MMR designs are akin to conducting quantitative and qualitative mini-studies within a single research study, where findings are often mixed at some point (Johnson & Onwuegbuzie, 2004). It is also important to note that MMR designs are influenced by the implementation approach and prioritisation of data collection methods (Tashakkori & Teddlie, 1998). According to Ngulube (2020), basic designs in MMR are concurrent and sequential. This is further expounded by Molina-Azorin and Font (2015), who note that if data are collected from both quantitative and qualitative methods simultaneously, the design will be concurrent, parallel, or simultaneous. On the other hand, if the information is gathered from quantitative and qualitative methods in phases, the design will be sequential or two-phase (Molina-Azorin & Font, 2015). As stated by Creswell

(2009), when adopting a sequential design, the researcher must first decide on the method to be used.

MMR designs also have either equivalent status or dominant-less-dominant (nested) designs. According to Molina-Azorin and Font (2015), the equivalent status design employs both quantitative and qualitative approaches equally throughout all stages of the study. In the dominant-less-dominant design, one approach, typically quantitative, dominates while the other is used to a lesser extent (Creswell, 1994). The researcher using the dominant-less-dominant design should give adequate justification for giving more weight to one approach over the other. Figure 3 illustrates and summarises the MMR designs.

Figure 3. MMR designs (Synthesised by the author)

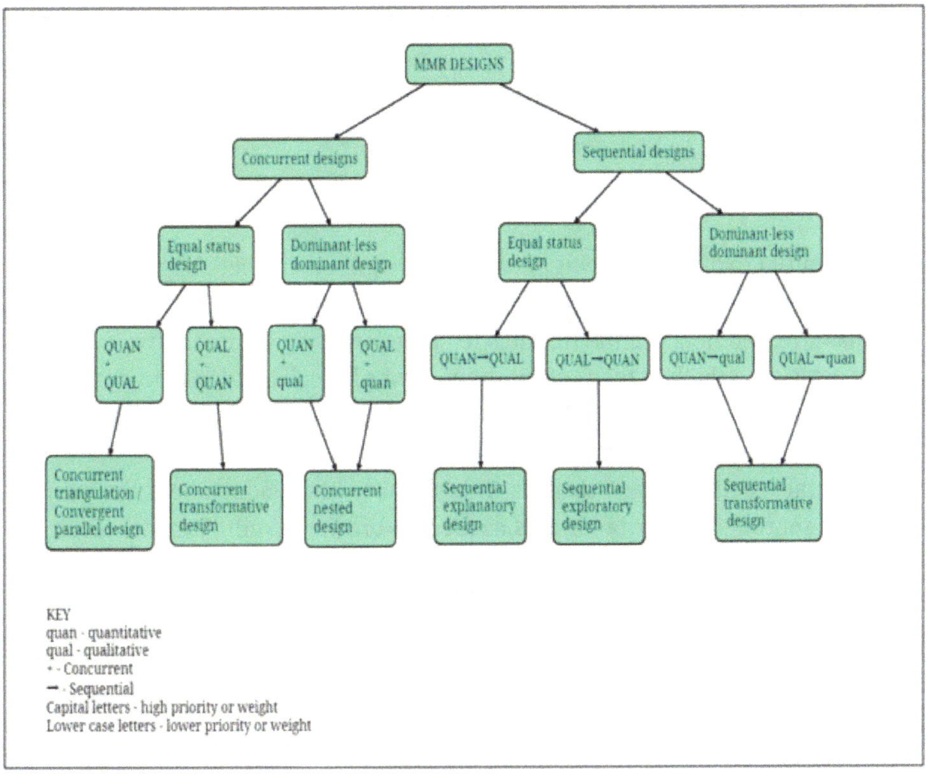

According to Dewasiri et al. (2018), the core MMR designs are explained as follows:

- Sequential explanatory design – This design consists of two distinct phases: initially, researchers gather and analyse quantitative data, followed by the collection and analysis of qualitative data to help interpret or contextualize the quantitative findings. This design is particularly valuable when seeking to understand the underlying reasons behind specific numerical outcomes, including unexpected patterns or trends. The qualitative phase builds on the quantitative results, offering deeper insights and elaboration. This approach is especially effective for studies driven primarily by quantitative questions that benefit from qualitative explanations to achieve a more comprehensive understanding. It is further illustrated in Figure 4.

Figure 4. Sequential explanatory design (Adopted from Creswell & Plano Clark, 2011)

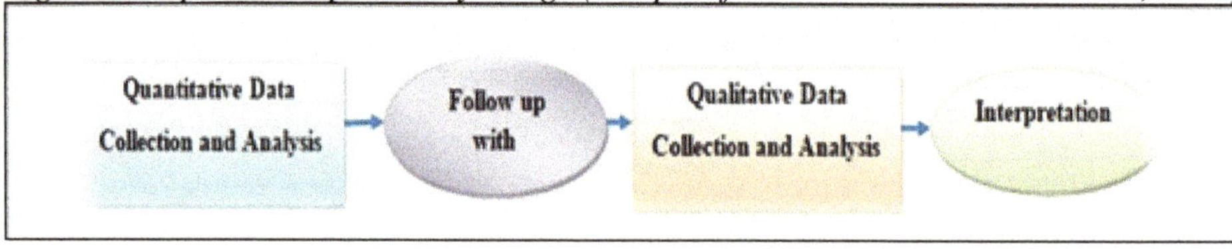

- Sequential exploratory design – In this design, researchers begin by collecting and analysing qualitative data to explore a topic deeply. After this initial phase, researchers use the findings to create or refine a quantitative tool, such as a survey, which is then tested in the next phase. The idea is to use the detailed insights from the qualitative data to shape the quantitative phase, which then measures or confirms these insights with a larger group. This design is especially helpful when there is little existing knowledge about the subject, and a thorough understanding is needed before performing broader testing or generalisation. The visual expression of this design is presented in Figure 5.

Figure 5. Sequential exploratory design (Adopted from Creswell & Plano Clark, 2011)

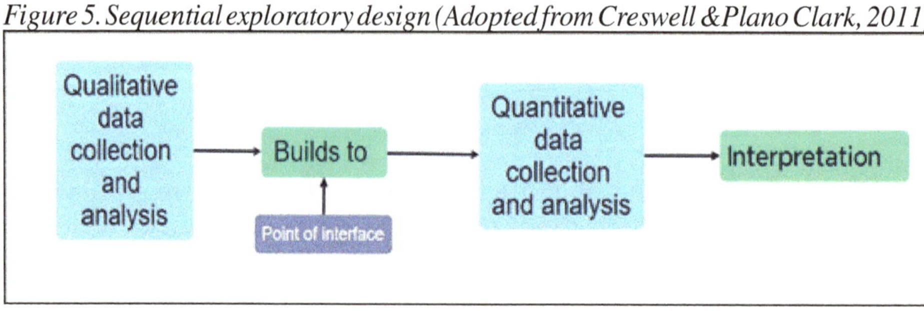

- Sequential transformative design - This MMR design involves collecting and analysing qualitative or quantitative data first, followed by the other type, in separate phases. This design is guided by a specific theoretical framework focused on social justice or advocacy, aiming to address issues affecting marginalized or underrepresented groups. The data from both phases are combined during interpretation to support the goal of promoting social change or empowerment. Researchers choose this design when they want to use both methods to advance a particular advocacy or theoretical perspective.
- Convergent parallel design - Uses two or more methods to confirm, cross-validate or corroborate findings within a study. A collection of both qualitative and quantitative data is done simultaneously. The collected data from the two approaches is treated as equally important. Each type of data is analysed separately in the usual way, and then the results are combined to form a comprehensive understanding of the bigger picture. This approach helps to understand the research question from different angles. This design best neutralises the weaknesses of one approach with the strengths of the other. Figure 6 further illustrates this design.

Figure 6. Convergent parallel design (Adopted from Creswell &Plano Clark, 2011)

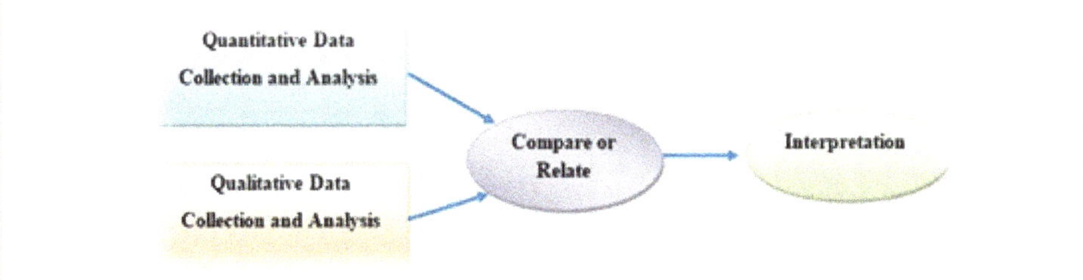

- Concurrent transformative design - In a concurrent transformative mixed methods research design, qualitative and quantitative data are collected simultaneously within a single phase of the study. This approach is grounded in a well-defined theoretical framework or a social justice lens, with the intention of promoting change or supporting advocacy efforts. Both data types are used in tandem to explore issues related to social transformation, frequently highlighting the experiences of marginalized populations or addressing systemic inequities. The integration of these data sets offers a more comprehensive understanding that advances the transformative objectives of the research.

- Concurrent nested design - Gives more weight and priority to one of the methods guiding the project. In a concurrent nested mixed methods research design, researchers collect both qualitative and quantitative data simultaneously. However, one type of data, either quantitative or qualitative, is given priority or emphasis over the other. In this design, the less dominant data type is "nested" or embedded within the dominant data collection process to provide additional context or support. This approach enables researchers to focus primarily on either qualitative or quantitative methods while still gaining insights from secondary data, all within a single phase of research. This design is visually presented in Figure 7.

Figure 7. Concurrent nested design (Adopted from Smith et al., 2023)

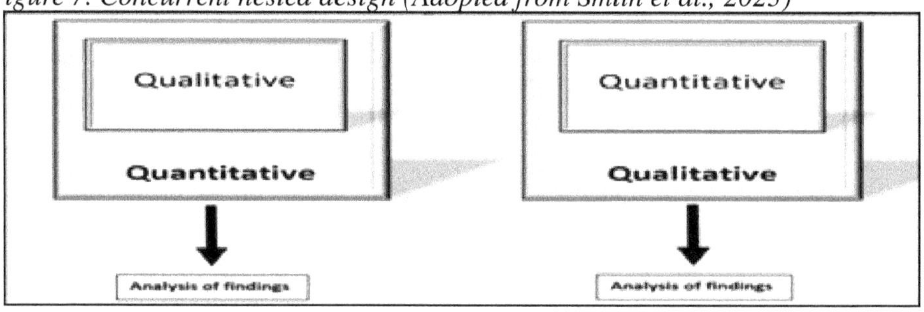

MMR SAMPLING

Sampling is the selection of study participants, which is essential in all research approaches. Elaborating further, Radhakrishnan (2014) says, "Sampling is the process of selecting a portion of the population to represent the entire population" (p. 24). Sampling is a crucial step in the research process, as it determines the quality of inferences a researcher can deduce from findings (Onwuegbuzie & Collins, 2007). By using the appropriate sampling technique, informatics researchers can generate findings that are both rich and inclusive in scope (Onwuegbuzie & Collins, 2007). Traditionally, two distinct sampling procedures dominated social and behavioural sciences. According to Teddlie and Yu (2007), these are probability and purposive sampling procedures. Probabilistic sampling schemes are typically associated with a quantitative approach, whereas non-probabilistic or purposive sampling schemes are commonly linked to a qualitative approach (Onwuegbuzie & Leech, 2006). However, advances in research studies have broken this dichotomy, increasing the options that quantitative and qualitative researchers can use in selecting samples (Onwuegbuzie & Collins, 2007). In fact, Onwuegbuzie and Leech (2006) conclude that both random and purposive sampling techniques are suitable for use in both quantitative and qualitative studies.

Probability (random) sampling involves choosing a significant number of units from a population or specific subgroups in a random way, where the likelihood of including each member of the population is determinable (Tashakkori & Teddlie, 2003b). The technique or sampling scheme aims to create a sample that accurately represents the entire population (Teddlie & Yu, 2007). Onwuegbuzie and Collins (2007) give the following examples of probability sampling techniques:

- Simple random sampling – Each member of a population has an equal chance of being selected.
- Stratified random sampling – The population is divided into sub-groups or strata, and random samples are drawn from each stratum.
- Cluster random sampling – The population is divided into naturally occurring groups or clusters. A random sample of these clusters is then selected, and data is collected from all or a randomly chosen subset of elements within the cluster.
- Systematic random sampling – Researchers select members of a population at regular intervals from a randomly chosen starting point.
- Multi-stage random sampling – At the first stage, large primary clusters are randomly selected from the overall population. In the following stages, the clusters are divided into smaller groups from which samples are randomly selected again.
- Multi-phase sampling - Is a sampling technique where information is collected in phases. A large sample is surveyed first to gather simple or preliminary data. In subsequent phases, detailed data is gathered from smaller subsamples drawn from the initial sample. Multi-phase sampling differs from multi-stage sampling, which involves sampling different units at different stages, whereas multi-phase sampling involves collecting different types of information from the same units over multiple phases.

Probability sampling approaches are ideal when the study aims to generalise findings to the population from which the sample is drawn (Onwuegbuzie & Collins, 2007).

Purposive sampling is also known as non-random, non-probability, purposeful or qualitative sampling (Collins, 2017). It refers to any sampling method where some elements of the population have no chance of selection or where the probability of selection cannot be accurately determined (Chaturvedi, 2016, p. 17). It involves selecting units, such as individuals, groups, or institutions, based on specific purposes related to answering the research questions (Teddlie & Yu, 2007). This technique deliberately chooses particular settings, documents, persons, or events for the valuable information they can provide that cannot be obtained from other available choices (Maxwell, 1997). This is supported by Patton (1990), who argues that the sample is rich in information. According to Onwuegbuzie and Collins (2007), purposive sampling techniques are ideal if the objective of the study is to obtain insights and maximise understanding of the underlying phenomenon. The subjects of the population are chosen arbitrarily by the researcher to be included in the sample. As presented by Teddlie and Yu (2007), examples of purposive sampling approaches include:

- Sampling to achieve representativeness or comparability – Involves the intentional selection of participants to either reflect the diversity proportions of the population or to enable meaningful group comparisons.
- Sampling special and unique cases – This is a purposive sampling approach that targets cases that stand out due to their rarity, significance, or illustrative value.
- Sequential sampling – With this technique, samples are taken one at a time or in successive groups, and data is analysed continuously or after each sample set.
- Sampling using multiple purposive techniques - Researchers combine different purposive sampling methods to enhance the quality and relevance of their sample.

MMR sampling involves a thoughtful combination and unification of qualitative and quantitative sampling techniques, customised to the study's objectives, to produce comprehensive and credible results. In MMR research, the choice of a sampling design is determined by whether the research design is concurrent or sequential, as well as the relationship between the quantitative and qualitative samples (Radhakrishnan, 2014). Consequently, we have two major categories of sampling designs: concurrent and sequential mixed methods sampling. With sequential mixed methods sampling, a sample for quantitative data can be drawn first, and later, the information is needed to draw the sample for qualitative data (Kemper, et al., 2003).

On the other hand, concurrent mixed methods sampling involves the simultaneous selection of units of analysis for quantitative and qualitative data (Teddlie & Yu, 2007). The relationship between the quantitative and qualitative samples determines the specific sampling approach for either concurrent or sequential sampling design. As Onwuegbuzie and Collins (2007) point out, these relationships can be identical, parallel, nested, or multi-level. The relationships are further explained in Table 1.

Table 1. MMR sample relationships (Source: Onwuegbuzie & Collins, 2007).

Sample relationship	Explanation
Identical	The same sample members participate in the study's quantitative and qualitative phases.
Parallel	The samples for the quantitative and qualitative components of the research are different, but they are drawn from the same population of interest.
Nested	Sample members selected for one phase of the study are a subset of those participants chosen for the other facet of the investigation.
Multi-level	Involves the use of two or more sets of samples that are extracted from different levels of the study

However, amateur researchers often face the challenge of balancing between quantitative and qualitative representation and intersection (Corrigan & Onwuegbuzie, 2020). Representation can be achieved in MMR by selecting a sample for each strand that aligns with the chosen design. According to Onwuegbuzie and Collins (2007), this is achievable through selecting samples that can generate thick descriptions and an adequate sample size, leading to a well-powered study to make valid statistical generalizations and enhance the meta-inferences. Since sampling is deeply interwoven in the integration of MMR studies, researchers must avoid sampling techniques that affect the integration of data collection, analysis, and findings in a mixed methods research study by ensuring that selected sampling designs are "realistic, efficient, practical, and ethical" (Onwuegbuzie & Collins, 2007, p. 305).

MMR DATA COLLECTION INSTRUMENTS

MMR studies use both quantitative and qualitative data collection techniques. Qualitative techniques yield rich, non-numerical data that facilitate an understanding of subjective meanings and social contexts (Neuman, 2014). Quantitative instruments yield measurable and statistically analysable data (Creswell, 2014). The strengths of each technique compensate for the weaknesses of the other. Data collected through these instruments supplement each other to enhance the validity and reliability of the findings. It is essential to note that the choice of instruments to use is influenced by the nature of the research problem to be addressed, as there is no one-size-fits-all approach (Ngulube, 2015). For the qualitative strand of an MMR study, examples of data collection techniques that can be used by informatics researchers include, but are not limited to:

- Interviews (semi-structured; unstructured).
- Focus groups.

- Observation.
- Open-ended surveys.
- Document and record analysis.
- Ethnographic research.
- Case studies.

On the other hand, examples of research instruments informatics researchers can use for the quantitative strand of the MMR study include, but are not limited to:

- Questionnaires.
- Experiments.
- Structured interviews.
- Longitudinal studies.
- Observation checklists.
- Existing data sets.

MMR DATA ANALYSIS

Data analysis entails converting gathered data into meaningful information (Taherdoost, 2020). It involves coding and sorting data to establish meaning (Blaikie, 2010). It also refers to the procedures followed in bringing order, structure, interpretation and presentation of gathered data (Creswell, 2009). Borrowing from Teddlie et al (2021) and Onwuegbuzie and Johnson (2021), data analysis in MMR may conform to the following typology:

- Data reduction through coding, thematic analysis.
- Data display, through the use of tables, graphs, matrices, and networks.
- Data transformation (converting qualitative data into a quantitative form, and vice versa).
- Data correlation, through mixed cross-tabulations.
- Crossover mixed methods analysis (using quantitative techniques to analyse qualitative data, and vice versa).
- Inherently mixed methods analysis.
- Data consolidation.
- Data comparison.

In simpler terms, data analysis typically involves three stages: preparing and organising data, coding the data, and presenting the data in the form of descriptive text, statistics, tables, and figures. Data analysis in MMR involves the analysis of quantitative data using quantitative methods and the analysis of qualitative data using qualitative methods (independent analysis); cross-over analysis; and joint analysis (Creswell & Plano Clark, 2007; Ngulube, 2023). The process of analysing data includes interpreting and summarising it to communicate its most important features (Creswell, 2014). Ngulube (2015) explains that data analysis reveals the hidden meaning that makes the data meaningful. Quantitative data in an MMR study can be analysed through descriptive statistics, inferential statistics, regression analysis, time series analysis, and cluster analysis. Qualitative data can also be thematically analysed.

The quantitative data can be prepared by converting it into a machine-readable numeric format. In analysing the quantitative data, the following steps that are in line with Bhattacherjee (2012)'s instructions on quantitative analysis can be followed:

- Data coding
- Data entry/recording
- Frequently checking the data for accuracy
- Empirical results presentation and analysis.

Qualitative data analysis may include:

- Transcribing (Hecker & Kalpokas, 2024)
- Importing data transcripts, records and documents into qualitative analysis software programs at once.
- Selecting quotations
- Data coding or categorisation based on content, themes and patterns (Charmaz, 2014).
- Empirical results presentation and analysis.

Although quantitative and qualitative data can be analysed manually, there are now a variety of software packages that informatics researchers can use to improve the quality of analysis, if used appropriately.

According to Oluwadiya (2022) and also Cooksey cited in Ngulube (2023), softwares that can accomplish quantitative data analysis include but are not limited to SPSS, SYSTAT, NCSS, STATGRAPHICS Centurion, R, Statistical Analysis System (SAS), STATA, Statistica, eViews, MPlus, UNISTA, XLSTAT, Microsoft Excel and AMOS. Among the top free statistics software packages are R, PSPP, JASP, and WinPEPI (Oluwadiya, 2022). SOCR, SciStatCal, SISA, and OpenEpi are among the free online (web-based) software packages informatics researchers can use. As observed by Abatan and Olayemi (2014), research involving quantitative methods cannot be done effectively without the aid of statistical software. Table 2 presents the following positive attributes of statistical software

Table 2. Positive attributes of statistical software (Source: Oluwadiya, 2022).

Key Statistical Package Attribute	**Example of Statistical Package(s) that can help**
Ease of use	SPSS and Minitab
Depth of point and click procedures	SPSS
Range, quality, and ease of use of statistical procedures offered	R, SAS, Stata, SPSS
Modifiability of analytical output specifications	SPSS
Range of graphical output offered	R
Speed of handling large data sets	SAS
Ease and flexibility of data importation	R, SAS, SPSS, and Stata
Ease of results exportation	SPSS
Ease and flexibility of data set manipulation	SPSS
Thoroughness of documentation	Stata

On the other hand, software programs known as CAQDAS (Computer-Assisted Qualitative Data Analysis Software), QDA (Qualitative Data Analysis), or QDAS (Qualitative Data Analysis Software) have gained increasing popularity in qualitative research over the past few years (Tani-Yildiz & Tam, 2023). Prominent QDA software includes MAXQDA, NVivo, and Atlas.ti (Tani-Yildiz & Tam, 2023). According to Oluwadiya (2022):

- Atlas.ti offers an intuitive interface with robust capabilities for data management, coding, and the addition of memos or comments to support data analysis and interpretation. It also features hyperlinking, various visualization tools (e.g., ad-hoc networks, Sankey diagrams, word clouds), advanced concept searching, and survey data import.
- MAXQDA excels in text analysis, data management, variable and survey data import, coding, memo writing, and mixed-methods analysis.
- NVivo is well-suited for text analysis, provides a multilingual interface, advanced text search functions, supports multiple output formats, and can import web pages and social media datasets.

Additional CAQDAS packages include Dedoose, Elan, F4 Analyse, HyperRESEARCH, QDA Miner, Quirkos, and Transana, among others (Paulus & Lester, 2020).

Regarding accessibility, Table 3 presents examples of freely accessible CAQDAS alongside proprietary software that requires licensing.

Table 3. Examples of free and proprietary CAQDAS (Source: Oluwadiya, 2022).

Examples of free CAQDAS	Examples of proprietary CAQDAS
Aquad	ATLAS.ti
Coding Analysis Toolkit (CAT)	MAXQDA
RQDA	NVvivo 10
QDA Miner Lite	QDA Miner
Open Code	Dedoose

These software programs are commonly employed by researchers to facilitate the analysis of data collected from interviews, focus groups, documents, field notes, and open-ended survey questions. As noted by Hoover and Koerber (2011), NVivo, Atlas.ti, and MAXQDA are the three leading qualitative data analysis software (QDAS) platforms, although other options are available. Within the field of media communication informatics, QDAS proves valuable for analysing diverse data types such as interview transcripts, blog or web pages, social media posts, images, audio and video recordings, media content, surveys, and articles (Marjaei et al., 2019). For instance, Magama and Ngoepe (2025), in their information science study titled "The proof of the pudding is in the eating: Archival services quality at the National Archives of Zimbabwe," utilized Atlas.ti to analyse interview transcripts and selected documents. Their analysis generated an ad hoc network of recommendations aimed at improving the quality of the National Archives of Zimbabwe's services, while significantly reducing the costs, time, and effort associated with manual coding. Additionally, QDAS enhances the transparency of analytical procedures, thereby facilitating the establishment of validity and trustworthiness (Woods et al., 2016).

MMR RESULTS REPORTING AND INTEGRATION

The way results are presented in mixed methods research varies according to the research design and the roles played by qualitative and quantitative data (Creswell & Plano Clark, 2018). Dewasiri et al. (2018) observe that:

- In convergent designs, the findings from each method are reported separately before being compared as illustrated in Figure 8.

- For explanatory sequential designs, quantitative results are presented first, followed by qualitative findings that help explain them.
- In exploratory sequential designs, qualitative results come first and inform the subsequent quantitative phase.
- In embedded designs, one type of data supports the other, shaping how the findings are combined in the results section.

Regardless of the chosen design, clear reporting, integration of data, and the strategic use of visual tools are essential. Utilizing joint displays, side-by-side comparisons, and direct qualitative quotes within the text can improve understanding and make mixed methods results more accessible to a broad audience (Creswell & Plano Clark, 2018). By thoughtfully organizing the results section, researchers can deliver mixed methods findings that are rich, insightful, and well-integrated, thereby advancing both scholarly knowledge and practical application

Figure 8. Mixing of results in the convergent design (Synthesised by the author)

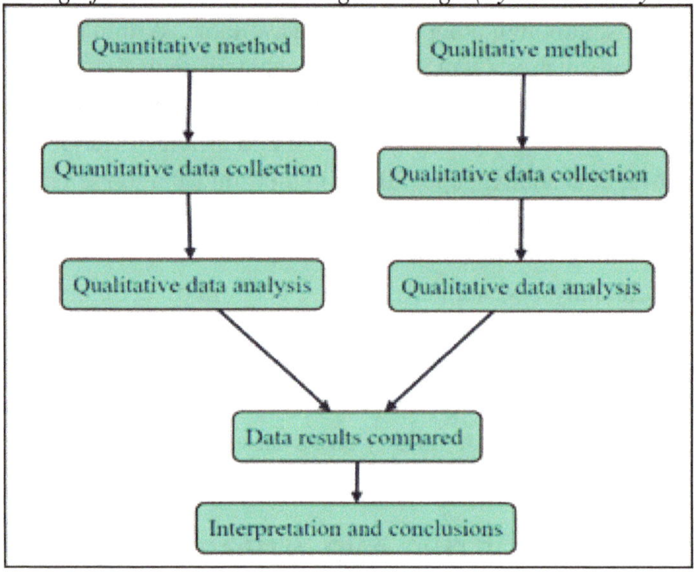

The choice of integration method in a mixed-methods study is shaped by its design. Regardless of the specific approach, the overarching goal remains that of generating insights that exceed what either qualitative or quantitative methods could achieve independently (Brannen, 2005). In a well-executed mixed-methods study, the connection between qualitative and quantitative data is clearly defined, supported by systematic analytical techniques and structured tools, such as joint displays. This integration enhances the interpretation of findings by exploring areas of convergence, divergence, and expansion, ensuring that each phase makes a meaningful contribution to the study's overall objectives (Creedon, 2016). Despite its importance, integration is often underdeveloped in mixed methods research. Researchers must be intentional in selecting integration strategies and transparent in reporting them (Morse & Niehaus, 2009). Such deliberate and clear practices enhance the rigour of the research and strengthen the capacity of mixed methods to address complex questions across various fields.

MMR VALIDITY AND RELIABILITY

Creswell (2014) argues that a valid and reliable study produces almost the same results if repeated using a different sample from the same population. According to Onwuegbuzie and Johnson (2006), validity in MMR can be termed legitimation (trustworthiness or defensibility or quality). As expounded further by Perez (2023), MMR researchers need to ascertain the validity of the quantitative strand, the trustworthiness of the qualitative strand, and the legitimation of the full MMR study. Failure to address validity issues in each strand spoils the credibility of the overall MMR study. Legitimation can be promoted by developing inferences drawn directly from the sample(s) to appropriately address legitimation types pertinent to the sample (Onwuegbuzie & Collins, 2007). Ensuring the validity of the study enhances the quality of research and the acceptability of the results. As Onwuegbuzie and Johnson (2006) argue, research needs to be defensible to the research and practice communities for which research is produced and used. MMR studies combined both quantitative and qualitative validity and reliability strategies to ensure the accuracy of data collection and analysis. As Terrel (2012) puts it, established rules exist for controlling validity in both standard quantitative and qualitative research, which must be followed when these methods are combined.

To achieve rigour, researchers can utilise the content validity strategy. Under this strategy, the researcher ensures that both quantitative and qualitative questions are carefully crafted and pretested. This is done to ensure that the questions reflect and are relevant to the phenomenon under investigation (Giddings & Grant, 2009). The questionnaire and interview guides can also be reviewed by experts. With this strategy, obscure questions are revised and reworded for clarity (Zohrabi, 2013). Pre-testing data collection instruments is crucial for ensuring that the data gathered is clear and reliable. This pre-testing process provides valuable insights, prompting the researcher to clarify technical terms, rephrase unclear questions, and eliminate redundant ones.

The researchers can also utilise the reliability and credibility strategy to enhance the validity and reliability. According to Nunan (1999), this strategy focuses on consistency, dependability and replicability of the results. This can be achieved by explicitly explaining the rationale, design, and subjects of the study, and integrating data obtained through quantitative and qualitative instruments to provide a sound interpretation and enhance the trustworthiness of the results. To ensure reliability, the researchers can take steps to guarantee that the findings are accurate and accurately reflect the participants' viewpoints. Additionally, the researchers can utilise multiple data sources to gain a more comprehensive and detailed understanding of the phenomenon being studied. This approach aligns with the principles of reliability and credibility (Yin, 2009) and helps to validate the findings, providing a more complete picture of the phenomenon (Billups, 2014).

The reliability and credibility of the results can also be enhanced by collecting similar data from individuals at different levels. The researcher should describe in detail how data were collected and analysed, and how results were obtained to establish dependability and auditability (Giddings & Grant, 2009). The actual data collection exercise must be carefully planned. To further enhance the validity and reliability of the results should conform Johnson and Christensen (2017), further recommend that researchers should:

- Ensure the research instruments are intended for the proposed sample (construct validity).
- Report what they saw or heard accurately for credible inferences to be generated from the data (descriptive validity).
- Provide accurate meanings to events, people, and behaviours that are being studied (interpretive validity).
- Conduct member checks with participants to ensure the researcher's interpretation is consistent with what the participant expressed (interpretive validity).

MMR ETHICAL/AXIOLOGICAL CONSIDERATIONS

Researchers have an obligation to safeguard their participants, build trust, uphold research integrity, and prevent misconduct (Creswell, 2009). Informatics researchers can accomplish this by adhering to research ethics. Coontz (2008) explains that research ethics encompass the manner in which research is conducted and whether researchers have acted responsibly to ensure the highest standards of excellence, integrity, and quality in scientific research. According to Terrel (2012) ethically sound research has the following attributes:

- Participants must participate voluntarily.
- Participants must understand the purpose and procedures of the study.

- Participants must understand that they have the right to a copy of the results.
- Participants must understand the potential benefits of the study and that their privacy will be respected.
- Researchers must understand the impact of their presence at research sites and ensure that these sites are left undisturbed at the end of the study.
- Care must be taken to identify and nullify any actual or perceived issues where power between the researcher and participant could be abused.
- Anonymity must be maintained during data analysis, and data must be kept for a reasonable period of time.
- Ensure that writing is free of bias towards any group (age, ethnicity, sexual orientation, race, gender, and so forth).
- The details of the study must be carefully explained in the report so that readers can judge its ethical quality for themselves.

In accordance with Terrel (2012)'s attributes of ethical research, informatics researchers must embrace internationally recognised moral principles of ethics, which are autonomy, beneficence, non-maleficence and justice.

Under autonomy, research should respect the participants' independence, rights, and dignity. Autonomy requires that individuals' participation is freely given, based on informed consent (UNISA, 2016). In conformity with the moral autonomy principle, direct or indirect coercion, undue influence, and inducement of participants through promises of benefits, such as material, financial, services, or opportunities, must be avoided. The study's overall objective must be clearly explained to participants, and the decision to participate must be made freely by them.

Informatics researchers should ensure respect for participants and respondents by obtaining their informed consent to participate in the study. The thorough pursuit of participants' consent aligns with Ruane (2016), who posits that informed consent is an ethical mandate that binds researchers to ensure that potential research participants are fully informed about the research and are given the opportunity to freely decide whether or not to participate. A schedule of dates for data collection should be drawn, diarised, and strictly followed as a blueprint for the data collection activities.

According to the moral principle of beneficence, the informatics researchers should foster a strong rapport with the informants to build trust. Additionally, the researchers should take time to explain the purpose and importance of the current study to the respondents and participants. The selection of research participants must be done scientifically. Furthermore, the researcher must be honest in all aspects of data collection, analysis, and interpretation. No preconceptions, prior conclusions, falsifications, or exaggerations should influence the study's results.

According to the principle of non-maleficence, research should be conducted in a manner that is free from harm to both participants and the public. Informatics researchers should respect and protect the dignity, privacy, and confidentiality of participants and institutions. This helps to ensure that participants' relevant social, cultural, and historical backgrounds are considered in the planning and conduct of research. In further pursuit of the non-maleficence moral principle, the researchers must ensure that the personal information of participants used for research purposes is adequately protected from unauthorised access.

Another important ethical principle is justice, which entails that the benefits and risks of research should be fairly distributed among individuals (UNISA, 2016). The principle further posits that participants and institutions should not be exposed to procedures or risks that are not directly attached to the research project or its methodology (UNISA, 2016). To conform to this principle, participants must be provided with adequate details about the research, including its purpose, objectives, and the questions it seeks to answer, so that they can be aware of the risks that may arise. Participants must be allowed to respond anonymously to protect their identity and privacy.

MMR ADVANTAGES AND DISADVANTAGES

Despite its many attractive features, MMR, like other approaches, has its fair share of weaknesses. According to Ngulube (2015), all research approaches have pros and cons. Consequently, Table 4 shows the advantages and disadvantages of the MMR approach.

Table 4. Advantages and disadvantages of the MMR approach

Advantages	Disadvantages
• It brings more clarity to the research problem and complex phenomena than a mono approach could (Cresswell & Plano Clark, 2011). "Quantitative tells us 'If'; and qualitative tells us 'how' and 'why'" (Terrel, 2012, p. 258). • Allows flexibility in examining problems and multiple insights into their solutions (Ngulube & Ngulube, 2015). • It increases the scope and level of possible analysis and substantial data interpretation (Johnson & Onwuegbuzie, 2004). • Achieves rigour, confidence, validity, and reliability of research findings through convergence and collaboration of results from different methods (Greene, et al., 1989). • Helps explore unanticipated findings and enhances generalisability (Mayoh & Onwuegbuzie, 2015). • Methodological integration offsets the weaknesses of one method with the strengths of the other (Caruth, 2013). • Knowledge produced is sufficient to inform theory and practice (Johnson & Onwuegbuzie, 2004). • Provides a holistic view of the research problem through allowing researchers to reflect on both the extensive, overall patterns and trends in the data (quantitative) and the meaning and context behind those patterns and trends (qualitative) (Sharma et al., 2023). • With MMR, researchers can develop, explain and validate a theory in one study. (Teddlie et al., 2021).	• Collecting and analysing data from multiple sources using multiple research questions makes MMR more complex (Sharma et al., 2023). • The researcher must be well-versed in using and applying both quantitative and qualitative approaches (McMillan & Schumacher, 2006). • Involves extensive data collection and expensive resources (Cameron & Miller, 2007). • Researchers may end up using MMR as a mere label for studies that only superficially mix the methods (Cameron & Miller, 2007). • Formulation of research questions is more difficult than in a mono method (Onwuegbuzie & Leech, 2006). • It is time-consuming (Molina-Azorin & Font, 2015). • Integrating and synthesizing the findings from both quantitative and qualitative data can be challenging as these two types of data are often analysed and presented in different ways (Sharma et al., 2023). • MMR is still evolving, and more work is needed to perfect philosophical positions, designs, data analysis, validity strategies, mixing, and integration of procedures (Johnson & Onwuegbuzie, 2004). • Limited generalizability and replicability (Sharma et al., 2023).

JUSTIFICATION FOR USING MMR IN INFORMATICS RESEARCH

Although MMR is not intrinsically superior to mono-method research, as Molina-Azorin and Font (2015) would argue, it can be adopted by informatics researchers when the overall aim and objectives of their studies are interdisciplinary, complex, and dynamic. Consequently, MMR can address informatics contemporary research phenomena in this ever-changing modern world characterised by *volatility* (rapid and unpredictable change), *uncertainty* (lack of predictability), *complexity* (interconnected and numerous factors) and *ambiguity* (unclear realities) (*VUCA*) framework. The MMR solutions to address VUCA elements are illustrated in Table 5. This aligns with Ngulube and Ngulube (2015), who argue that a method is not used merely for its own sake, but rather for its ability to address research problems and questions. The quantitative dimension can provide general patterns and width, while the qualitative dimension reflects upon experience and depth to increase confidence in the results (Newby, 2014). Johnson and Onwuegbuzie (2004) add that the quantitative method provides hard generalisable data, while the qualitative method provides deep and rich exploratory data. Additionally, the qualitative data can be employed to provide a deeper understanding of the quantitative data during the analysis and interpretation process. According to Greene, Caracelli, and Graham (1989), the convergence and corroboration of results from different methods achieve scientific rigour and avoid inconsistent and incomplete results. Consequently, MMR research output is of higher value than quantitative or qualitative research alone can produce (Muskat et al., 2013). In summary form, Molina-Azorin and Font (2015) presented the rationale for using the MMR approach, which can also benefit informatics researchers as follows:

- Complementarity and utility – providing clarifications of findings from one method with the other.
- Development – using results from one method to help design or inform the other.
- Integration – cross-validating results through multiple data sources and methods to enhance the research credibility.
- Expansion – extending the scope or magnitude of inquiry through using different methods for different components of the study.
- Initiation – discovering incongruity or new perspectives that challenge research questions and open new avenues for inquiry.

Table 5. MMR solutions to address VUCA elements (Synthesised by author).

VUCA element	MMR Solution
Volatility	Quantitative techniques enable tracking of trends and detection of patterns in swiftly shifting data, while qualitative methods delve into the user experiences and behaviours driving those changes. This combined approach enables researchers to quantify change and understand its underlying causes.
Uncertainty	Using both methods is well-suited for investigating unpredictable future outcomes. Quantitative data can forecast possible scenarios, whereas qualitative insights explore the perceptions, beliefs, and contexts shaping those uncertainties. Together, they offer a stronger foundation for decision-making than either approach alone.
Complexity	Informatics systems often involve numerous interconnected variables, creating high complexity. Mixed methods help unravel this by using quantitative data to chart relationships among variables and qualitative data to reveal hidden or non-linear links that numbers alone cannot expose.
Ambiguity	Qualitative data adds depth and context to interpret unclear situations. For instance, while a quantitative survey may indicate user dissatisfaction with a software update, qualitative interviews can provide insight into the reasons behind that dissatisfaction. Such rich, contextual detail is essential for understanding ambiguity in complex informatics environments.

FURTHER RESEARCH DIRECTIONS

Additional MMR is necessary to further streamline the approach by utilising summary tables, diagrams, clearer section headings, joint displays, and side-by-side comparisons. Future studies should capitalise on the flexible nature of MMR by incorporating a broader range of indigenous perspectives and methodologies within diverse cultural contexts. This integration will enhance researchers' understanding and application of MMR, particularly in the Global South and Africa. Alignment with cultural differences fosters the creation of contextually relevant knowledge and the design of solutions and interventions grounded in local traditions and values (Garcia & Ramirez, 2021). Moreover, future research should emphasise the incorporation of indigenous theories and knowledge systems alongside Western-based intervention frameworks.

CONCLUSION

Mixed-methods research provides a strong framework for informatics studies by delivering a richer, more comprehensive understanding of complex issues and overcoming the drawbacks inherent in relying on a single method. It allows investigators to tackle complex issues with both adaptability and thoroughness, answering questions about the mechanisms and reasons behind human behaviour and social phenomena, while simultaneously advancing theoretical understanding and practical solutions. This approach entails the deliberate combination of data collection, analysis, and interpretation phases, creating a dynamic interaction between numerical data and narrative insights. It is a practical method driven primarily by the research question rather than rigid philosophical distinctions between different methodologies. This combined approach broadens and deepens analysis, resulting in more meaningful and practical insights. Important elements of using mixed methods in informatics include tackling complex research questions. Since informatics often involves complex systems and human factors, mixed methods enable researchers to examine these intricacies by collecting both quantitative and qualitative data.

REFERENCES

Abatan, S. M., & Olayemi, M. S. (2014). The role of statistical software in data analysis. *International Journal of Applied Research and Studies*, *3*(8), 1–15.

Adeleye, J. O. (2017). *Pragmatism and its implications on the teaching and learning in Nigerian schools*. Retrieved on 10 October 2025 from https://www.isres.org/pragmatism-and-its-implications-on-teaching-and-learning-in-nigerian-schools-51-s.html

Babbie, E., & Mouton, J. (2001). *The practice of social research: An introduction to reading research*. Oxford University Press.

Bans-Akutey, A., & Tiimub, B. M. (2021). Triangulation in research. *Academia Letters. Article*, *3392*, 1–6. DOI: 10.20935/AL3392

Bartlett, C., Marshall, M., & Marshall, A. (2012). Two-eyed seeing and other lessons learned within a co-learning journey of bringing together indigenous and mainstream knowledges and ways of knowing. *Journal of Environmental Studies and Sciences*, *2*(4), 331–340. DOI: 10.1007/s13412-012-0086-8

Bazeley, P. (2004). Issues in mixing qualitative and quantitative approaches to research. In Buber, R., Gadner, J., & Richards, L. (Eds.), *Applying qualitative methods to marketing management research* (pp. 141–156). Palgrave Macmillan.

Bhattacherjee, A. (2012). *Social Science Research: Principles, Methods, and Practices*. Retrieved on 10 October 2025 from https://digitalcommons.usf.edu/oa_textbooks/3

Billups, F. (2014). The quest for rigor in qualitative studies; strategies for institutional research. *The NERA Researcher*, *52*, 10–12.

Brannen, J. (2005). Working qualitatively and quantitatively. In Seale, C., Gobo, G., Gubrium, J. F., & Silverman, D. (Eds.), *Qualitative Research Practice*. Sage.

Bryman, A. (2012). *Social research methods* (4th ed.). Oxford University Press.

Cameron, R., Dwyer, T., Richardson, S., Ahmed, E., & Sukumaran, A. (2013). Lessons from the field: Applying the Good Reporting of a Mixed Methods Study (GRAMMS) framework. *Electronic Journal of Business Research Methods*, *11*(2), 55–66.

Cameron, R., & Miller, P. (2007). Mixed methods research: Phoenix of the paradigm wars. In R. Chapman (Ed.), *Managing our intellectual and social capital: Proceedings of the 21st ANZAM Conference, Sydney, 4–7 December 2007.*

Caruth, G. D. (2013). Demystifying mixed methods research design: A review of the literature. *Mevlana International Journal of Education, 3*(2), 112–122.

Charmaz, K. (2014). *Constructing grounded theory: A practical guide through qualitative analysis* (2nd ed.). Sage.

Chaturvedi, K. (2016). Sampling methods. Retrieved on 10 October 2025 from https://www.pitt.edu/~super7/43011-44001/43911.pdf

Chilisa, B., & Tsheko, G. N. (2014). Mixed methods in indigenous research: Building relationships for sustainable intervention outcomes. *Journal of Mixed Methods Research, 8*(3), 222–233. DOI: 10.1177/1558689814527878

Collins, K. M. T. (2017). Sampling designs in educational research. In Wyse, D., Smith, E., Suter, L., & Selwyn, N. (Eds.), *SAGE Handbook of Educational Research* (pp. 280–292). Sage.

Coontz, P. (2008). The responsible conduct of social research. In Yang, K., & Miller, G. J. (Eds.), *Handbook of research methods in public administration*. Taylor and Francis Group.

Creedon, R. (2016). *A mixed method study to explore the impact of nurse prescribing in clinical practice* (Unpublished doctoral dissertation). University College Cork, Cork, Ireland. Retrieved on 10 October 2025 from https://cora.ucc.ie/server/api/core/bitstreams/d8c1d35a-4e3b-4978-8518-95f7334c1828/content

Creswell, J. W. (1994). Research designs: Qualitative and quantitative approaches. *Sage (Atlanta, Ga.)*.

Creswell, J. W. (2009). *Research design: Qualitative, quantitative and mixed methods approaches* (3rd ed.). Sage.

Creswell, J. W. (2014). *Research design: Qualitative, quantitative and mixed methods approaches* (4th ed.). Sage.

Creswell, J. W., & Plano Clark, V. L. (2007). Designing and conducting mixed methods research. *Sage (Atlanta, Ga.)*.

Creswell, J. W., & Plano Clark, V. L. (2011). *Designing and conducting mixed methods research* (2nd ed.). Sage.

Creswell, J. W., & Plano Clark, V. L. (2018). *Designing and conducting mixed methods research* (3rd ed.). Sage.

Dawadi, S., Shrestha, S., & Giri, R. A. (2021). Mixed methods research: A discussion on its types, challenges, and criticisms. *Journal of Practical Studies in Education, 2*(2), 25–36.

de Waal, C. (2001). *On Peirce*. Wadsworth.

Dewasiri, N. J., Weerakoon, Y. K. B., & Azeez, A. A. (2018). Mixed methods in finance research: The rationale and research designs. *International Journal of Qualitative Methods, 17*, 1–13.

Dovetail Editorial Team. (2023). Mixed methods research guide with examples. https://dovetail.com/research/mixed-methods-research/

Garcia, G. A., & Ramirez, J. J. (2021). Proposing a methodological borderland: Combining Chicana feminist theory with transformative mixed methods research. *Journal of Mixed Methods Research, 15*(2), 240–260. DOI: 10.1177/1558689820954023

Giddings, L., & Grant, B. (2009). From rigour to trustworthiness: Validating mixed methods. In Andrew, S., & Halcomb, E. J. (Eds.), *Mixed methods research for nursing and health science*. Blackwell Publishing Ltd.

Goundar, S. (2012). *Research methodology and research method*. Retrieved on 10 October 2025 from https://www.researchgate.net/publication/333015026_Chapter_3_Research_Methodology_and_Research_Method

Granikova, V., Hong, Q. N., Crist, E. C., & Pluye, P. (2020). Mixed methods research in library and information science: A methodological review. *Library & Information Science Research, 42*(1), 1–10. DOI: 10.1016/j.lisr.2020.101003

Greene, J. C., & Caracelli, V. J. (Eds.). (1997). *Advances in mixed-method evaluation: The challenges and benefits of integrating diverse paradigms: New directions for evaluation*. Jossey-Bass.

Greene, J. C., Caracelli, V. J., & Graham, W. F. (1989). Toward a conceptual framework for mixed method evaluation designs. *Educational Evaluation and Policy Analysis, 11*, 255–274.

Gunasekare, U. L. T. P. (2015). Mixed research method as the third research paradigm: A literature review. *International Journal of Scientific Research, 4*(8), 361–367.

Harris, J. (2021). Mixed methods research in developing country contexts: Lessons from field research in six countries across Africa and the Caribbean. *Journal of Mixed Methods Research*. Advance online publication. DOI: 10.1177/15586898211032825

Harvey, L. (2012). Phenomenological approach to validity. *Quality Research International.* Retrieved on 10 October 2025 from https://www.qualityresearchinternational.com/methodology/RRW1pt8pt3Phenvalidity.php

Hecker, J., & Kalpokas, N. (2024). The ultimate guide to qualitative research - Part 2: Handling qualitative data. Retrieved on 10 October 2025 from https://atlas.ti.com/guides/qualitative-research-guide-part-2/qualitative-data-analysis

Hoover, R. S., & Koerber, A. L. (2011). Using NVivo to answer the challenges of qualitative research in professional communication: Benefits and best practices tutorial. *Transactions on Professional Communication, 54*(1), 68–82. DOI: 10.1109/TPC.2009.2036896

Hunter, A. D., & Brewer, J. (2015). Designing multimethod research. In Hesse-Biber, S. N., & Johnson, R. B. (Eds.), *The Oxford Handbook of Multimethod and Mixed Methods Research Inquiry* (pp. 185–205). Oxford University Press.

Johnson, R. B., & Christensen, L. B. (2017). *Educational research: Quantitative, qualitative, and mixed approaches* (6th ed.). Sage.

Johnson, R. B., & Onwuegbuzie, A. J. (2004). Mixed methods research: A research paradigm whose time has come. *Educational Researcher, 33*(7), 14–26.

Kandiero, A., & Makuwatsine, C. (2022). Exploring determinants of internet service provider customer switching barriers using an exploratory sequential mixed methods research design. In Ngulube, P. (Ed.), *Handbook of research on mixed methods research in information science* (pp. 293–323). IGI Global.

Kemper, E. A., Stringfield, S., & Teddlie, C. (2003). Mixed methods sampling strategies in social science research. In Tashakkori, A., & Teddlie, C. (Eds.), *Sage handbook of mixed methods in social and behavioural research* (pp. 1–20). Sage.

Leedy, P., & Ormrod, J. (2005). *Practical research: Planning and design* (7th ed.). Pearson.

Lincoln, Y. S., Lynham, S. A., & Guba, E. G. (2011). Paradigmatic controversies, contradictions, and emerging confluences, revisited. In Denzin, N. K., & Lincoln, Y. S. (Eds.), *The Sage handbook of qualitative research* (4th ed., pp. 97–128). Sage.

Magama, B., & Ngoepe, M. (2025). The proof of the pudding is in the eating: Archival services quality at the national archives of Zimbabwe. *Information Development, 0*(0), 1–14. DOI: 10.1177/02666669251375096

Mamorobela, S. P. (2022). Understanding a social media-enabled knowledge management adoption model for small and medium enterprises in South Africa: An explanatory sequential mixed methods research design study. In Ngulube, P. (Ed.), *Handbook of research on mixed methods research in information science* (pp. 324–339). IGI Global.

Marjaei, S., Yazdi, F. A., & Chandrashekara, M. (2019). MAXQDA and its application to LIS research. *Library Philosophy and Practice*, (e-journal) 2325, 1–10. https://digitalcommons.unl.edu/libphilprac/2325/

Maxwell, J. (1997). Designing a qualitative study. In Bickman, L., & Rog, D. J. (Eds.), *Handbook of applied social research methods* (pp. 69–100). Sage.

May, E. M., Hunter, B. A., & Jason, L. A. (2017). Methodological pluralism and mixed methods to strengthen community psychology research: An example from Oxford House. *Journal of Community Psychology*, *45*, 100–116. PMID: 28839344

Mayoh, J., & Onwuegbuzie, A. J. (2015). Toward a conceptualization of mixed methods phenomenological research. *Journal of Mixed Methods Research*, *9*, 91–107.

McMillan, J., & Schumacher, S. (2006). *Research in education: Evidence-based inquiry* (6th ed.). Pearson.

Melnikovas, A. (2018). Towards an explicit research methodology: Adapting research onion model for futures studies. *Journal of Futures Studies*, *23*(2), 29–44. DOI: 10.6531/JFS.201812_23(2).0003

Mertens, D. M. (2010). Research and evaluation in education and psychology: Integrating diversity with quantitative, qualitative and mixed methods. *Sage (Atlanta, Ga.)*.

Molina-Azorín, J. F., & Font, X. (2015). Mixed methods in sustainable tourism research: An analysis of prevalence, designs and application in JOST 2005–2014. *Journal of Sustainable Tourism*, *2015*, 1–25.

Morse, J. M., & Niehaus, L. (2009). *Mixed method design: Principles and procedures*. Retrieved on 20 October from https://www.researchgate.net/publication/290746053_Janice_M_Morse_Linda_Niehaus_2009_mixed_method_design_Principles_and_procedures

Muskat, M., Blackman, D., & Muskat, B. (2013). *Mixed Methods: Combining expert interviews, cross-impact analysis and scenario development*. University of Canberra.

Mutsagondo, S. (2022). Application of the convergent mixed methods research design in the study of email records management in Zimbabwe's central government. In Ngulube, P. (Ed.), *Handbook of research on mixed methods research in information science* (pp. 558–574). IGI Global.

Nakaziba, S., & Namuguzi, M. (2022). Application of mixed methods in information science research in Africa: A methodological review. In Ngulube, P. (Ed.), *Handbook of research on mixed methods research in information science* (pp. 1–19). IGI Global.

Neuman, W. L. (2014). *Social research methods: Qualitative and quantitative approaches* (7th ed.). Pearson.

Newby, P. (2014). *Research methods for education* (2nd ed.). Routledge.

Ngulube, P. (2015). Trends in research methodological procedures used in knowledge management studies 2009–2013. *African Journal of Library Archives and Information Science*, *25*(2), 1–30.

Ngulube, P. (2020). The movement of mixed methods research and the role of information science professionals. In Ngulube, P. (Ed.), *Handbook of research on connecting research methods for information science research* (pp. 425–455). IGI Global., DOI: 10.4018/978-1-7998-1471-9.ch022

Ngulube, P. (2023). Improving the quality of reporting findings using computer data analysis applications in educational research in context. *Heliyon*, *9*, e19683. DOI: 10.1016/j.heliyon.2023.e19683 PMID: 37810136

Ngulube, P., & Ngulube, B. (2015). Mixed methods research in the South African Journal of Economic and Management Sciences: An investigation of trends in the literature. *Suid-Afrikaanse Tydskrif vir Ekonomiese en Bestuurswetenskappe*, *18*(1), 1–13.

Ngulube, P., & Ukwoma, S. C. (2021). Prevalence of methodological transparency in the use of mixed methods research in library and information science research in South Africa and Nigeria, 2009–2015. *Library & Information Science Research*, *43*(4), 101124. DOI: 10.1016/j.lisr.2021.101124

Nunan, D. (1999). *Research methods in language learning*. Cambridge University Press.

O'Cathain, A., Murphy, E., & Nicholl, J. (2008). The quality of mixed methods studies in health services research. *Journal of Health Services Research & Policy*, *13*(2), 92–98. DOI: 10.1258/jhsrp.2007.007074 PMID: 18416914

Oluwadiya, K. S. (2022). Software packages for quantitative and qualitative research. Retrieved on 20 October from https://oluwadiya.com/wp-content/uploads/2022/07/Software-Packages-for-Quantitative-and-Qualitative-Research.pdf

Onwuegbuzie, A. J., & Collins, K. M. T. (2007). A typology of mixed methods sampling designs in social science research. *Qualitative Report*, *12*(2), 281–316.

Onwuegbuzie, A. J., & Johnson, R. B. (2006). The validity issue in mixed research. *Research in the Schools*, *13*(1), 48–63.

Onwuegbuzie, A. J., & Johnson, R. R. (2021). *The reviewer's guide for mixed methods research analysis*. Routledge.

Onwuegbuzie, A. J., & Leech, N. L. (2006). Linking research questions to mixed methods data analysis procedures. *The Qualitative Report*, *11*, 474–498.

Patton, M. (1990). *Qualitative evaluation and research methods* (2nd ed.). Sage.

Perez, A. (2023). Practical recommendations for mixed methods sampling in psychological intervention research: A mixed methods case study. Retrieved on 20 October from https://digitalcommons.unl.edu/cehsdiss/427/

Phillips, D. C. (2004). *Is mixed methods research an epistemological oxymoron?* [Paper presentation]. American Educational Research Association.

Poth, C., & Munce, S. E. (2020). Commentary—Preparing today's researchers for a yet unknown tomorrow: Promising practices for a synergistic and sustainable mentoring approach to mixed methods research learning. *International Journal of Multiple Research Approaches*, *12*(1), 56–64. DOI: 10.29034/ijmra.v12n1commentary

Radhakrishnan, G. (2014). Sampling in mixed methods research. *International Journal of Advances in Nursing Management*, *2*(1), 24–27.

Rahman, S. (2017). The advantages and disadvantages of using qualitative and quantitative approaches and methods in language "testing and assessment" research: A literature review. *Journal of Education and Learning*, *6*(1), 102–112.

Ruane, J. M. (2016). *Introducing social research methods*. Wiley.

Sharma, L. R., Bidari, S., Bidari, D., Neupane, S., & Sapkota, R. (2023). Exploring the mixed methods research design: Types, purposes, strengths, challenges, and criticisms. *Global Academic Journal of Linguistics and Literature*, *5*(1), 3–12. DOI: 10.36348/gajll.2023.v05i01.002

Smith, B., Paton, C., & Ramaraj, P. (2023). Teaching basic surgical skills using a more frugal, near-peer, and environmentally sustainable way: Mixed methods study. *JMIR Perioperative Medicine*, *6*(e50212), 1–10. DOI: 10.2196/50212 PMID: 37966886

Taherdoost, H. (2020). Different types of data analysis; data analysis methods and techniques in research projects. *International Journal of Academic Research in Management*, *9*(1), 1–9.

Tani-Yıldız, E., & Tam, M. S. (2023). Using qualitative data analysis software (QDAS) in communication studies: A systematic review. *Akdeniz İletişim, (20.Yıl Özel Sayısı - 42)*, 51–68. https://doi.org/DOI: 10.31123/akil.1330172

Tashakkori, A., & Creswell, J. W. (2007). Editorial: The new era of mixed methods. *Journal of Mixed Methods Research*, *1*(1), 3–7.

Tashakkori, A., & Teddlie, C. (1998). Mixed methodology: Combining qualitative and quantitative approaches. *Sage (Atlanta, Ga.)*.

Tashakkori, A., & Teddlie, C. (2003a). Major issues and controversies in the use of mixed methods in social and behavioural sciences. In Tashakkori, A., & Teddlie, C. (Eds.), *Mixed methods in social and behavioural research*. Sage.

Tashakkori, A., & Teddlie, C. (2003b). *Handbook of mixed methods in social and behavioural research*. Sage.

Teddlie, C., Johnson, R. B., & Tashakkori, A. (2021). *Foundations of mixed methods research: Integrating quantitative and qualitative techniques in the social and behavioral sciences* (2nd ed.). Sage.

Teddlie, C., & Tashakkori, A. (2009). Foundations of mixed methods research: Integrating quantitative and qualitative approaches in the social and behavioural sciences. *Sage (Atlanta, Ga.)*.

Teddlie, C., & Yu, F. (2007). Mixed methods sampling: A typology with examples. *Journal of Mixed Methods Research*, *1*(1), 77–100.

Woods, M., Paulus, T., Atkins, D. P., & Macklin, R. (2016). Advancing qualitative research using qualitative data analysis software (QDAS): Reviewing potential versus practice in published studies using Atlas.ti and NVivo, 1994–2013. *Social Science Computer Review*, *34*(5), 597–617. DOI: 10.1177/0894439315519631

Yin, R. K. (1994). Case study research: Design and methods. *Sage (Atlanta, Ga.)*.

Yin, R. K. (2009). *Case study research: Design and methods* (4th ed.). Sage.

Zohrabi, M. (2013). Mixed method research: Instruments, validity, reliability and reporting findings. *Theory and Practice in Language Studies*, *3*(2), 254–262.

ADDITIONAL READING

Chilisa, B. (2019). *Indigenous research methodologies* (2nd ed.). Sage.

Creswell, J. W., & Creswell, J. D. (2018). *Research design: Qualitative, quantitative, and mixed methods approaches.* Sage. Bazeley, P. (2019). Integrative analysis for mixed methods research. In Atkinson, P., Delamont, S., Cernat, A., Sakshaug, J. W., & Williams, R. A. (Eds.), *SAGE research methods foundations*. Sage.

Hirose, M., & Creswell, J. W. (2022). Applying core quality criteria of mixed methods research to an empirical study. *Journal of Mixed Methods Research*, *17*(1), 12–28. DOI: 10.1177/15586898221086346

Ngulube, P. (Ed.). (2022). *Handbook of research on mixed methods research in information science*. IGI Global.

KEY TERMS AND DEFINITIONS

Abductive reasoning: Starts with observing an empirical phenomenon and seeks the most plausible explanation based on available evidence. It involves alternating between inductive and deductive processes to develop the best possible conclusion. It is invaluable for explaining surprising phenomena that do not align with existing theories.

Axiology: Means what counts as fundamental values and what is consciousness (moral choices, ethics and normative judgements).

Deductive reasoning: Begins with an established theory, from which a research question or hypothesis is formulated. Data is then collected to test and either confirm or refute the hypothesis. This approach primarily guides the quantitative research.

Doxology: Is a philosophical concept that emphasizes the distinction between subjective beliefs and objective, verifiable knowledge, and functions as a foundational premise for the research process.

Epistemology: Is the branch of philosophy concerned with examining the processes through which individuals acquire knowledge about the world, the definition of knowledge itself, and the limitations associated with it. It investigates how humans come to understand phenomena and assesses the reliability of that knowledge.

Induction reasoning: Starts with observing and gathering data, followed by describing and analysing it to develop a new theory. The approach is commonly used to guide the qualitative research.

Mixed Methods Research: Is the combining of qualitative and quantitative research methods in one or more phases of a single study to thoroughly address the research question by gathering both types of data either simultaneously, sequentially, or in multiple stages, aiming to leverage their strengths and reduce their limitations.

Ontology: Refers to the assumptions regarding the nature of reality. It seeks to understand what exists independently of human perception and to develop a comprehensive classification or framework of the fundamental categories of entities that constitute reality.

Chapter 6
Designing Robust Sampling Frameworks for Mixed Methods Research: Challenges and Practical Guidelines

Martha Lyaka
http://orcid.org/0000-0002-7993-613X
University of South Africa, South Africa

Isaac Mukungu
http://orcid.org/0000-0003-3259-2701
Busitema University, Uganda

Nina Olivia Rugambwa
http://orcid.org/0000-0002-5127-965X
Kyambogo University, Uganda

Patrick Ngulube
http://orcid.org/0000-0002-7676-3931
University of South Africa, South Africa

ABSTRACT

In recent years, research in mixed methods has received considerable attention in a wide range of fields, including computer science and informatics. This methodology encourages the integration of both qualitative and quantitative data to provide a comprehensive picture of complex phenomena. However, one of the main problems faced by researchers participating in mixed method projects is that of sampling. Sampling, as a cornerstone of methodological rigour, has an impact on the validity

DOI: 10.4018/979-8-3693-4018-9.ch006

and application of research findings. The aim of this chapter is to analyse the nuances of sampling in the context of mixed-method research, specifically to address the complexity involved and to provide practical suggestions for computational and informatics researchers.

INTRODUCTION

In the rapidly evolving field of research methodology, mixed-methods research (MMR) has emerged as a powerful third methodology, combining qualitative and quantitative approaches to gain a more comprehensive understanding of complex phenomena (Zhou et al., 2023). Mixed method research recognises that relying on only qualitative or quantitative data can lead to a limited view of the research questions (Creswell & Creswell, 2018), therefore, integrating both methods gives a more comprehensive and nuanced understanding of the issues at stake (Fetters, 2020; Ivankova & Plano Clark, 2018). This integrated approach is particularly important in areas such as health, social sciences, education, and, more recently, computer science and informatics, where the convergence of different user experiences, technological environments, and perspectives of stakeholders requires a multidisciplinary approach to research topics (Akhai, 2024; Guetterman, 2020; Lukito & Pruden, 2023).

Sampling is an essential component of mixed-methods research, presenting both opportunities and challenges. This is the mechanism by which researchers select participants for their studies, and it is essential for ensuring that the results of research are valid, reliable, and applicable to the wider world. The sampling process is particularly complex in MMR, requiring the careful integration of both qualitative and quantitative sampling strategies to capture the diversity of experiences and achieve a consistent outcome (Ngulube, 2022; Onwuegbuzie & Collins, 2017). This duality of purpose, balancing the need for depth in qualitative data with the rigour of quantitative analysis, requires researchers to navigate a complex landscape of sampling decisions that have a direct impact on the outcome of their research.

Various sampling techniques have been developed over the years to meet the unique requirements of MMR. Traditional methods, such as purposive and random sampling, have been adapted to the specific situation of the MMR, and innovative approaches, including stratification and parallel sampling (Schoonenboom, 2024), have been developed. These advances provide researchers with a range of tools to strategically select participants and ensure the relevance of their studies to the diverse populations they seek to study. Despite these developments, important challenges remain. Issues such as sampling bias, representation, and integration of findings from different sampling approaches can undermine the reliability of research. Researchers must be aware of these challenges and prioritise transparency, flexibility, and methodological rigour in their sampling strategies (Etikan & Bala, 2017).

The importance of addressing the complexity of sampling is emphasised by the increasing recognition that the integrity of MMR lies not only in its ability to extract insights from quantitative and qualitative data, but also in its methodology, in how it selects, implements, and reports on the sampling (Tashakkori et al., 2021; Teddlie & Tashakkori, 2010). There is a wealth of literature where inadequate sampling strategies have led to contradictory interpretations and conclusions, particularly in rapidly evolving areas such as informatics and computing, where the adoption of technologies and user experience are constantly changing.

This chapter, therefore, examines the complexity of MMR sampling, with a particular focus on integrating different sampling strategies and their impact on data quality and reliability. This chapter outlines the various sampling methods available, analyses potential problems, and provides practical recommendations to enhance the robustness and relevance of MMR. Moreover, this chapter is of value to researchers from various fields who seek to deal with complicated sampling pathways in a holistic manner, ensuring that their findings contribute meaningfully to their field of expertise. By comprehensively addressing the complexity of MMR sampling, this chapter provides a valuable resource for both novice and experienced researchers. Ultimately, our focus on effective sampling procedures will enable researchers to make informed decisions, thereby enhancing MMR's credibility and integrity, enriching discussions in their respective areas, and fostering a common understanding of multiple human experiences and technological interactions.

THE IMPORTANCE OF SAMPLING IN MIXED METHODS RESEARCH

Sampling lies at the intersection of the qualitative and quantitative paradigms of MMR, playing a crucial role in demonstrating the consistency, validity, and universality of integrated data. A coherent sampling strategy for MMR will address (a) the selection of participants or cases for the qualitative and quantitative strands, (b) the integration of both strands at the level of the sampling design (e.g., concurrent versus sequential, identical versus sequential), and (c) the limits of sampling to support the intended conclusions of the inter-method interaction. Foundational discussions and literature metadata highlight the importance of careful sampling, robust data integration, and overall robustness of mixed-method studies (Palinkas et al., 2010; Ramanujan et al., 2022; Tashakkori et al., 2021). Researchers have also demonstrated that the explicit articulation of the sampling design and rationale enhances transparency, allowing for replication or comparison across MMR studies (Fetters,

2020; Salmona et al., 2019). Moreover, the literature shows that explicit sampling frameworks for MMR (Creswell & Creswell, 2018; Creswell & Plano Clark, 2018; Kettles et al., 2011) are useful in various practice contexts (information science, computer science, healthcare, applied sciences, education and nursing) to ensure that data types are consistent with each other.

Mixed-method research sampling does not involve a simple transfer of sampling methods; it requires the integration of probabilistic and purposive approaches within a single study. Early and influential discussions identified a continuum and the need to explain how sampling frameworks (random) and sampling limits (proportional) contribute to the mixed methods of the study (Guetterman, 2020; Teddlie & Yu, 2007). This conceptual framework is reinforced by meta-frameworks and synthesis papers that explain how the choice of MMR sampling design affects the mixing dimension (partial versus total integration), timing (concurrent versus sequential data collection), and the weighting of different methods (dominant versus non-dominant). In practice, researchers must explicitly specify (a) the population or frame to which each method is to be addressed, (b) how sampling units for the qualitative and quantitative strands are to be linked, and (c) how the proposed design supports the research questions and the intended conclusions. Empirical reviews highlight the impact of sampling on the consistency of conclusions drawn from different data sets and on the overall interpretation of MMR studies (Ramanujan et al., 2022).

Several canonical MMR sampling patterns have been identified and developed in the literature, including sequential and concurrent sampling patterns, as well as strategic sampling patterns that support triangulation, complementarity, or tool development. For example, the MMR design reviews emphasise the sequencing and weighting of methods (e.g., the use of a Morse-inspired taxonomy) to inform the collection and analysis of data across the strands (Palinkas et al., 2010). Other syntheses map the different sequence and integration functions of MMR designs and show how sampling decisions interact with the rationale for mixing and integrating results (Onwuegbuzie & Collins, 2007). The parallel convergent design is highlighted as a widely used design that explicitly seeks complementary data from both methods on the same subject to improve understanding (Creswell & Creswell, 2018; Tashakkori et al., 2021). Sampling in the context of MMR should be carefully aligned with the selected design to maximise integration and interpretive consistency (Dawadi et al., 2021; Teddlie & Tashakkori, 2010).

The focus on qualitative components remains a cornerstone of MMR, as these components aim to illuminate contexts, meanings, and processes that quantitative measures may not fully capture. Representative synthesis data show that purposeful sampling is often used to select cases with rich information, informant perspectives, or settings with the highest variability, thereby enriching the qualitative component of the data and reconciling it with the quantitative component of the integration

(Guetterman, 2020; Palinkas et al., 2010). Insufficient justification for mixed methods or poor integration of qualitative and quantitative sampling may undermine rigour, thus underlining the need for clear justification and transparent reporting (Creswell & Creswell, 2018; Guetterman, 2020; Ngulube & Ukwoma, 2022; Tashakkori et al., 2021). Mixed-methods research literature also shows that mixed-method sampling promotes greater generalizability when combined with careful design but requires a clear understanding of how sampling boundaries and frameworks relate to the research questions and the intended conclusions of different methods (Guetterman, 2020; Onwuegbuzie & Collins, 2017; Ramanujan et al., 2022; Teddlie & Yu, 2007).

In different fields and designs, sampling is a critical determinant of the robustness and interpretability of mixed-method studies. Thoughtful sampling procedures enable the coherent integration of qualitative and quantitative findings, enhance the reliability of conclusions, and, where appropriate, improve the transferability and generalizability of results. TExisting scholarship identifies the following key claims: (a) MMR samples must be carefully designed to fit MMR concepts and research questions; (b) explicit articulation and justification of the sampling decisions improves robustness and transparency; (c) the sampling frame and the relationship between qualitative and quantitative samples shape MMR integration and the validity of mixed conclusions (Onwuegbuzie & Collins, 2007); (d) the methodological guidelines provide practical guidance on the interaction between methods (Palinkas et al., 2010); and (e) by combining these strands, researchers can better plan, conduct, and report MMR studies that use sampling strategies to support robust, reliable, and actionable conclusions.

SAMPLING TECHNIQUES IN MIXED METHODS RESEARCH

The MMR sampling involves coordinating qualitative and quantitative strands to support consistent conclusions. The four techniques described each have a distinct role in MMR and are often combined during the study to ensure that they are as comprehensive as possible. The following summary includes key terms from the literature and references provided to illustrate how each technique works, when to use it, and how to combine it with other techniques (sequential, integral, and triangulation) to maximise integration and reliability.

Purposive Sampling

Purposive sampling enables the collection of in-depth qualitative insights and theory-building that supports quantitative goals by purposefully selecting participants or cases based on their information-rich attributes pertinent to the research questions (Atkins et al., 2012; Collins, 2017; Palinkas et al., 2013). This method is a common one in MMR, where the qualitative components seek to provide depth and contextual nuance, clarifying processes, meanings, and contexts that may be absent in the quantitative approach.

In the qualitative strand, a qualitative sample is used to collect a range of views, expertise, or typologies that illuminate mechanisms or provide a basis for developing tools in the quantitative field (Creswell & Creswell, 2018; Ivankova et al., 2006; Ngulube, 2015; O'Cathain et al., 2008; Ramanujan et al., 2022). In the exploratory phase, where sampling is less common for quantitative generalised estimates, point-of-care sampling may be used to ensure coverage of relevant conditions or subgroups before introducing more extensive probability sampling (Atkins et al., 2012; O'Cathain et al., 2008).

The purposeful sampling, in combination with stratified or random sampling, offers explanatory depth, which helps to optimise measurement constructs and to interpret quantitative patterns. To ensure consistency, it is essential to clearly explain the rationale behind the project selection and its relationship to the research questions and the intended integration (Collins et al., 2007; Powell et al., 2008; Tashakkori et al., 2021).

Random Sampling

Random sampling selects participants from the population with a similar probability, maximizing representativeness for quantitative estimation and supporting the generalizability of findings when the population is well-defined (Etikan & Bala, 2017; Guetterman, 2020; Teddlie & Yu, 2007). In MMR, the quantitative part is often reinforced by random sampling, while the qualitative part can use purposive sampling to explain or extend random findings (Collins, 2017; Palinkas et al., 2013; Teddlie & Yu, 2007).

Quantitative strands favour random or probabilistic sampling to test hypotheses, estimate population parameters, and allow for generalizable findings across subsets based on important variables (Guetterman, 2020). Sequential design enables the characterisation of large samples by first sampling randomly, followed by purposive qualitative sampling to explain observed variations or anomalies in subgroups identified during the quantitative phase (Corrigan & Onwuegbuzie, 2020; Ivankova & Plano Clark, 2018).

The integrated interpretation is strengthened by ensuring that the qualitative samples are linked to the quantitative framework. For example, nested or parallel sampling schemes help to clarify and validate the cross-methodologies (Collins et al., 2007; Onwuegbuzie & Collins, 2017).

Stratified Sampling

Stratified sampling divides the population into subgroups based on common characteristics and then draws samples from each subgroup to represent the key variables, allowing for segmentation (Mason et al., 2021). This approach enhances the accuracy and power of quantitative subgroup differences while maintaining representativeness for the population.

Stratified random sampling enables the proportionate or ideal representation of significant covariates in the quantitative strands, facilitating reliable quantitative analyses and subgroup comparisons (Etikan & Bala, 2017; Mason et al., 2021). In the qualitative strands, stratified purposive sampling can be used to identify cases with a lot of information within each stratum to guarantee contextual depth in various contexts (e.g., libraries of different sizes or educational institutions in both urban and rural areas) (Atkins et al., 2012; Collins et al., 2007).

By examining the similarities and differences between strata, cross-stratum integration allows for richer triangulation and a more nuanced meta-inference. The logical interpretation and transferability of results are supported by the explicit alignment of the strata between the qualitative and quantitative strands (Åkerblad et al., 2020; Creswell & Creswell, 2018; Hands, 2022).

Sequential Sampling

Sequential sampling involves several stages, and the selection of the samples made in one stage affects the subsequent stages. This method is often employed in sequential explanatory or exploratory designs, facilitating the iterative improvement of tools, investigations, or theoretical frameworks in response to new information (Etikan & Bala, 2017; Ivankova et al., 2006; Izgar & Aktürk, 2018; Sharp et al., 2011).

Typically, sequential design involves the collection and evaluation of quantitative data first (e.g., experiment or survey), followed by a qualitative sampling phase to clarify or further clarify the findings; alternatively, the pre-qualification phase may guide the development of a quantitative measurement or tool later (Creswell & Creswell, 2018; Fetters, 2020). To highlight specific elements (e.g., embedded patterns), a secondary strand is collected during or after the primary phase. while maintaining overall design coherence; in sequential triangulation, initial findings lead to targeted replication or extension in subsequent samples (Schoonenboom, 2024; Sharp et al., 2011).

Regarding integration, define how the findings of the first stage will inform the second-stage sampling (e.g., which subgroups will be sampled more extensively) and specify the criteria for maintaining consistency (Izgar & Aktürk, 2018). Then clearly indicate how sampling decisions developed over time and how these decisions influenced the interpretation of the cross-methods, including any deviations and the reasons for them (Collins, 2017).

MIXED METHODS RESEARCH EMPLOYS UNIQUE SAMPLING TECHNIQUES

Mixed methods studies employ both qualitative and quantitative methods, thereby requiring unique sampling plans to meet two competing goals, namely depth and generalizability. In mixed-methods sampling, the underlying probability emphasis of quantitative research is combined with the purposefulness characteristic of qualitative inquiry (Onwuegbuzie & Collins, 2017). Quantitative components typically employ probability-based approaches; various sampling procedures may utilise simple random, stratified, or cluster techniques to obtain representative samples, which in turn facilitate statistical generalisations (Kalpokas & Hecker, 2023). Qualitative, on the other hand, focuses on obtaining diverse and rich perspectives through a purposive approach, such as maximum variation, snowball, or theoretical sampling, to achieve depth of understanding about the phenomenon (Teddlie & Yu, 2007).

To integrate various research paradigms, researchers commonly use mixed methods sampling typologies, which include basic, sequential, concurrent (parallel), and multilevel designs. Basic mixed-methods sampling involves selecting separate samples for qualitative and quantitative components. In contrast, sequential sampling leverages findings from one phase to inform participant selection in the next phase; this approach is often seen in explanatory or exploratory sequential designs (Creswell & Plano Clark, 2018). Concurrent or parallel sampling collects qualitative and quantitative data simultaneously, either from the same group of participants or different samples. This method promotes triangulation and complementarity (Teddlie & Yu, 2007). Multilevel sampling expands on this idea by incorporating hierarchical structures. For example, researchers may quantify data at the school level and then conduct qualitative interviews with teachers within those selected schools (Onwuegbuzie & Collins, 2017).

These sampling strategies are not merely technical choices; they have a significant impact on the quality of inferences and the consistency of interpretations. Validating findings in mixed-methods research is crucial (Onwuegbuzie & Collins, 2017).

Ultimately, successful mixed methods sampling requires clear design, justification of sampling choices, and transparency in reporting, as highlighted by the A*PA's Journal Article Reporting Standards for Mixed Methods* (APA, 2024). By thoughtfully integrating these sampling strategies, researchers can harness the strengths of both qualitative and quantitative paradigms, yielding findings that are both statistically robust and contextually rich—characteristics that define high-quality mixed-methods research. Validating findings in mixed-methods research is crucial (Onwuegbuzie & Collins, 2017). The Appendix illustrates all the unique characteristics of mixed methods research sampling.

INTEGRATING SAMPLING ACROSS METHODS IN MIXED METHODS RESEARCH

Consequently, integrating sampling across the qualitative and quantitative strands is essential for achieving consistent conclusions in MMR. Effective integration requires explicit design decisions on how the sampling of each method relates to the overall design of the MMR, how the selection of the sampling in one part informs the selection of the other part, and how the chosen integration mode (sequential, embedded, or triangulation) shapes the interpretation of the results (Palinkas et al., 2010; Teddlie & Yu, 2007). The following section summarises the guidance provided and demonstrates how sequential design integration, embedded design integration, and triangulation integration can be implemented through coordinated sampling decisions, drawing on the broader literature on MMR sampling and representative empirical work.

Sequential Design Integration

Sequential integration involves the collection and analysis of data in stages, where the findings of the first stage guide the selection and collection of data in the subsequent stages. This approach is often used to refine questions, develop tools, or elaborate mechanisms before extending the study to the next methodological phase (Fàbregues et al., 2024; Ivankova & Plano Clark, 2018; O'Cathain et al., 2007a). In sequential design, the decision on the selection of the sample is iterative; the qualitative phase can identify relevant constructs or inform the sampling framework for the quantitative phase, or the quantitative results can be used to guide the subsequent qualitative study (Guetterman et al., 2015; López-Fernández & Molina-Azorín, 2011; Nanthagopan, 2021). In informatics, this approach supports the development of tools, the collection of user-driven requirements, and the iterative assessment of interfaces, data models, and algorithms (Petter & Gallivan, 2004; Turner et al., 2016).

The following should be taken into consideration when sequentially sampling:

1. Phase 1 informs Phase 2 sampling frames, such as qualitative purposive sampling in the first phase, which may specify the target characteristics, ranges, or contexts to be represented in the subsequent quantitative sample (Johnson & Onwuegbuzie, 2004; O'Cathain et al., 2007a; Tashakkori et al., 2021).

2. Establish clear boundaries to ensure that populations sampled in phases 1 and 2 are logically linked (e.g., qualitative cases nested in a quantitative cohort and vice versa) to support consistent conclusions (Corrigan & Onwuegbuzie, 2020; Johnson & Onwuegbuzie, 2004).
3. Start with a small enough sample to test depth or instrument development, then expand to a larger or different composition of quantitative samples to test or generalise findings; explicitly justify the migration criteria (e.g., saturation thresholds, pilot estimates) (Younas et al., 2019).
4. Documenting the integration decision by reporting on how the results of the phase 1 sampling and the expected integration path (triangulation, development, or complementarity) shaped the phase 2 sampling to increase transparency and replicability (O'Cathain et al., 2007a; Younas et al., 2019).

Sequential explanatory designs often employ an initial quantitative phase followed by a qualitative phase, where the qualitative phase utilises qualitative sampling to explain or refine the quantitative results (Ivankova et al., 2006; James et al., 2022; López-Fernández & Molina-Azorín, 2011). If the initial results indicate different subgroups or contextual differences, a purposive sampling in the subsequent qualitative phase may ensure that these differences are investigated in greater depth and strengthen the coherence between methods (Fàbregues et al., 2024; Guetterman, 2020). For example, in cloud computing adoption studies at a medical university, researchers initially employed stratified random sampling to collect quantitative data across departments (Hosseini-Ashlaghi et al., 2023). The patterns that emerged then guided the purposive selection of interview participants, allowing the qualitative phase to investigate organisational dynamics and contextual factors that the numbers had only hinted at.

Embedded Design Integration

Embedded designs involve one data type within a larger design that is dominated by another; the embedded data type plays a secondary, supporting role (e.g., qualitative data embedded in a predominantly quantitative study, and vice versa) (Schoonenboom, 2024). Sampling in embedded designs typically follows the logic of the dominant design; however, the auxiliary strand utilises the sampling method that best serves its specific analysis.

The following should be taken into consideration when using embedded sampling:

1. Align the purpose of the additional sample with the needs of analysis; for example, in a predominantly quantitative study, include qualitative cases which reveal key patterns, abnormalities, or mechanisms; sampling for the embedded part

may be more useful than for representativeness (Collins et al., 2007; Fàbregues et al., 2024).

2. The embedded strand is collected in parallel or in the same time frame as the dominant strand, but the sampling framework should be selected in such a way as to optimise the effectiveness and information output in relation to the primary objectives of the study (Barbour & Barbour, 2003; Kadushin et al., 2008; Nanthagopan, 2021).
3. Clearly distinguish which findings are derived from the dominant versus embedded strand and justify how the embedded data contribute to the interpretation, relevance, or development of a theory; transparency in the selection criteria for embedded cases increases the robustness of the data (Barbour & Barbour, 2003; Schoonenboom, 2024).
4. Sample size considerations: as the embedded strand is a secondary strand, its sample size may be smaller or targeted at specific issues that the primary strand addresses, with the rationale being the analytical contribution of the embedded data (Fàbregues et al., 2024; Schoonenboom, 2024).

Embedded designs are used in various fields to confirm or contextualise quantitative results with qualitative details (Barbour & Barbour, 2003; Schoonenboom, 2024). Where the embedded strand focuses on mechanisms or explanations for observed patterns, a purposive sampling of cases rich in information is common and justified by the need to improve understanding rather than to achieve statistical generalisation (Schoonenboom, 2024).

Triangulation (Convergent Design)

Triangulation in MMR refers to the use of independent data sources or sampling units to validate findings, assess convergence and enhance the reliability of inferences. Triangulation often involves collecting complementary data from different samples or environments to investigate whether patterns are convergent across methods and contexts (Ivankova et al., 2006; Kadushin et al., 2008; Kelle et al., 2019).

The following should be taken into consideration when applying triangulation:

1. Triangulation benefits from sampling in different contexts, populations, or settings to test whether observed associations are maintained across changes in variables; this supports robustness of the conclusions and improves transferability where appropriate (Fàbregues et al., 2024; Ivankova et al., 2006; Kadushin et al., 2008).
2. Design an analysis comparing the results of different samples or data sources to assess convergence or divergence and define criteria for what constitutes a

reasonable convergence to support a conclusion (Ivankova et al., 2006; Kadushin et al., 2008; Kelle et al., 2019).

3. Explicitly document the selection of samples, the degree of independence of the samples and how the approximation will be assessed, thereby reducing ambiguities in the claims of validity, and enhancing the reliability of the interpretation (Ivankova et al., 2006; Kelle et al., 2019; O'Cathain et al., 2007b).

Kadushin et al. (2008) illustrate triangulation as part of the assessment of mixed methods and highlight how correlations between methods can be interpreted within a triangulated framework; this work highlights the value and pitfalls of considering triangulation as an epistemic strategy rather than a mechanical requirement. Ivankova et al. (2006) state that to explain the sequence of triangulation considerations that arise in interpreting how qualitative and quantitative lines converge or diverge, and to reinforce the need for careful planning of sampling to allow meaningful triangulation conclusions.

Therefore, one should start with the overall design of the MMR and explicit integration objectives (triangulation, development, or complementarity) to guide the selection of the sample (O'Cathain et al., 2007a; Palinkas et al., 2010; Teddlie & Yu, 2007). For sequential designs, it is necessary to define how the results of Phase 1 sampling will inform the Phase 2 sampling, including the candidate subgroups, contexts, or tools to be targeted for subsequent data collection (Guetterman et al., 2015). For embedded designs, specify the dominant design and justify the sampling strategy for the embedded design according to its analytical function (Barbour & Barbour, 2003; Schoonenboom, 2024) while maintaining consistency with the main sampling framework. For triangulation, propose a sampling pattern across multiple sources or populations that maximises the convergence potential and allows for reliable cross-method validation; articulate criteria for assessing convergence and how discrepancies are to be interpreted (Ivankova et al., 2006; Kadushin et al., 2008). Ensure transparent reporting of sampling boundaries, frameworks, and cross-sampling between methods to support replication, robustness assessment, and comparability between studies (Guetterman, 2020; O'Cathain et al., 2007b; Younas et al., 2019). The type of sampling integration discussed is shown in Table 1 and Appendix I.

Table 1. The level of sampling integration in MMR

Integration Pattern	Purpose	Sampling Relationship Across Strands	Sampling Decision Focus`	Illustrative Examples / Applications
Sequential Design Integration	Build upon or elaborate on results from one strand to the next.	The Phase 1 sample informs Phase 2 (e.g., qualitative → quantitative or vice versa).	Define linkage between phases, justify migration criteria, and ensure logical population overlap.	Qualitative findings refine the quantitative instrument or target subgroups; quantitative patterns guide purposive qualitative follow-ups.
Embedded Design Integration	Supplement a dominant method with a supporting strand.	A secondary strand is sampled within or parallel to the main sample, but for different analytical goals.	Align the embedded sample purpose with the dominant study aims; clarify the analytical contribution.	Qualitative interviews embedded in randomised control trials (RCTs) to explain mechanisms; survey data embedded in ethnography to quantify observed patterns.
Triangulation (Convergent) Integration	Validate findings across independent samples or contexts.	Multiple independent samples are used to cross-check consistency and convergence.	Ensure diversity of contexts or populations; predefine criteria for convergence/divergence.	Parallel qualitative and quantitative samples testing the same hypotheses in different contexts.

Coordinated sampling supports stronger meta-indications across sequential, embedded, and triangulation integrations by harmonising who is being studied, how they are selected, and how the findings are synthesised across strands. Careful alignment of the sampling design to the chosen integration pattern is crucial to the validity, interpretive consistency, and reliability of mixed-method claims (Kadushin et al., 2008; Teddlie & Yu, 2007; Younas et al., 2019). When sampling is explicitly designed to facilitate cross-method integration, researchers can triangulate results, develop tools and generalise across contexts while preserving the distinct strengths of each methodological area (Johnson & Onwuegbuzie, 2004; O'Cathain et al., 2007a; Tashakkori et al., 2021).

COMMON PITFALLS TO AVOID IN MIXED METHODS SAMPLING

Researchers using mixed methods of sampling encounter several recurring problems that may compromise the consistency, validity, and integrative power of their studies. The use of post-hoc or ad-hoc sampling decisions, where participants are selected after the start of data collection without a theoretical basis or design justification, is a frequent problem. These practices disrupt the internal logic of mixed methodologies and undermine cross-cutting conclusions, highlighting the need for predetermined sampling grounds, limits, and progression criteria, as well as transparent documentation (Bryman, 2007; Johnson & Onwuegbuzie, 2004; O'Cathain et al., 2009). The lack of consistency between sampling frameworks across the qualitative and quantitative areas is closely related. If populations, units, or samples are poorly matched, meaningful integration becomes difficult or even impossible. Scholars recommend creating explicit links between samples, either through nested, parallel, or cross-classification schemes, and utilising visual mapping tools such as sampling matrices and common displays to ensure traceability (Bryman, 2007; Hirose & Creswell, 2023; O'Cathain et al., 2009).

Another major weakness is the imbalance between methodological strands, where one method has a privileged degree of scope or rigour, creating asymmetric evidence and hindering triangulation. To mitigate this, clarity on the purpose of the mixing whether triangulation, evolution, or extension and justification for the relative weight and sample size of the strands (Johnson & Onwuegbuzie, 2004; Sanscartier, 2018; Taherdoost, 2022) should be provided. Weak reporting on sampling procedures further undermines the credibility of the study, as omissions concerning inclusion criteria, recruitment strategies, and crossline relationships obscure the conclusions (Bazeley, 2009; O'Cathain et al., 2009). Sampling bias and non-response bias are additional threats, especially when the differences between participants and non-participants affect both strands of the analysis. The combination of probabilistic strategies in quantitative work and purposive approaches in qualitative investigations, including documentation of non-response and sensitivity analysis, offers partial solutions (Hands, 2022; Hirose & Creswell, 2023).

Mixed-method studies also risk underestimating the contextual diversity, which would limit the depth and portability of the interpretation. Without stratified or multilayer sampling to capture the variation between roles, contexts and cultural groups, the two lines of inquiry can converge to overly narrow conclusions (O'Cathain et al., 2009; Tunarosa & Glynn, 2016). In addition, the excessive reliance on single data sources introduces bias in the methodologies and limits triangulation. The integration of multiple sources and clarification of how sampling supports convergence between data sources enhances interpretative reliability (Hirose & Creswell, 2023; Sanscartier, 2018). Poor planning of integration artefacts such as joint displays, matrices, and other tools used to aggregate qualitative and quantitative findings often leads to a shallow or fragmented integration. Researchers are encouraged to design these integration procedures at the sampling phase and not to repeat them after analysis (Hirose & Creswell, 2023; Sanscartier, 2018).

Other complications arise from ethical and practical constraints. If logistical barriers, restrictions imposed by gatekeepers, or ethical constraints constrain the possible sample, bias may occur unless these restrictions are explicitly agreed to and documented (Bryman, 2007; O'Cathain et al., 2009). Another persistent problem is the overgeneralization of non-representative samples, especially when researchers confuse statistical universality with qualitative transferability. Accurate interpretation requires that the claims be consistent with the limitations inherent in the sampling logic of each strand (Bryman, 2007; O'Cathain et al., 2009). Finally, many studies lack predefined convergence criteria, leaving researchers without clear guidelines to determine whether the findings are consistent, complementary, or contradictory. The introduction of these criteria, and their linkage with the overall rationale for the mixed method approach, enhances the transparency and consistency of the interpretation (Johnson & Onwuegbuzie, 2004; Nanthagopan, 2021; Taherdoost, 2022). These common challenges are summarised in Figure 1.

Figure 1. Common pitfalls in MMR sampling (Designed by Authors)

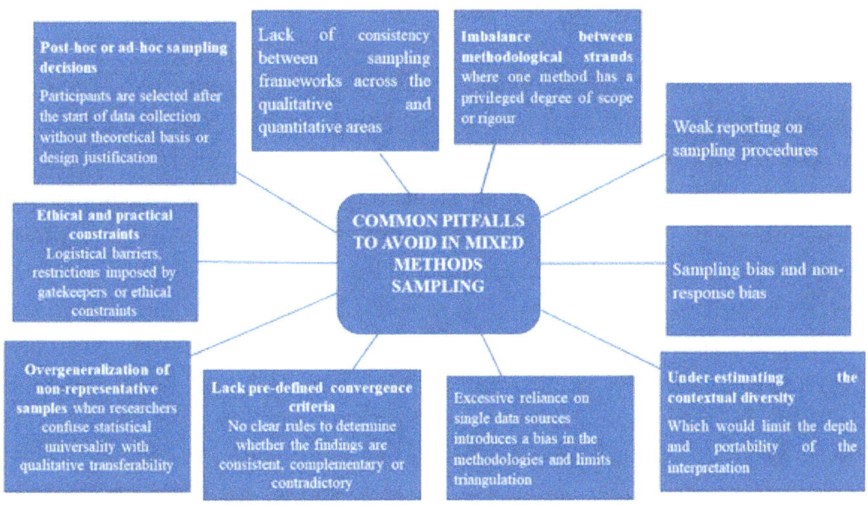

PRACTICAL RECOMMENDATIONS FOR RESEARCHERS

The practical implications for researchers in all fields underscore the significant implications of MMR sampling. Palinkas et al. (2010) provide a framework for evaluating MMR design, including the mixing, sequencing and data integration functions, all of which are closely linked to the way samples are selected and inter-linked across methods. Corrigan and Onwuegbuzie (2020) provide a meta-framework for MMR analysis, which explicitly prescribes sampling design, sample size and inter-sampling for optimal meta-induction. The literature reviewed highlights the central challenge of integrating qualitative and quantitative findings in cases where sampling decisions are misaligned or underreported; therefore, the robust reporting of MMR data is crucial for credibility (Fàbregues & Paré, 2018; Fàbregues et al., 2024; Younas et al., 2019). The foundational work of Johnson and Onwuegbuzie situates MMR within a pragmatic paradigm, offering multiple design options for integrating methods, which underscores the importance of carefully matching the sample to philosophical attitudes and research questions (Johnson & Onwuegbuzie, 2004).

For effective sampling management in the MMR, researchers must clearly define their sampling objectives, have flexibility in the sampling design, thoroughly document the process, and, in some cases, engage in cross-disciplinary collaborations, as outlined below.

Define clear sampling objectives

- Articulate precise sampling objectives that align with your mixed methods design (sequential, embedded, or triangulation) and the specific inferences you aim to make across strands (Schoonenboom, 2024; Sharp et al., 2011). This clarity helps ensure that both qualitative and quantitative samples are purposefully chosen to support the intended meta-inferences and integration goals (Schoonenboom, 2024; Sharp et al., 2011).
- Specify population boundaries, units of analysis, and the planned linkage between qualitative cases and quantitative participants (e.g., nested or parallel samples) to guide transparent sampling decisions from the outset (Sharp et al., 2011; Taherdoost, 2022; Teddlie & Yu, 2007).
- Predefine how sampling frames will support integration artefacts (joint displays, data matrices, meta-inference schemas) and what constitutes adequate coverage of key dimensions (e.g., contexts, roles, configurations) for robust cross-method synthesis (Schoonenboom, 2024; Teddlie & Yu, 2007).

Flexibility in sampling design

- Build in deliberate flexibility to adapt sampling in response to emerging findings while maintaining a documented rationale for any deviations. Recognise that "mess" and emergence are inherent in mixed methods research and provide criteria for when and how to adjust sampling boundaries or unit definitions considering new insights (Johnson & Onwuegbuzie, 2004; Sanscartier, 2018; Taherdoost, 2022).
- Use design-pattern awareness to accommodate multiple integration patterns. For instance, in sequential designs, be prepared to transition to a broader or

differently weighted sampling plan post-phase if phase 1 results reveal unanticipated subgroups or mechanisms (Teddlie & Yu, 2007).

- Plan for hybrid sampling, if necessary, combining probabilistic methods for the quantitative part with purposive or maximum variation approaches for the qualitative part, to balance generalisation and depth and justify the hybrid method in the protocol (Bryman, 2007; O'Cathain et al., 2009; Schoonenboom, 2024).
- Consider domain-agnostic or cross-domain applicability by proposing sampling strategies that can be transposed across environments by focusing on general principles (representation, boundary management, integration preparedness) rather than on specific domain-specificities (Bryman, 2007; O'Cathain et al., 2009).

Thorough documentation

- Document the sampling decisions in a comprehensive manner, including inclusion and exclusion criteria, selection methods, sampling frameworks, boundaries and explicit mapping between qualitative cases and quantitative units. Transparent reporting improves replicability and critical assessment of cross-method inferences (Bazeley, 2009; Etikan & Bala, 2017; O'Cathain et al., 2009; Schoonenboom, 2024).
- Document the progression criteria for sampling in the sequential designs, specifying what constitutes an adequate phase progression (e.g., saturation, minimum viable sample size, pilot estimates), and how the results will trigger changes in the next sampling (Johnson & Onwuegbuzie, 2004; O'Cathain et al., 2007a; O'Cathain et al., 2009).
- Capture the integration plan in detail, define the integration artefacts (joint displays, matrices, data-fusion protocols) and show how the selection of samples will interact with these artefacts to support convergence or explain differences (Hirose & Creswell, 2023; Jackson-Gordon & Clark, 2023; James et al., 2022).
- Use standardised reporting templates, adopt structured sections for sampling design, boundaries and cross-method links to facilitate comparison between studies and meta-analysis (James et al., 2022; O'Cathain et al., 2009).

In summary, before data collection, create a sample design sketch or an appendix that specifies the overall research design of the mixed methods (sequential, integral, or triangulation) and the intended integration method. Sampling frames for each strand, their boundary, and explicit linkages between samples (nested, parallel, identical or cross-classified). Phase-by-phase sampling criteria for the sequential designs, including predefined triggers for the step-by-step progress. The role and scope of each embedded or auxiliary strand, with explicit justification linked to the objectives of the analysis. Planned integration artefacts (images, matrices) and how the selection of the samples will allow for their construction and interpretation. Pilot the sampling plan with a small mock-up or pilot study to test whether the proposed samples provide the expected integration opportunities and whether any biases or gaps are present. As the study progresses, document any changes to the sample, their rationale, and their impact on integration. Regularly audit the consistency of the sampling decisions with the integration results.

FURTHER RESEARCH DIRECTIONS

Future developments in mixed-methods sampling are likely to focus on enhancing methodological coherence, transparency, and responsiveness to the increasingly complex research environment. One area of progress is adaptive sampling designs, which enable sampling procedures to adjust to new findings while still adhering to predefined rules of thumb. These approaches can avoid ad hoc adjustments and are particularly useful in fast-moving areas, such as information science and the digital research environment (Henrys et al., 2024; Poth et al., 2023, 2024).

The second focuses on improving the integration of multi-modal and cross-platform sampling. As studies increasingly combine qualitative data with digital trace data, computational records and other diverse sources, more robust frameworks are needed to harmonise the units of analysis across the different modalities. Future methodological work should develop methods to ensure compatibility between sample levels at the participant level, system-level data, and context information, thus enhancing the validity of the interpretation (Sardana et al., 2021; Shin, 2016).

There is also an unmet need for equity-aware sampling strategies. Traditional sampling often reproduces inequalities in access and visibility, which limits the representativeness of mixed-method studies. Future research should prioritise sampling frameworks that capture structural, cultural, and technological forms of diversity, which extend beyond demographic differences to encompass differences in access to digital systems and information environments.

Additionally, the field would benefit from a more formal integration of diagnostic tools to assess the consistency of qualitative and quantitative samples. Existing practices, such as joint displays, often lack clear evaluation criteria. New analytical frameworks capable of identifying integration biases and of testing the integrity of the sample could greatly enhance the reliability of the conclusions.

Finally, the increasing momentum for open and pre-registered sampling protocols is likely to shape future practice. Transparency of sampling rationale, contingencies and justified deviations may improve both the reproducibility and the interpretability of the data, even where qualitative flexibility is required. Together, these trends indicate the need for more consistent, fair, and analytically reliable mixed-method sampling procedures.

CONCLUSION

This chapter reviewed the sampling in MMR in detail, focusing on how the choice of the sample influences the design, integration and conclusions in both qualitative and quantitative terms. Across the different sections, several key themes emerge, which together provide insights into good practice, recurring challenges and practical ways forward for researchers from different fields. Sampling is not a minor methodological detail, but a fundamental design decision that shapes the consistency and reliability of the conclusions drawn from cross-methods. The well-articulated sampling plan should be consistent with the overall MMR design (sequential, embedded or triangulated) and the intended mode of integration (triangulation, embedded, complementarity or sequential). A deliberate articulation of population frameworks, sampling boundaries, and relationships between qualitative and quantitative cases enhances transparency, facilitates replication, and improves comparability between studies.

For qualitative strands, the qualitative sampling remains essential to provide depth, context and meaning; its purposeful match to quantitative issues reinforces integration. Random and stratified sampling are valuable for generating generalizable and representative quantitative findings, as well as for ensuring that cross-method comparisons across key subsets are meaningful. Sequential and integrated sampling designs require explicit planning to ensure that sampling in one phase or strand informs the other in a meaningful way, and to clearly specify the criteria for sequencing, iteration, and integration. Triangulation as an integration objective benefits from sampling across different contexts, populations and data modalities, and the rigour is enhanced by predefined convergence criteria.

Sequential design integration benefits from a consistent stepwise sampling approach, with qualitative findings guiding the quantitative sampling boundaries and the development of the instruments. Embedded design integration requires alignment of the helper strand to its analytical role, with a selection that optimizes the explanatory yield while maintaining coherence with the dominant strand. Triangulation requires sampling across multiple contexts and data flows, with explicit convergence criteria and transparent rules for interpretation.

The principles discussed are deliberately framed to have a general scope, applicable to education, health, the social sciences, engineering, the humanities, informatics, and other fields. Emphasis is placed on representation, clarity of boundaries, readiness for integration and transparent reporting rather than on discipline-specific conventions. However, specific sampling matrices and integration artefacts should be informed by domain-specific considerations (e.g. user roles, settings, configurations, or artefact lifecycle).

REFERENCES

Åkerblad, L., Seppänen-Järvelä, R., & Haapakoski, K. (2020). Integrative strategies in mixed methods research. *Journal of Mixed Methods Research*, *15*(2), 152–170. DOI: 10.1177/1558689820957125

Akhai, S. (2024). A review on optimizations in μ-edm machining of the biomedical material Ti6Al4V using the Taguchi method: Recent advances since 2020. *Latest Trends in Engineering and Technology*, 395-402.

American Psychological Association. (2024). Journal article reporting standards for mixed methods research (JARS). Retrieved on 21 November from https://apastyle.apa.org/jars/mixed-methods

Atkins, S., Launiala, A., Kagaha, A., & Smith, H. (2012). Including mixed methods research in systematic reviews: Examples from qualitative syntheses in TB and malaria control. *BMC Medical Research Methodology*, *12*(1). Advance online publication. DOI: 10.1186/1471-2288-12-62 PMID: 22545681

Barbour, R. S., & Barbour, M. K. (2003). Evaluating and synthesizing qualitative research: The need to develop a distinctive approach. *Journal of Evaluation in Clinical Practice*, *9*(2), 179–186. DOI: 10.1046/j.1365-2753.2003.00371.x PMID: 12787181

Bazeley, P. (2009). Editorial: Integrating data analyses in mixed methods research. *Journal of Mixed Methods Research*, *3*(3), 203–207. DOI: 10.1177/1558689809334443

Bordens, K. S., & Abbott, B. B. (2018). *Research design and methods: A process approach* (10th ed. ed.). McGraw-Hill.

Bryman, A. (2007). Barriers to integrating quantitative and qualitative research. *Journal of Mixed Methods Research*, *1*(1), 8–22. DOI: 10.1177/2345678906290531

Collins, K. M. T. (2017). Sampling decisions in educational research. In Wyse, D., Smith, E., Suter, L., & Selwyn, N. (Eds.), *British Educational Research Association (BERA/SAGE handbook of educational research)* (Vol. 1, pp. 280–292). Sage.

Collins, K. M. T., Onwuegbuzie, A. J., & Jiao, Q. G. (2007). A mixed methods investigation of mixed methods sampling designs in social and health science research. *Journal of Mixed Methods Research*, *1*(3), 267–294. DOI: 10.1177/1558689807299526

Corrigan, J. A., & Onwuegbuzie, A. J. (2020). Toward a meta-framework for conducting mixed methods representation analyses to optimize meta-inferences. *The Qualitative Report*. Advance online publication. DOI: 10.46743/2160-3715/2020.3579

Creswell, J. W., & Creswell, J. D. (2018). Research design: Qualitative, quantitative, and mixed methods approaches. *Sage (Atlanta, Ga.)*.

Creswell, J. W., & Plano Clark, V. L. (2018). Designing and conducting mixed methods research. *Sage (Atlanta, Ga.)*.

Dawadi, S., Shrestha, S., & Giri, R. A. (2021). Mixed-methods research: A discussion on its types, challenges, and criticisms. *Journal of Practical Studies in Education, 2*(2), 25–36. https://doi.org/DOI:10.46809/jpse.v2i2.20

Etikan, I., & Bala, K. (2017). Sampling and sampling methods. *Biometrics & Biostatistics International Journal, 5*(6), 00149.

Fàbregues, S., & Paré, M. H. (2018). Appraising the quality of mixed methods research in nursing: A qualitative case study of nurse researchers' views. *Nursing Inquiry, 25*(4). Advance online publication. DOI: 10.1111/nin.12247 PMID: 29927008

Fàbregues, S., Younas, A., Escalante-Barrios, E. L., Molina-Azorin, J. F., & Vázquez-Miraz, P. (2024). Toward a framework for appraising the quality of integration in mixed methods research. *Journal of Mixed Methods Research, 18*(3), 270–280. DOI: 10.1177/15586898241257555

Fetters, M. D. (2020). The mixed methods research workbook: Activities for designing, implementing, and publishing projects. *Sage (Atlanta, Ga.)*.

Guetterman, T. C. (2020). Qualitative, quantitative, and mixed methods research sampling strategies. https://doi.org/DOI: 10.1093/obo/9780199756810-0241

Guetterman, T. C., Fetters, M. D., & Creswell, J. W. (2015). Integrating quantitative and qualitative results in health science mixed methods research through joint displays. *Annals of Family Medicine, 13*(6), 554–561. DOI: 10.1370/afm.1865 PMID: 26553895

Hands, A. S. (2022). Integrating quantitative and qualitative data in mixed methods research: An illustration. *Canadian Journal of Information and Library Science, 45*(1), 1–20. DOI: 10.5206/cjilsrcsib.v45i1.10645

Henrys, P. A., Mondain-Monval, T. O., & Jarvis, S. G. (2024). Adaptive sampling in ecology: Key challenges and future opportunities. *Methods in Ecology and Evolution, 15*(9), 1483–1496. https://doi.org/https://doi.org/10.1111/2041-210X.14393

Hirose, M., & Creswell, J. W. (2023). Applying core quality criteria of mixed methods research to an empirical study. *Journal of Mixed Methods Research, 17*(1), 12–28. DOI: 10.1177/15586898221086346

Hosseini-Ashlaghi, M., Gholamhosseinzadeh, M., Zameni, F., & Siamian, H. (2023). Identifying and ranking critical success factors for cloud computing deployment at Mazandaran University of Medical Sciences. *Educational Research in Medical Sciences, 12*(2).

Ivankova, N., & Plano Clark, V. (2018). Teaching mixed methods research: Using a socio-ecological framework as a pedagogical approach for addressing the complexity of the field. *International Journal of Social Research Methodology, 21*(4), 409–424. https://doi.org/DOI:10.1080/13645579.2018.1427604

Ivankova, N. V., Creswell, J. W., & Stick, S. L. (2006). Using mixed-methods sequential explanatory design: From Theory to practice. *Field Methods, 18*(1), 3–20. DOI: 10.1177/1525822x05282260

Izgar, G., & Aktürk, A. O. (2018). A mixed method research on peer assessment. [IJERE]. *International Journal of Evaluation and Research in Education, 7*(2), 118. DOI: 10.11591/ijere.v7i2.12770

Jackson-Gordon, R., & Clark, V. L. P. (2023). Using a joint display for building integration in a sequential study: Informing data collection for a participatory second phase. *Journal of Mixed Methods Research, 18*(2), 137–146. DOI: 10.1177/15586898231179848

James, T. G., DeJonckheere, M., & Guetterman, T. C. (2022). Integrating transformative considerations and quantitative results through a participant selection joint display in explanatory sequential mixed methods studies. *Journal of Mixed Methods Research, 18*(1), 14–30. DOI: 10.1177/15586898221149470 PMID: 38344136

Johnson, R. B., & Onwuegbuzie, A. J. (2004). Mixed methods research: A research paradigm whose time has come. *Educational Researcher, 33*(7), 14–26. DOI: 10.3102/0013189x033007014

Kadushin, C., Hecht, S., Sasson, T., & Saxe, L. (2008). Triangulation and mixed methods designs: Practicing what we preach in the evaluation of an Israel experience educational program. *Field Methods, 20*(1), 46–65. DOI: 10.1177/1525822x07307426

Kalpokas, N., & Hecker, J. (2023). The guide to mixed methods research. ATLAS. ti Research Hub. Retrieved on 20 October from https://atlasti.com/guides/the-guide-to-mixed-methods-research

Kelle, U., Kühberger, C., & Bernhard, R. (2019). How to use mixed-methods and triangulation designs: An introduction to history education research. *History Education Research Journal, 16*(1). Advance online publication. DOI: 10.18546/herj.16.1.02

Kettles, A., Creswell, J. W., & Zhang, W. (2011). Mixed methods research in mental health nursing. *Journal of Psychiatric and Mental Health Nursing*, *18*(6), 535–542. DOI: 10.1111/j.1365-2850.2011.01701.x PMID: 21749560

López-Fernández, O., & Molina-Azorín, J. F. (2011). The use of mixed methods research in interdisciplinary educational journals. *International Journal of Multiple Research Approaches*, *5*(2), 269–283. DOI: 10.5172/mra.2011.5.2.269

Lukito, J., & Pruden, M. L. (2023). Critical computation: Mixed-methods approaches to big language data analysis. *The Review of Communication*, *23*(1), 62–78. DOI: 10.1080/15358593.2022.2125821

Mason, P., Augustyn, M. M., & Seakhoa-King, A. (2021). Mixed methods research in tourism: A systematic sequential approach. *Folia Turistica, 56*, 0-0. https://doi .org/DOI: 10.5604/01.3001.0014.8956

McNabb, D. E. (2021). *Research methods for political science: Quantitative and qualitative methods* (3rd ed.). Routledge.

Nair, S. S., & Prem, S. S. (2020). A framework for mixed-method research. *Shanlax International Journal of Management*, *8*(2), 45–53. DOI: 10.34293/management. v8i2.3220

Nanthagopan, Y. (2021). Review and comparison of multi-method and mixed method application in research studies. *Journal of Advanced Research*, *2*(3), 55–78.

Ngulube, P. (2015). Qualitative data analysis and interpretation: systematic search for meaning. In Mathipa, E. R., & Gumbo, M. T. (Eds.), *Addressing research challenges: making headway for developing researchers* (pp. 131–156). Mosala-MASEDI Publishers & Booksellers.

Ngulube, P. (2022). Using simple and complex mixed methods research designs to understand research in information science. In *Handbook of research on mixed methods research in information science* (pp. 20–46). IGI Global Scientific Publishing., DOI: 10.4018/978-1-7998-8844-4.ch002

Ngulube, P., & Ukwoma, S. C. (2022). Mixed methods research in open distance learning: A content analysis of the literature. *Open Learning*, •••, 1–17. DOI: 10.1080/02680513.2022.2149321

O'Cathain, A., Murphy, E., & Nicholl, J. (2007a). Integration and publications as indicators of "yield" from mixed methods studies. *Journal of Mixed Methods Research*, *1*(2), 147–163. DOI: 10.1177/1558689806299094

O'Cathain, A., Murphy, E., & Nicholl, J. (2007b). Why, and how, mixed methods research is undertaken in health services research in England: A mixed methods study. *BMC Health Services Research*, *7*(1). Advance online publication. DOI: 10.1186/1472-6963-7-85 PMID: 17570838

O'Cathain, A., Murphy, E., & Nicholl, J. (2008). The quality of mixed methods studies in health services research. *Journal of Health Services Research & Policy*, *13*(2), 92–98. DOI: 10.1258/jhsrp.2007.007074 PMID: 18416914

O'Cathain, A., Nicholl, J., & Murphy, E. (2009). Structural issues affecting mixed methods studies in health research: A qualitative study. *BMC Medical Research Methodology*, *9*(1). Advance online publication. DOI: 10.1186/1471-2288-9-82 PMID: 20003210

Onwuegbuzie, A. J., & Collins, K. M. (2007). A typology of mixed methods sampling designs in social science research. *The Qualitative Report*, *12*(2), 281–316. DOI: 10.46743/2160-3715/2007.1638

Onwuegbuzie, A. J., & Collins, K. M. (2017). The role of sampling in mixed methods research. *KZfSS Kölner Zeitschrift Für Soziologie Und Sozialpsychologie*, *69*(2), 133–156.

Palinkas, L. A., Aarons, G. A., Horwitz, S. M., Chamberlain, P., Hurlburt, M. S., & Landsverk, J. (2010). Mixed method designs in implementation research. *Administration and Policy in Mental Health*, *38*(1), 44–53. DOI: 10.1007/s10488-010-0314-z PMID: 20967495

Palinkas, L. A., Horwitz, S. M., Green, C. A., Wisdom, J. P., Duan, N., & Hoagwood, K. (2013). Purposeful sampling for qualitative data collection and analysis in mixed method implementation research. *Administration and Policy in Mental Health*, *42*(5), 533–544. DOI: 10.1007/s10488-013-0528-y PMID: 24193818

Petter, S., & Gallivan, M. (2004). Toward a framework for classifying and guiding mixed method research in information systems. *37th Annual Hawaii International Conference on System Sciences, 2004. Proceedings of the, Big Island, HI, USA*, 2004, pp. 10 pp. https://doi.org/DOI: 10.1109/hicss.2004.1265614

Poth, C. N., Bullock, E. P., Eppel, E., Cameron, R., & Golenko, X. (2023). Adaptive mixed methods research design practices to address complexity in business and management research. In *Handbook of mixed methods research in business and management* (pp. 329–347). Edward Elgar Publishing., DOI: 10.4337/9781800887954.00031

Poth, C. N., Wongvorachan, T., Bulut, O., & Otto, S. J. G. (2024). Adaptive case study-mixed methods design practices for researchers studying complex phenomena. *Journal of Mixed Methods Research*, *18*(3), 292–303. DOI: 10.1177/15586898241250217

Powell, H., Mihalas, S. T., Onwuegbuzie, A. J., Suldo, S. M., & Daley, C. E. (2008). Mixed methods research in school psychology: A mixed methods investigation of trends in the literature. *Psychology in the Schools*, *45*(4), 291–309. DOI: 10.1002/pits.20296

Ramanujan, P., Bhattacharjea, S., & Alcott, B. (2022). A multi-stage approach to qualitative sampling within a mixed methods evaluation: Some reflections on purpose and process. *The Canadian Journal of Program Evaluation*, *36*(3), 355–364. DOI: 10.3138/cjpe.71237

Salmona, M., Lieber, E., & Kaczynski, D. (2019). Qualitative and mixed methods data analysis using Dedoose: A practical approach for research across the social sciences. *Sage (Atlanta, Ga.)*.

Sanscartier, M. D. (2018). The craft attitude: Navigating mess in mixed methods research. *Journal of Mixed Methods Research*, *14*(1), 47–62. DOI: 10.1177/1558689818816248

Sardana, D., Kahu, S. Y., Gračanin, D., & Matković, K. (2021). Multi-modal data exploration in a mixed reality environment using coordinated multiple views. In Yamamoto, S., & Mori, H. (Eds.), *Human interface and the management of information. information presentation and visualization*.

Schoonenboom, J. (2024). Design patterns in mixed methods research: Embedding mixed methods matched comparisons sampling in previous design decisions. *Journal of Mixed Methods Research*, *18*(3), 383–392. DOI: 10.1177/15586898241257284

Sharp, J. L., Mobley, C., Hammond, C., Withington, C., Drew, S. F., Stringfield, S., & Stipanovic, N. (2011). A mixed methods sampling methodology for a multisite case study. *Journal of Mixed Methods Research*, *6*(1), 34–54. DOI: 10.1177/1558689811417133

Shin, D.-H. (2016). Cross-platform users' experiences toward designing interusable systems. *International Journal of Human-Computer Interaction*, *32*(7), 503–514. DOI: 10.1080/10447318.2016.1177277

Straits, B. C., & Singleton, R. A. (2018). *Social research: Approaches and fundamentals*. Oxford University Press.

Taherdoost, H. (2022). What are different research approaches? Comprehensive review of qualitative, quantitative, and mixed method research, their applications, types, and limitations. *Journal of Management Science & Engineering Research, 5*(1), 53–63. https://doi.org/https://doi.org/10.30564/jmser.v5i1.4538

Tashakkori, A., Johnson, R. B., & Teddlie, C. (2021). *Foundations of mixed methods research: Integrating quantitative and qualitative approaches in the social and behavioral sciences* (2nd ed.). Sage.

Teddlie, C., & Tashakkori, A. (2010). *Overview of contemporary issues in mixed methods research* (Vol. 2). Sage. https://doi.org/DOI:10.4135/9781506335193

Teddlie, C., & Yu, F. (2007). Mixed methods sampling. *Journal of Mixed Methods Research, 1*(1), 77–100. DOI: 10.1177/1558689806292430

Tunarosa, A., & Glynn, M. A. (2016). Strategies of integration in mixed methods research. *Organizational Research Methods, 20*(2), 224–242. DOI: 10.1177/1094428116637197

Turner, S. F., Cardinal, L. B., & Burton, R. M. (2016). Research design for mixed methods. *Organizational Research Methods, 20*(2), 243–267. DOI: 10.1177/1094428115610808

Younas, A., Pedersen, M., & Tayaben, J. L. (2019). Review of mixed-methods research in nursing. *Nursing Research, 68*(6), 464–472. DOI: 10.1097/nnr.0000000000000372 PMID: 31693552

Zhou, Y., Yuchun, Z., & Machtmes, K. (2023). Mixed methods integration strategies used in education: A systematic review. *Methodological Innovations, 17*(1), 41–49. DOI: 10.1177/20597991231217937

ADDITIONAL READING

Eisele, G., Vachon, H., Lafit, G., Tuyaerts, D., Houben, M., Kuppens, P., Myin-Germeys, I., & Viechtbauer, W. (2022). A mixed-method investigation into measurement reactivity to the experience sampling method: The role of sampling protocol and individual characteristics. *Psychological Assessment, 35*(1), 68–81. DOI: 10.1037/pas0001177 PMID: 36174163

Fradelos, E. C., Alikari, V., Artemi, S., Missouridou, E., Mangoulia, P., Kyranou, M., Saridi, M., Toska, A., Tsaras, K., & Tzavella, F. (2024). A mixed-method study on the assessment of factors influencing nurses' provision of spiritual care. *Health Care*, *12*(8), 854–854. DOI: 10.3390/healthcare12080854 PMID: 38667616

Jackson, S. D., Mohr, J. J., & Kindahl, A. M. (2021). Intersectional experiences: A mixed methods experience sampling approach to studying an elusive phenomenon. *Journal of Counseling Psychology*, *68*(3), 299–315. DOI: 10.1037/cou0000537 PMID: 34043376

Storey, M.-A., Hoda, R., Milani, A. M. P., & Baldassarre, M. T. (2025). Guiding principles for mixed methods research in software engineering. *Empirical Software Engineering*, *30*(5). Advance online publication. DOI: 10.1007/s10664-025-10629-x

KEY TERMS AND DEFINITIONS

A sample: is the actual population taken from the study's target population for the purpose of gathering research data (McNabb, 2021) or generalising the findings to a larger population (Bordens & Abbott, 2018).

Mixed methods sampling: involves the intentional and coordinated selection of participants or data sources for both the quantitative and qualitative components of a study. This approach ensures that numerical data and rich descriptive insights inform, compare, and complement each other. It also highlights the distinct nature of mixed methods research sampling compared to traditional probability and non-probability sampling techniques and underscores the relationship between samples.

Sampling: is the process of locating and choosing individuals, things, or events to be included in a research study (Straits & Singleton, 2018).

The sampling framework: is a structured plan that specifies who or what is eligible for sampling, how the list of samples is drawn up, and how the selected sample will represent a broader population or phenomenon.

APPENDIX

A Mixed Method Study Illustrating Unique Sampling Techniques (Collins, 2017; Creswell & Plano Clark, 2018; Nair & Prem, 2020; Onwuegbuzie & Collins, 2007; Sharp Et Al., 2011)

Table 2. Proposed study: Adoption and impact of artificial intelligence (AI) in selected public higher education institutions in South Africa

Technique	Meaning	Application		Benefit
Parallel sampling	Quantitative and qualitative samples are drawn independently from the same population	Quantitative (QUANT)	Survey 500 lecturers and students across 6 public universities using stratified random sampling	Allows comparison of findings without one sample influencing the other
		Qualitative (QUAL)	Independently select 25-60 participants (lecturers, students, administrators) using purposive sampling for interviews	
Sequential sampling	One sample informs the selection of the other (common in explanatory designs)	Phase 1 (QUANT)	Conduct a survey to identify high AI adopters and low adopters	Qualitative data explains quantitative results
		Phase 2 (QUAL)	Select interview participants from these groups to explore reasons behind adoption patterns	
Multilevel sampling	Samples are taken from various levels within a hierarchy, such as institution, faculty, and individual.	Level 1	Select universities (clusters)	Captures institutional, departmental, and individual perspectives
		Level 2	Within each university, select faculties (strata)	
		Level 3	Within faculties, sample lecturers and students	
Nested sampling	The qualitative sample is a smaller group taken from the larger quantitative sample.	Quantitative	Survey 500 respondents	Ensures direct linkage between quantitative and qualitative data
		Qualitative	Select 30 for interviews from the 500 respondents based on interesting patterns (e.g., extreme adopters or sceptics)	

continued on following page

Table 2. Continued

Technique	Meaning	Application		Benefit
Parallel multilevel sampling	Both quantitative and qualitative samples are drawn from multiple levels, but independently	Quantitative	Survey students and lecturers at multiple universities	Provides a holistic view across levels without dependency
		Qualitative	Conduct interviews with administrators, lecturers, and students at different universities (not necessarily the same individuals as in the survey)	

Chapter 7

Document Analysis as a Robust Qualitative Research Method in the Context of Computing and Information Systems

Tlou Maggie Masenya
http://orcid.org/0000-0001-5400-0549
Durban University of Technology, South Africa

Collence T. Chisita
http://orcid.org/0000-0002-7375-8627
Durban University of Technology, South Africa

ABSTRACT

This narrative review aimed to track the adoption of the document analysis approach and to understand the extent to which it has been used as a robust stand-alone qualitative research method in the context of computing and Information Systems (IS). Although document analysis research method is seen as a valuable and beneficial research approach, a thematic synthesis of the studies revealed limited or scant use and adoption of this method in the field of computing and Information Systems (IS). This study is intended to bridge the disparity in literature by focusing on how document analysis can be used as a robust stand-alone approach through determining the importance of conducting qualitative document analysis in computing and IS research, the criteria to effectively select the relevant documents for qualitative research as well as the strengths and limitations of using document analysis as stand-alone qualitative research in the context of Computing and IS.

DOI: 10.4018/979-8-3693-4018-9.ch007

The chapter thus mapped and provided insights into the use of document analysis as an independent qualitative research method in order to understand the deeper meaning and underlying patterns within the documents.

INTRODUCTION AND BACKGROUND

Mettler (2025) described qualitative analysis as an essential component of the dynamic process of sensemaking, where researchers sift through data to extract innovative insights that can contribute to new theoretical perspectives. As noted by Morgan (2022), qualitative studies can be conducted at the descriptive level, where researchers seek explicit meaning in the data rather than focusing on interpreting latent meaning. In most cases, this involves analysing primary data (gathered from naturalistic inquiries or direct sources) and secondary data (gathered from existing data), and this can be done through document analysis. Document analysis is a valuable research approach that can be employed independently or to complement other qualitative methods, such as interviews or observations, thereby helping to triangulate findings and strengthen the research. Blaxter, Hughes, and Tight (2006) assert that document analysis is a fundamental component of most research, regardless of the specific field or methodological approach. Although this approach has been primarily used to complement other data collection methods, such as interviews, focus groups, and case studies, it has also been employed as a stand-alone approach and a robust qualitative research method (Bowen, 2009). This method is also a special form of qualitative research that relies solely on the analysis of documents. Documentary analysis or review refers to the study of documents that detail the acts, policies, procedures and standards as requirements for the proper functioning of an organisation (Patton, 2002). This implies a critical review of procedures, policies, memos, guidelines, reports, newspapers, institutional/organisational files, and other relevant documents concerning the phenomenon. Atkinson and Coffey (1997:47) emphasised that:

> *We should not use documentary sources as surrogates for other kinds of data and for instance, we cannot learn through records alone and how an organization operates daily. Equally, we cannot treat records as firm evidence of what they report and therefore this strong reservation does not mean that we should ignore or downgrade documentary data.*

Mackieson, Shlonsky, and Connolly (2019) argue that document analysis, content analysis, and thematic analysis are distinct research methodologies that focus on textual content. As stated above, document analysis encompasses the process of analysing documents in general, which may include content and theme analysis. In contrast, content analysis involves categorising and quantifying the content of the data into predetermined groups relevant to the primary study topic or questions in a methodical, reliable, and scientific manner (Mackieson, Shlonsky & Connolly, 2019). Written documents enable modern societies to function and are essential components of bureaucracies across various disciplines (Hashimov, 2025). These documents allow researchers to describe the content or categorise approaches to specific problems in existing procedures and policies (Adebiyi et al., 2019). For example, in other fields, such as public health, this approach has facilitated the comprehension and interpretation of South African health policy responses to foetal alcohol spectrum disorder (Katchmarchi et al., 2018; Adebiyi et al., 2019). This is due to the fact that document analysis is a systematic procedure for reviewing or interpreting documents, utilising both primary and secondary data, to supplement other types of research data and corroborate other sources (Bowen, 2009).

Therefore, document analysis, as a research method, is particularly applicable to qualitative intensive studies that produce rich descriptions of a single phenomenon, event, organisation, or program (Stake, 1995; Yin, 1994). In qualitative research, the process of reviewing and analysing documents to investigate and discuss a research issue is known as documentary analysis. This process involves identifying, evaluating, incorporating, and drawing conclusions from reliable sources, such as academic papers, official reports, and guidelines (Bowen, 2009). This method can also be applicable in Computing and IS qualitative research. Numerous authors have used document analysis to unpack the impact of ICTs on urban spaces (Afradi & Nourian, 2022). The primary goal of documentary analysis is to re-examine documents to investigate and analyse a research problem, and it serves as a qualitative research approach that relies on researchers to locate, assess, and combine data derived from trustworthy sources, i.e., official reports and guidelines, in order to draw conclusions (Fitzgerald, 2012). The document analysis approach was employed in a comparative study to develop a comprehensive synthesis of the official documentation of five mobile testing tools: Appium, Espresso, Robotium, Frank, and Earl Grey (da Silva & de Souza Santos, 2023). According to da Silva and de Souza Santos (2023), the six elements that were considered for the research were platform support, automated testing, gadget simulation, troubleshooting, tool manuals, and the CI/CD process.

This chapter demonstrated the importance of properly maintained documentation in facilitating empirical and transparent comparisons, confirmed document analysis as a method for independently assessing software tools, and urged its use for initial instrument evaluations, functionality audits, and comparative analyses. Another study by Nguyen et al. (2024), which examined "Fairness Testing Tools for the Software Industry Adoption," involved a document analysis of user manuals, readme files, educational materials, and research articles that first suggested or assessed the tools. Through the extraction and synthesis of pertinent data, the methodology provided a thorough understanding of the potential and constraints of each instrument.

The study conducted by Nguyen et al. (2024) revealed that document analysis is a dependable and efficient method of evaluating software solutions, especially in situations when empirical investigation may not be practical. It also highlights the importance of thorough documentation in promoting tool adoption and facilitating empirical evaluations. Based on Nurmaidah, Kustiningsih, and Rahayu's (2024) study of electronic medical/health record systems in the digital transformation era, document analysis is a reliable and effective method for investigating organisational shifts and intricate systems in ICT and healthcare, especially when there are limitations on physical access to participants and systems.

As observed by Bowen (2009), documents may be the only necessary data source for studies designed within an interpretive paradigm, as in hermeneutic inquiry, or it may simply be the only reliable or trustworthy source, as in cross-cultural and historical research. However, Blaxter, Hughes, and Tight (2008) argued that virtually all research projects involve to a greater or lesser extent, the use and analysis of documents. When conducting a document analysis, researchers typically choose between using qualitative research methods that focus on measurements facilitating comparison and statistical aggregation of data or using numbers and statistics to make sense of the data (Patton, 2015). Qualitative researchers often prefer using other data collection methods, such as interviews, over analysing documents because of their desire to participate more actively in field research (Morgan, 2022). Merriam and Tisdell (2016) further stated that some researchers may avoid analysing documents because they are unaware of their value, and as a result, analysing documents has frequently been an underutilised approach in qualitative research. Many researchers, particularly novice researchers, struggle to comprehend qualitative research in document analysis, including its approaches, document selection criteria, and data analysis strategies. Although there are few studies and articles relating to the use of document analysis as a robust stand-alone qualitative research method, novice researchers sometimes lose interest in adopting this qualitative approach due to a lack of understanding of the critical approaches (Cleland, 2017). Furthermore, a limited literature review also sometimes demotivates researchers from adopting a document analysis approach independently. Therefore, this study aimed to address the challenges related to the adoption of document analysis as a stand-alone approach guided by the following objectives:

- Examine the criteria used in selecting the relevant documents for qualitative research.
- Assess the strengths and limitations of analysing documents for qualitative research.
- Determine the implications of document analysis as a qualitative research method in Computing and IS research.

LITERATURE REVIEW

The following section reviews the literature with specific reference to sources that explore the importance of employing qualitative document analysis in Computing and IS research.

There are numerous examples of distinct techniques centred on different viewpoints in philosophy, rendering it challenging to determine what constitutes worthwhile qualitative research (Yadav, 2022). Qualitative research emphasises interpretative, naturalistic, and holistic inquiry (Anderson, 2017). Despite the wide variety of perspectives, beliefs, and methods used by qualitative researchers, they all agree that multiple realities exist, which are considered intangible, contextually located, and shaped, maintained, or transformed by the experiences and meanings of participants (Henry, 2015). While the assumptions of quantitative research are positivist or post-positivist, the methodologies of qualitative researchers are placed within a range of paradigmatic perspectives, including critical theory, interpretivism, feminism, constructivism, and critical realism, among others (Anderson, 2017). Citing Kitto et al. (2008), Patias and Hohendorff (2019) outline seven criteria for qualitative research that centre on the following:

a) elucidation regarding clarity of the research problem and its objectives;
b) justification with a focus on the choice of the design and the rationale for choosing the qualitative method as the ideal method of tackling the research problem;
c) Methodological rigour and transparency;
d) Clarity on the representativeness nature of the sample and the technique used to define the sum of participants and how the chosen technique maintains abstract generalisation;
e) interpretation of the theoretical argument of results related to the existing theory or the development of a new theory;
f) reflexivity and evaluative rigour in considering possible biases of the researcher in choosing the theme, method, relationship with research participants and interpretation of results; and
g) Ethical approval and generalizability.

THE IMPORTANCE OF USING DOCUMENTS AS VALUABLE DATA SOURCES

Although document analysis has been underused in qualitative research (Tisdell, Merriam, & Stuckey-Peyrot, 2025), this approach can be valuable for many reasons. Merriam and Tisdell (2016) contended that, given qualitative researchers' preference for more involvement in field research, they often choose to use alternative methodologies, including interviews, over document analysis. Some scholars or postgraduate students may not have the time or resources needed to conduct interviews, participant observation and survey questionnaires. As noted by Morgan (2022), this method enables researchers to conduct studies that might otherwise be impossible to complete when used to analyse pre-existing documents. Postgraduate students, scholars, and participants may lack the technical skills or knowledge necessary for using technology, such as video conferencing, for data collection, or may not have the means to afford or access it. Adopting document analysis as an independent method in research can also reduce some of the ethical concerns associated with other qualitative research methods (Morgan, 2022). As pointed out by Merriam and Tisdell (2016), using pre-existing documents as a form of data also raises fewer ethical concerns than using other qualitative methods.

Document analysis also enables researchers to access data that would otherwise require more time and significant effort to collect (Merriam & Tisdell, 2016). For example, although public documents are available for anyone to analyse, the accessibility of these documents usually reduces the ethical concerns associated with their usage. Researchers thus need to ask themselves if the authors of the content to be analysed will be harmed by their research (Braun & Clarke, 2013). It is essential to examine the various definitions of documents, as there are multiple interpretations that should be taken into consideration. According to Cicero, Sutton, and Rackham (1977), the etymology of the term "documentation" is derived from the Latin word docere, meaning "teaching" or "instructing." A document is defined by Scott (2014: 5) as an artefact that has an inscribed text as its central feature. A document is a source of facts as it is produced, disseminated, and used in a socially ordered manner (Atkinson & Coffey, 1997, p. 58). The definition of a document provided by Karppinen and Moe (2012) is more comprehensive since it emphasises documents as texts and media, including audio-visual sources.

Documents encompass a wide variety of sources, such as diaries, letters, written reports, speeches, photographs, films, and more, when interpreted broadly (Karppinen & Moe, 2019). A further method for categorising a document is to determine whether it was created intentionally or unintentionally, whether it was intended for public consumption, and whether it serves as a source or a study object (Karppinen & Moe, 2019). Like historical research, documents can serve as factual or contextual sources that reveal the writer's objectives and interests or offer insight into the policy-making process (Karppinen & Moe, 2019). Although implementing multi-methods, such as interviews, participant observation, and survey questionnaires (triangulation), to collect data increases the trustworthiness of a study, however, introducing one method of data collection to beginners can also make qualitative research easier to conduct, which is a complex process (Green & Chian, 2018). The use of document analysis in conjunction with other qualitative research approaches is a common method for triangulating research methods in studying similar subjects (Denzin, 2017: 48).

This approach is more useful when researchers or scholars are unable to complete studies using other qualitative methods. Using document analysis as a data collection method can thus be more appropriate and convenient for researchers and postgraduate students, while other data collection methods may require extensive travel to a site to conduct interviews and observations. For instance, the COVID-19 pandemic has had wide-ranging implications for the academic and research community, affecting academic research, scholars, and their research participants in various aspects and at multiple levels (Meiljering et al., 2024). This pandemic significantly impacted how qualitative researchers conducted data collection effectively and safely (Hangel et al., 2022; Howlett, 2022), as they faced substantial restrictions on travelling to study sites during lockdowns, which hindered their ability to conduct fieldwork and gather data in person. Additionally, the time and money required to travel to the sites may hinder researchers or scholars from conducting interviews and observations. Furthermore, field researchers may be required to do more work after data collection because they need to transcribe their interviews, which is unnecessary when analysing pre-existing data (Braun et al., 2016). Merriam and Tisdell (2016) identified several factors that hinder the adoption of this approach, despite its reputation as a valuable and beneficial research method. These include researchers who prefer creating their own data and seeing situations and people in person; they feel they lack the skills needed to analyse documents, and they underestimate the potential of documents for yielding data.

METHODOLOGY

The study employed a hermeneutic review, which differs from systematic reviews in that it is not subject to rigorous protocols and statistical quantification of the articles reviewed. The study employed a qualitative methodology, focusing on situational and textual analysis, and adopted hermeneutics as an interpretive method. As a result of the qualitative need to interpret texts in consideration of their social context, interpretation-focused hermeneutics emerged (Muganga, 2015). A hermeneutic approach employs understanding as a point of reference to produce an interpretive understanding (Boell & Cecez-Kecmanovic, 2014). The study thus employed a qualitative research methodology to investigate how document analysis might be applied as a qualitative research method in the context of information systems and computing. Qualitative research is a research methodology that primarily enables the exploration or investigation of the quality of relationships, activities, circumstances, or materials (Fraenkel & Wallen, 2019). The crucial approaches for producing a qualitative study article, thesis, or dissertation have been identified by Creswell and Poth (2024) as narrative, phenomenology, grounded theory, ethnography, and case studies. The "how" and "why" research questions are addressed through qualitative research, which also enables researchers to gain a deeper understanding of the situation, phenomena, and experiences (Islam & Aldaihani, 2022).

Narrative evaluations are recursive in nature, usually involving several cycles of searching, analysing, and interpreting. Typically, high-quality narrative reviews include articles that deal with the phenomenon of interest or other relevant manuscripts (Sukhera, 2022). A significant number of narrative reviews adhere to subjectivist and interpretivist paradigms, which are prevalent in various social science and humanities disciplines (Aaron, Webb, & Luhanga, 2020). According to these worldviews, reality is dynamic and subjective, and they stand in opposition to positivist and post-positivist worldviews, which form the basis of systematic reviews and hold that experimental research may reveal a single reality. Given that the word "narrative review" is often misinterpreted, misused, and unfairly disregarded, the systematic review approach has been incorrectly regarded as the international standard (Greenhalgh, Thorne & Malterud, 2018). Narrative reviews are a specialised form of knowledge synthesis that has a strong foundation in a specific research tradition. Narrative reviews are often presented as non-systematic, suggesting that there is a hierarchy of evidence that places them below other review formats (Greenhalgh, Thorne, & Malterud, 2018).

Nonetheless, medical researchers and educators find narrative reviews to be very helpful. A narrative review can incorporate a wide range of studies and offer an overall overview, along with interpretation and critique, whereas a systematic review typically concentrates on a specific subject in a particular context and follows a predetermined way to integrate findings from related studies (Greenhalgh, Thorne & Malterud, 2018). One can continuously improve their understanding by critically analysing dataset components, such as individual primary studies, as part of a wider body of work (Boell & Cecez-Kecmanovic, 2014). The narrative review is a distinct and possibly complementary type of scholarship rather than a poor cousin of the systematic review (Greenhalgh, Thorne & Malterud, 2018). Concepts and definitions crucial to the topic and research question should be identified, along with the limits and scope of the topic (Sukhera, 2022). There are usually no rigid selection or exclusion criteria in narrative reviews. Narrative reviews should describe the search terms and databases used in the synthesis, as well as the reasoning behind their selection (Sukhera, 2022). While the narrative review did not adhere to the strict procedures of a systematic literature review, the researchers used sources from Google Scholar, EBSCOhost, Scopus, Web of Science and JSTOR databases. Grey literature sources were accessed from National libraries and Digital repositories.

The selection criteria included empirical studies and other reviews published in peer-reviewed journals that focused on the use of document analysis as an approach. Although many articles related to the study were retrieved, a thorough review of all the articles revealed that some were found to be irrelevant and were excluded.

FINDINGS AND DISCUSSION

The findings were presented in three themes: document selection criteria for qualitative research, strengths and limitations of analysing documents for qualitative research and the implications of document analysis as a stand-alone qualitative research method in Computing and IS research.

DOCUMENTS SELECTION CRITERIA FOR QUALITATIVE RESEARCH

Documents of all types can help researchers develop an understanding, uncover meaning, and gain insights relevant to the research problem, as stated by Merriam (1988). Flick (2018) further noted that textual documents and visual materials can be a potential source for qualitative document analysis. Patton (2015) added that any textual document is a potential source for qualitative data analysis. The first crucial step in the process of conducting document analysis is selecting appropriate documents for the study. Flick (2018) identified four factors to consider when selecting documents to analyse for the study, namely authenticity, credibility, representativeness, and meaning. The authenticity of the documents to be analysed is a foundational element of any research (Mogalakwe, 2009), and this involves the extent to which the analysed document is genuine (Dunne et al., 2016). For instance, researchers or scholars are required to select the appropriate interview participants, and similarly, they must ensure that the selected documents for analysis are authentic (Mogalakwe, 2009).

Determining whether a document is a primary or secondary source is one of the most effective ways to assess its authenticity. Credibility involves the extent to which the source is free from error and distortion (Dunne et al., 2016). Researchers thus need to determine whether document creators are reliable sources to assess the credibility of documents (Flick, 2018). However, certain documents are more likely to be less credible than others. For instance, the formal documents produced by commercial organisations are usually intended to create a good impression. As noted by Payne and Payne (2004), employees at these organisations often conceal their mistakes, as they are typically interested in protecting themselves, leading to a discrepancy between what a document contains and what actually happens within their organisations. Representativeness refers to the typicality of a document and its degree of representativeness or standardisation within its category or type. For instance, a document lacks representativeness if it contains idiosyncratic content rather than information or data that reflects the content of a collection of other documents about the same topic, as stated by Kridel (2015). However, according to Payne and Payne (2004), assessing the representativeness of any organisation's internal documents is difficult for external researchers, as senior management can restrict access only to those they wish to be investigated. Meaning involves the significance of the document's content and whether the information or data is clear and understandable. Therefore, to assess the meaning of a text as a whole, researchers and data analysts must connect the literal meaning to the context in which the document was created (Mogalakwe, 2009).

STRENGTHS AND LIMITATIONS OF ANALYSING DOCUMENTS FOR QUALITATIVE RESEARCH

Some postgraduate students and researchers, including emerging researchers, use document analysis as their primary research method, although this approach is often employed to complement other research methods. In relation to other qualitative research methods, document analysis has both strengths and limitations. However, it offers advantages that clearly outweigh the limitations, given its efficiency and cost-effectiveness. In some instances, using pre-existing documents allows scholars, emerging researchers and postgraduate students to gain access to the best source of data for completing their research or studies. For instance, interviewing people who lived many years ago or conducting direct observations on how they lived is impossible, especially when conducting historical research. Therefore, documents such as books and articles can be an important source for investigating this subject since it may be difficult to observe their relationships (Merriam & Tisdell, 2016). In highly violent settings, researchers and scholars are at higher risk of being harmed, and therefore, analysing documents may be a safer approach (Goldstein, 2014). Analysing documents, such as journal articles and books, can also be instrumental because the data is stable, as researchers may influence participants during interviews or observations (Merriam & Tisdell, 2016). Since the advent of the Internet, many documents have entered the public domain and are now obtainable without the authors' permission, making document analysis a suitable approach for qualitative researchers. The analysed documents are unobtrusive, non-reactive, and unaffected by the research process. Merriam and Tisdell (2016) further noted that data is unaffected when conducting a document analysis with pre-existing texts. Other advantages of analysing documents in qualitative research are as follows:

- **Efficient method:** Document analysis requires data selection, instead of data collection, and it is thus less time-consuming and more efficient than other research methods.
- **Cost-effectiveness:** Document analysis is less costly than other research methods, and it is considered a suitable method, especially when the collection of new data is infeasible. Researchers can only evaluate the content and quality of the data contained in documents, as it has already been gathered.
- **Stability:** Documents are stable, and the researcher or data analyst's presence does not alter what is being studied (Merriam, 1988).
- **Coverage:** Documents provide broad coverage, and they cover many events and settings (Yin, 1994).

Several limitations inherent in analysing documents are also identified during narrative review. Although other forms of thematic analysis often applied to document analysis may be valuable for conducting positivist and interpretive studies, these methods have been criticised for failing to produce the insights that reflexive thematic analysis can yield (Braun et al., 2019). Documents may not include important information that other methods may uncover; for instance, an organisation can provide access to only the content aligning with its values when allowing outsiders to examine its documents. Investigating public records and personal documents are other examples of potential biases that could result from analysing documents, as these documents could be biased and deceptive (Merriam & Tisdell, 2016). Researchers also need to be concerned about the authenticity of analysed documents and the data provided by sources who wish to remain anonymous (Merriam & Tisdell, 2016). Additionally, working with limited data and researchers is another weakness of relying on pre-existing texts, as researchers may need to adjust their research interests or questions based on the available data (Blackstone, 2019). The use of pre-existing documents as the sole source of data may not provide the useful information needed to conduct a study, as many documents are not produced or created primarily for research purposes (Merriam & Tisdell, 2016). Group members from marginalised communities sometimes distrust researchers due to a fear that the research findings may be useless or harmful to their communities (Ellard-Grey et al., 2015). Some of the limitations of analysing documents in qualitative research are as follows:

- **Insufficient detail:** Documents are produced independently for some purposes other than research.

- **Low retrievability:** Documentation retrievability is sometimes difficult, and as stated by Yin (1994), access to relevant and useful documents may be deliberately blocked.
- **Biased selectivity:** Incomplete collection of documents suggests 'biased selectivity' (Yin, 1994). In an organisational context, the available and suitable documents are likely to be aligned with the corporate policies and procedures, reflecting the organisation's principles.

IMPLICATIONS OF DOCUMENT ANALYSIS AS A STAND-ALONE QUALITATIVE RESEARCH METHOD IN COMPUTING AND IS RESEARCH

Document analysis is applicable in computing and Information Systems in various areas ranging from software engineering, ICT Policy and Governance, Digital Transformation and Change Management, Human-Computer Interaction (HCI), Knowledge Management and Collaboration Systems, Open Source and Community-Driven Development, Cybersecurity and Risk Management, among others. An entire software project encompasses many documents, including the project vision document, the project plan document, the software requirements specification (SRS), the software design specification (SDD), and the software test specification (STD) (ISO, 2019). Triangulating results from various data sources, such as survey results, focus group or interview transcripts, and observations, is a common approach in document analysis (Gross, 2018). It can be conducted independently or in conjunction with a more comprehensive mixed-methods or phenomenological investigation (Gross, 2018). By employing documents in triangulation, researchers can mitigate bias, verify or invalidate, clarify, or build on insights from different sources of data (Gross, 2018). Cardno (2018) extolled policy document analysis as a means for exploring the nature of a policy document to gain insight into what lies beneath it and what is contained within it. Additionally, it is particularly well-suited for use as a method in qualitative study projects. Research that focuses on educational issues can use policy documents to understand the origins and causes of convoluted issues.

Cardno (2018) emphasised that document analysis serves as a rigorous qualitative tool that helps educational leaders analyse and assess policy texts while also serving as a means of addressing praxis-oriented leadership demands and academic research goals. The study by Cardno (2018) confirmed the importance of policy documents as rich data sources that represent institutional values, priorities, and decision-making. By analysing official texts, investigators can determine their significance, syntax, and intent, promote inquiry, and reveal power relationships and organisational approaches. Cardno's (2018) study has implications for educational leadership; however, it also has beneficial implications in ICT and governance. Furthermore, when guided by accepted norms and regulatory research, analysing documents serves as a potent and valuable technique for creating cybersecurity plans for cutting-edge technologies like blockchain (Kabanda, 2021).

Analysing source documents, processed transactions, and the connections between them are all examples of how document analysis can be used to examine and understand audit trails in computer and information systems. An audit trail is a documentation of activities in a system of accounting that validates the system's operation (Nzechukwu, 2016, p. 590). A thorough audit trail enables auditors to track transactions in a client's accounting files, from source documents through subsidiary ledgers, to the general ledger, and ultimately to basic financial reports, payments, and receipts. Using document analysis, trends, variations, or patterns in system or use changes can be identified; logs can be correlated with regulations and policies to verify compliance with processes and correct compliance; the movement and transformation of data between systems can be examined by inspecting time stamps, user identifiers, and actions, and events can be evaluated to reconstruct the incidents that led to system security breaches, glitches, or unauthorized access to data.

Document analysis was reinterpreted as a flexible and participative technique in the Human-Document Interaction Systems study by Karatzas et al. (2016), which enabled more complex and organic connections between individuals and documents in both online and offline environments. In their study of an interaction-oriented approach to document analysis, Karatzas et al. (2016) highlight that document analysis is not simply about analysing static images; it incorporates understanding of how people use both physical and digital documents. Williams (n.d.) highlighted how document analysis can provide a framework for better understanding people's social conditions and experiences, as academics become more aware of how multiple document resources can enhance qualitative research. Such a method provides a multifaceted perspective that quantitative methods may overlook, cultivating a comprehension that extends beyond conventional data collection (Williams, n.d.). Therefore, conducting a qualitative document analysis on existing documents is a desirable method of study, as it is particularly advantageous in circumstances where researchers lack the ability to conduct qualitative research using other methodologies (Morgan, 2022). Morgan (2022) cited the COVID-19 pandemic as a classic example of the advantages of document analysis. However, researchers risk conducting invalid studies if they fail to validate their findings for integrity, trustworthy proportionality, and significance (Morgan, 2022). Creative thinking, discreteness, reliability, accessibility, affordability, and usefulness are just a few of the many reasons that document analysis has been praised (Bowen, 2009; Prior, 2012).

FURTHER RESEARCH DIRECTIONS

Although document analysis, a qualitative research method, has been regarded as a beneficial and valuable research approach, the lack of literature on its adoption is unfortunate, considering its considerable value. Numerous students and emerging scholars continue to struggle with adopting this approach in postgraduate research due to a lack of guidelines on how to employ this method. The conventional approach to document analysis is often too narrow in postgraduate research because it frequently oversimplifies the role of documents and underutilises their full potential, partly due to a lack of awareness of this approach's value and a preference for more active data collection methods such as interviews. As a result, students, emerging researchers, instructors and supervisors are failing to effectively adopt this approach in postgraduate research. This challenge also provokes more general concerns over the value of supplementary or complementary qualitative research methods in postgraduate scholarship in the field of computing and IS. There is therefore a need for a paradigm shift in the adoption of qualitative research methods, such as document analysis, in computing and IS postgraduate research, and to overcome the perception that this approach is the "difficult option". Future research should thus focus more on:

- discussing document analysis procedures.
- implementing guidelines to overcome the difficulties associated with employing document analysis as a robust stand-alone qualitative research method in postgraduate research, specifically in the context of Computing and IS.
- justifying its use to postgraduate students and emerging researchers by emphasising the value of document analysis in postgraduate research and its unique ability to capture the complexity of lived experience and human behaviour.

CONCLUSION AND RECOMMENDATIONS

This chapter contributes to strengthening the knowledge base and advancing understanding of document analysis as a viable qualitative research method. It has discussed the importance of using document analysis as a stand-alone qualitative research method and delineated the strengths and limitations of this approach. After increasing the awareness of how beneficial and valuable this approach can be, researchers could be more inclined to use this approach for qualitative studies in the context of Computing and IS research. However, the increased understanding of document analysis is crucial if this research approach is to be used effectively.

The researcher or scholars also need to determine the authenticity and usefulness of particular documents, and not only their suitability and accessibility, considering the original purpose of each document, the context in which it was produced and the intended audience. Researchers or data analysts should make the process of data analysis as rigorous and as transparent as possible. Document analysis should therefore not be viewed as a supplementary or complementary research method but rather as a valid, reliable and authentic research method that can be used as a stand-alone or independent research method.

REFERENCES

Afradi, K., & Nourian, F. (2022). Understanding ICT's impacts on urban spaces: A qualitative content analysis of literature. *GeoJournal, 87*(2), 701–731. DOI: 10.1007/s10708-020-10277-2

Anderson, V. (2017). Criteria for evaluating qualitative research. *Human Resource Development Quarterly, 28*(2), 125–133.

Atkinson, P. A., & Coffey, A. (1997). Analysing documentary realities. *Qualitative Research: Theory, Method and Practice,* 56–75.

Blackstone, A. (2019). *Social research: Qualitative and quantitative methods*. Flat World.

Blaxter, L., Hughes, C., & Tight, M. (2006). *How to research* (3rd ed.). Open University Press.

Boell, S. K., & Cecez-Kecmanovic, D. (2014). A hermeneutic approach for conducting literature reviews and literature searches. *Communications of the Association for Information Systems, 34*(1), 12.

Bowen, G. (2009). Document Analysis as a Qualitative Research Method. *Qualitative Research Journal, 9*(2), 27–40.

Braun, V., & Clarke, V. (2013). *Successful qualitative research: A practical guide for beginners*. Sage Publishers.

Braun, V., Clarke, V., Hayfield, N., & Terry, G. (2019). Thematic analysis. *In P. Liamputtong (Ed.), Handbook of research methods in health social sciences* (pp. 843-860). Springer. California State Department of Education. 10 quick ways to analyze children's books for racism and sexism. Retrieved on 16 October 2025 fromhttps://cmascanada.ca/wp-content/uploads/2011/11/article-10-ways-to-analyze-childrens-books-for-sexism-and racism.pdf

Cardno, C. (2018). Policy Document Analysis: A practical educational leadership tool and a qualitative research method. *Educational Administration: Theory & Practice, 24*(4), 623–640.

Cicero, M. T., Sutton, E. W., & Rackham, H. (1977). *De Oratore; De Fato; Paradoxa Stoicorum; De Partitione Oratoria*. Harvard University Press.

Cleland, J. A. (2017). The qualitative orientation in medical education research. *Korean Journal of Medical Education*, (2), 61–71. PMID: 28597869

Creswell, J. W., & Poth, C. N. (2024). *Qualitative inquiry and research design: Choosing among five approaches* (5th ed.). SAGE Publications.

da Silva, G., & de Souza Santos, R. (2023, October). Comparing mobile testing tools using documentary analysis. In *2023 ACM/IEEE International Symposium on Empirical Software Engineering and Measurement (ESEM)* (pp. 1-6). IEEE.'

Denzin, N. K. (2017). *The research act: A theoretical introduction to sociological methods*. Routledge.

Dunne, B., Pettigrew, J., & Robinson, K. (2016). Using historical documentary methods to explore the history of occupational therapy. *British Journal of Occupational Therapy*, *79*(6), 376–384.

Ellard-Gray, A., Jeffrey, N. K., Choubak, M., & Crann, S. E. (2015). Finding the hidden participant: Solutions for recruiting hidden, hard-to-reach, and vulnerable populations. *International Journal of Qualitative Methods*, *14*(5), 1–10.

Fitzgerald, T. (2012). Documents and documentary analysis. *Research Methods in Educational Leadership and Management*, *3*, 296–308.

Flick, U. (2018). *An introduction to qualitative research.* Sage Publishers.

Goldstein, D. M. (2014). *Qualitative research in dangerous places: Becoming an "ethnographer" of violence and personal safety.* Social Science Research Council.

Green, J. L., & Chian, M. M. (2018). Triangulation. In Frey, B. (Ed.), *The Sage encyclopedia of educational research, measurement, and evaluation* (pp. 1718–1720). Sage.

Greenhalgh, T., Thorne, S., & Malterud, K. (2018). Time to challenge the spurious hierarchy of systematic over narrative reviews? *European Journal of Clinical Investigation*, *48*(6), e12931. PMID: 29578574

Gross, J. M. (2018). Document analysis. *The SAGE encyclopedia of educational research, measurement, and evaluation,* 1, 545-548.

Hangel, N. F., Schönweitz, F., McLennan, S., & Fiske, A. (2022). Solidarity behavior and its limits: A qualitative study about German and Swiss residents' behaviors towards public health measures during COVID-19 lockdown in April 2020. *SSM. Qualitative Research in Health*, *2*(2), 100051. PMID: 35975169

Hashimov, A. (2025). Max Weber's Bureaucratic Theory: Its Strengths and Weaknesses, and Adaptation to the Modern Era - A Theoretical and Empirical Analysis. *Acta Globalis Humanitatis et Linguarum*, *2*(4), 128–166.

Howlett, M. (2022). Looking at the 'field' through a Zoom lens: Methodological reflections on conducting online research during a global pandemic. *Qualitative Research, 22*(3), 387–402. PMID: 35663097

ISO, I. (2019). I. 21839, "*Iso/IEC/IEEE International Standard–Systems and Software Engineering–System of Systems (SOS) Considerations in Life Cycle Stages of a System,*" (pp. 1-40). ISO/IEC/IEEE 21839: 2019, Kabanda, G. (2021). Cybersecurity risk management plan for a blockchain application model. *Trans Eng Comput Sci, 2*(1), 221.

Karatzas, D., ... (2016, April). Human-document interaction systems-A new frontier for document image analysis. *In 2016 12th IAPR Workshop on Document Analysis Systems (DAS)* (pp. 369-374). IEEE.

Karppinen, K., & Moe, H. (2019). Texts as data I: Document analysis. In *The Palgrave handbook of methods for media policy research* (pp. 249–262). Springer International Publishing.

Katchmarchi, A. B., Taliaferro, A. R., & Kipfer, H. (2018). A Document Analysis of Drowning Prevention Education Resources in the United States. *International Journal of Injury Control and Safety Promotion, 25*(3). PMID: 28701104

Kitto, S. C., Chesters, J., & Grbich, C. (2008). Quality in qualitative research. *The Medical Journal of Australia, 188*(4), 243–246. PMID: 18279135

Kridel, C. (2015). The biographical and documentary milieu. In He, M. F., Schultz, B. D., & Schubert, W. H. (Eds.), *The Sage guide to curriculum in education* (pp. 311–318). Sage Publishers.

Mackieson, P., Shlonsky, A., & Connolly, M. (2019). Increasing rigor and reducing bias in qualitative research: A document analysis of parliamentary debates using applied thematic analysis. *Qualitative Social Work: Research and Practice, 18*(6), 965–980.

Merriam, S. B. (1988). *Case study research in education: A qualitative approach.* Jossey-Bass.

Merriam, S. B., & Tisdell, E. J. (2016). *Qualitative research: A guide to design and implementation* (4th ed.). Jossey Bass.

Mettler, T. (2025). Computational Text Analysis for Qualitative IS Research: A Methodological Reflection. *Communications of the Association for Information Systems, 56*, 349–378.

Mogalakwe, M. (2009). The documentary research method: Using documentary sources in social research. *Eastern Africa Social Science Research Review*, *25*(1), 43–58.

Morgan, H. (2022). Conducting a qualitative document analysis. *The Qualitative Report*, *27*(1), 64–77.

Muganga, L. (2015). The importance of hermeneutic theory in understanding and appreciating interpretive inquiry as a methodology. *Journal of Social Research & Policy*, *6*(1), 65.

Nguyen, T., Baldassarre, M. T., de Lima, L. F., & de Souza Santos, R. (2024, October). From literature to practice: Exploring fairness testing tools for the software industry's adoption. *In Proceedings of the 18th ACM/IEEE International Symposium on Empirical Software Engineering and Measurement* (pp. 549-555).

Nurmaidah, N., Kustiningsih, N., & Rahayu, S. (2024). Change management strategy in the implementation of electronic medical record system in the era of digital transformation: Case study at Rahman Rahim Hospital. *Gema Wiralodra*, *15*(1), 228–238.

Nzechukwu, P. O. (2016). *Internal Audit Practice from A to Z*. CRC Press.

Patton, M. Q. (2015). *Qualitative research & evaluation methods: integrating theory and practice*. SAGE Publishers.

Payne, G., & Payne, J. (2004). *Key concepts in social research*. Sage Publishers.

Prior, L. (2012). The role of documents in social research. In *Handbook of qualitative research in education*. Edward Elgar Publishing.

Scott, J. (2014). *A matter of record: documentary sources in social research*. John Wiley & Sons.

Stake, R. E. (1995). *The art of case study research*. Sage Publishers.

Sukhera, J. (2022). Narrative reviews: Flexible, rigorous, and practical. *Journal of Graduate Medical Education*, *14*(4), 414–417. PMID: 35991099

Tisdell, E. J., Merriam, S. B., & Stuckey-Peyrot, H. L. (2025). *Qualitative research: A guide to design and implementation*. John Wiley & Sons. Williams, B. (n.d). *Role of document analysis in qualitative research methods. Insight*. Retrieved on 16 October, 2025 from https://insight7.io/role-of-document-analysis-in-qualitative-research-methods/

Yadav, D. (2022). Criteria for good qualitative research: A comprehensive review. *The Asia-Pacific Education Researcher, 31*(6), 679–689.

Yin, R. K. (1994). *Case study research: Design and methods* (2nd ed.). Sage Publishers.

ADDITIONAL READING

Aaron, M., Webb, A., & Luhanga, U. (2020). A narrative review of strategies to increase patient safety event reporting by residents. *Journal of Graduate Medical Education, 12*(4), 415–424. PMID: 32879681

Cardno, C. (2018). Policy document analysis: A practical educational leadership tool and a qualitative research method. *Educational Administration: Theory & Practice, 24*(4), 623–640.

Fraenkel, J. R., & Wallen, N. E. (2019). *How to design and evaluate research in education* (7th ed.). McGraw Hill Higher Education.

Henry, P. (2015). Rigor in qualitative research: Promoting quality in social science research. *Research Journal of Recent Sciences, 2277*, 2502.

Islam, M. A., & Aldaihani, F. M. F. (2022). Justification for adopting qualitative research methods, research approaches, sampling strategy, sample size, interview method, saturation, and data analysis. *Journal of International Business and Management*, 5(1), 01-11.

Wong, G. (2015). Literature reviews: who is the audience? *Researching Medical Education*, 25-34.

KEY TERMS AND DEFINITIONS

Computing: is the use of computers that perform calculations and other logical operations to solve problems which are machines. It encompasses a wide range of activities including software development, the design of hardware and the study of algorithms.

Data analysis: is the process of collecting, cleaning, transforming, and interpreting or modelling data to discover useful information, identify patterns, and support business decision-making.

Data collection method: are techniques used to gather information which can be categorized as primary data (original data collected directly from the source) or secondary data (using existing data).

Document analysis: is a systematic procedure for reviewing or evaluating documents/materials both physical and electronic (computer-based and Internet-transmitted) to extract meaning and gain a deeper understanding of a phenomenon.

Information systems: are an integrated set of components or computer-based solutions which include hardware, software and networks that gather, store, analyse and process information and knowledge to support organisational goals and decision-making.

Qualitative research method: is an approach used to collect and analyze non-numerical data to understand the concepts, the underlying motivations, opinions and experiences, focusing on answering the "why" and "how" questions by exploring complex human behavior in natural settings.

Chapter 8
Understanding Phenomenological Qualitative Inquiry in Information Science and Informatics Research

Isaac Mukungu
http://orcid.org/0000-0003-3259-2701
Busitema University, Uganda

Martha Lyaka
http://orcid.org/0000-0002-7993-613X
University of South Africa, South Africa

ABSTRACT

Phenomenology is a philosophical approach that focuses on the lived experiences of people in different situations and circumstances. Phenomenology provides a window for conducting qualitative research that explores human experiences by engaging with study participants in detail to collect rich data about phenomena. Phenomenology can be applied in various fields of study, including information science and informatics, to examine the experiences of different individuals and inform theoretical, policy, and practical dimensions. In this chapter, we discuss the definition of phenomenology, its history, the various types of phenomenology, and the justification for adopting phenomenological approaches. We also examine sampling, data collection, and data analysis in phenomenological studies, as well as the limitations of the approach.

DOI: 10.4018/979-8-3693-4018-9.ch008

INTRODUCTION

Information science research relates to studies and investigations on how information is applied and manifested to support and facilitate organisational and personal activities through information acquisition, organisation, management, including storage, retrieval and use. Information science research aims to encompass key aspects of information, including retrieval, preservation, and access, as well as security in information retrieval and storage. It also addresses information behaviour, community outreach, and emerging areas, including artificial intelligence, the Internet of Things, data literacy, open access, and open science (Hawamdeh & Madali, 2023; Järvelin & Vakkari, 2022; Togia & Malliari, 2017). Specifically, this chapter provides guidance to qualitative informatics researchers when adopting phenomenology to study phenomena in the fields of informatics and information science. Clearly developing parameters for a research helps to focus the study and avoid vagueness in the research process. Qualitative phenomenology provides distinct parameters that guide researchers in refining their study aims and objectives, developing appropriate research methods, including analysis and ethics, and presenting findings that are meaningful to the readers.

DEFINING AND CONTEXUALISING PHENOMENOLOGY

Phenomenology was first introduced as a philosophical study by German philosopher Edmund Husserl (1859-1938), who is considered the father of Phenomenology (Briegel & Müller, 2025; Gallagher, 2022; Oluka, 2025). Several attempts have been made by researchers to define and explain phenomenology in research, with one common occurrence: the concept of "lived experience". Phenomenology takes several forms, and it has been portrayed in the literature as a philosophy and a methodology for research or study (Sloan & Bowe, 2014). As a research approach or methodology, phenomenology is a qualitative research method that encompasses the study of individuals' lived experiences within the world. (Neubauer et al., 2019). Phenomenology is an approach to research that seeks to describe the essence of a phenomenon by exploring it from the perspective of those who have experienced it (Neubauer et al., 2019). The major objective of phenomenology is to identify and illuminate the problem or phenomenon in a way that shows and explains how it is perceived by study participants in a given context and how they experience it (Dangal & Joshi, 2020).

Phenomenology is a research design for qualitative inquiry, focusing on the in-depth study of results. Qualitative inquiry focuses on comprehending a phenomenon by employing a suitable process of in-depth study that involves interaction between the phenomenon being studied and the researcher (Daruhadi, 2024). Qualitative research is an essential methodological approach for several disciplines, including information science research. Qualitative approaches to research are inclined towards acknowledging people's subjective and experiential life-world, involving a detailed description of these experiences (Sloan & Bowe, 2014). Experiential perceptions and emotions form the basis for measuring phenomena in qualitative studies, as qualitative researchers focus on the subjective experiences of participants and real-life practices (Ravn, 2023). Information science and informatics researchers are increasingly adopting qualitative research approaches to explore and discover behaviour patterns of stakeholders in information science studies, including students and practitioners.

In informatics research, phenomenology aims to comprehend the experiences of participants by describing and/or interpreting these subjective experiences, thereby gaining a better understanding of the phenomenon under study from the participants' perspectives. Adopting phenomenology in computing and informatics offers a philosophical, theoretical, methodological and unifying framework that guides the research process of in-depth investigation of the lived experiences of participants on computing and informatics topics, in which participants' realities of life are represented (Mwadulo & Odoyo, 2020; Ndame, 2023). Phenomenology is a recognised qualitative research approach devoted to the exploration as well as interpretation of participants' lived experiences with the aim of unearthing inherent meanings that individuals assign and attribute to a phenomenon (Oluka, 2025). It is a research approach that aims at describing the essence of a phenomenon from the point of view of the people who experience it, with the objective of describing what this experience means in terms of "what and how it was experienced (Neubauer et al., 2019). According to Ngulube and Ngulube (2017), phenomenology focuses on people's lived experiences from a subjective point of view. It is a theoretical approach involving the study of an individual's experience, as the behaviour of people is determined by experience.

Phenomenology is an essential approach to philosophy as it influences the interpretivist notion that people interpret phenomena (Williamson & Johanson, 2018). In informatics research, it is crucial to understand and measure the subjective meanings people attribute to their information environments, thereby comprehending their lived experiences as a basis for the meaning-making process of data. In general, phenomenology examines the structure of various forms of experience, encompassing perception, memory, thought, emotion, imagination, volition, desire, bodily awareness, social activity, personified action, and linguistic activity (Smith, 2018). Research in informatics using phenomenology requires a thorough understanding of the study participants' experiences in all their forms. Qualitative phenomenology illuminates people's experiences with phenomena, making the approach ideal for studies that require an understanding of individuals' perceptions and viewpoints on the study phenomenon as they live it. Phenomenology is the science of describing what a person receives, feels, and knows in his immediate consciousness and experience. Informatics researchers employ phenomenology to get deeper insights and the meaning of phenomena being studied. For example, Keshavarz (2020) applied interpretive phenomenology to explore the lived experiences of participants regarding research support and information services, and described the participants' views on information behaviour, knowledge organisation, ICTs, and the role of librarians in information science. Similarly, (2024) espoused phenomenology to explore the lived experiences of stroke survivors on their perception of digital technologies in their daily lives, with resultant in-depth data supporting rehabilitation processes using digital technologies, among other studies. Phenomenology is applicable in various fields where human instincts, perceptions, emotions, desires, feelings, and understanding are critical (Dangal & Joshi, 2020).

TYPES OF PHENOMENOLOGY

Information science and informatics researchers can adopt either of the two major types of phenomenology expounded on in the literature, depending on the aim and objectives of the study being explored. The literature distinguishes between hermeneutic and transcendental phenomenology as the two primary types of phenomenology. Hermeneutic phenomenology, also known as interpretive phenomenology, was founded by philosopher Martin Heidegger, who was a colleague of Edmund Husserl, who founded transcendental or descriptive phenomenology (Creswell, 2014; Creswell & Poth, 2018; Dangal & Joshi, 2020; Ngulube & Ngulube, 2017; Sloan & Bowe, 2014).

Descriptive Phenomenology (Transcendental Phenomenology)

Transcendental phenomenology, also referred to as descriptive phenomenology, was developed by Edmund Husserl and examines how an individual perceives a given phenomenon without the researcher's interpretation (Tavakol & Sandars, 2025). In transcendental phenomenology, the researcher's intuitions, thoughts, and opinions are set aside during the research process through bracketing, ensuring that the researcher's subjectivity does not influence participants' descriptions (Neubauer et al., 2019). Bracketing as a concept is explained as an attempt by the researcher to stand apart from the study in as unprejudiced an approach as possible, so that participants' lived experiences can be adjudged without the researcher's opinions or thoughts (Dangal & Joshi, 2020; Gill, 2020; Neubauer et al., 2019; Ngulube &

Ngulube, 2017; Sloan & Bowe, 2014). The aim is to present the individual's lived experience as it is by describing their feelings, thoughts, memories and how they perceive a given phenomenon.

Bracketing implies that information science researchers ought to put aside their preconceptions, assumptions and other biases in their investigation so as to gain objectivity in the study findings by focusing on the depth of the participant's lived experiences (Daruhadi, 2024; Squires, 2023; Thomas & Sohn, 2023). Bracketing is crucial in avoiding the researcher's biases from contaminating and influencing study results, thereby ensuring objectivity. Bracketing is important given the close bond that naturally builds between the study phenomenon and the researcher before and during the research processes in qualitative studies, to ensure the researcher is shielded from actions that can cumulatively contaminate the study (Habibullah et al., 2023). It is therefore necessary for information science researchers to establish bracketing approaches, such as self-reflection and continuous reflexivity, in their studies to enable them to approach study participants with an open mind and allow for objectivity in data gathering and analysis. Researchers should bracket their preconceptions, needs, and beliefs, as well as their personal feelings and previous knowledge of the phenomenon, and any biases they may hold about the study participants.

Descriptive phenomenology is concerned with establishing and realising the objective reality of the problem under study (Ngulube & Ngulube, 2017). Researchers in information science are guided by the desire to describe situations in their research areas, adopting a transcendental phenomenological approach. Therefore, transcendental phenomenology focuses on a thorough comprehension of participants' explanations, which provide a clear understanding of their lived experiences shared through stories, thoughts, emotions, and views about their lives in specific situations. This understanding serves as a basis on which researchers can describe participants' lived experiences in detail. This phenomenological approach emphasises establishing the essence of a problem under investigation by comprehending how study participants think, feel, remember, and perceive particular experiences in life (Tavakol & Sandars, 2025). Descriptive phenomenology aims to reveal the perception of the world with consciousness, excluding any personal beliefs and preconceptions, thereby presenting a clear and unfiltered understanding of social phenomena.

The researcher who places himself or herself within the phenomenological paradigm must set aside all preconceptions, judgments or prejudices towards a particular topic in order to make an objective analysis of the information participants bring to an investigation (Padilla-Díaz, 2015). Descriptive phenomenology was propagated by Edmund Husserl with the aim of overcoming researcher biases, which limit and minimise the achievement of pure consciousness (Ngulube & Ngulube, 2017; Tavakol

& Sandars, 2025). It therefore focuses on how events and practices related to the research phenomenon are described without the researcher's bias, of whatever kind.

Interpretive Phenomenology (Hermeneutic Phenomenology)

Interpretive phenomenology, also sometimes referred to as hermeneutic phenomenology, also known as interpretive phenomenology was propagated by Martin Heidegger, whose philosophical inclination of phenomenology was based on Husserl's work, which he built on to provide the Heideggerian phenomenology with distinctions from transcendental phenomenology (Neubauer et al., 2019). Hermeneutic phenomenology examines the nature of being, an ontological dimension, and is concerned with individuals and their lifeworld. Hermeneutic phenomenology directs information science researchers to comprehend their study environments and phenomena by focusing on the realities of individuals experiencing a given phenomenon, by going beyond description of the phenomenon to making deductions and interpretations of the lived experiences. Hermeneutic phenomenology aims to realise the depth of people's lived experiences by providing a detailed account and interpretation of their experiences (Dangal & Joshi, 2020). It focuses on identifying themes for comprehending experiences rather than searching for essence, which is tagged to descriptive or transcendental phenomenology (Ngulube & Ngulube, 2017).

In contrast to transcendental phenomenology, hermeneutic phenomenology emphasises reflexivity, as opposed to bracketing, which is untenable in interpretive phenomenology, where researchers directly use their prior knowledge and experience of a phenomenon to interpret the participants' lived experience. Interpretive phenomenology rejects and disregards bracketing, but advances reflexivity, which tasks the researcher with continuously demonstrating transparency, self-reflection, and self-awareness of their likely biases that can contaminate the study findings (Frechette et al., 2020; Ngulube & Ngulube, 2017; Tavakol & Sandars, 2025; Urcia, 2021). In this phenomenological approach, informatics researchers should interpret the participants' lived experiences using their own experiences and knowledge (Tavakol & Sandars, 2025; Urcia, 2021). Researchers are expected to reflect on the entire research process, rather than adopting bracketing of their preconceptions, so as to engage with participants and unfold contextual elements and fundamental meanings that influence a phenomenon (Dangal & Joshi, 2020; Urcia, 2021). The aim is to understand the participant's subjectivity on a phenomenon and make interpretations of those subjective views.

Interpretation denotes attaching and attributing significance to something, especially what is discovered, making and constructing a sense of what is found, providing explanations and making conclusions of what was found, picking out lessons, making inferences, making considerations for meanings and putting order on what was found out in the context of the study(Gabriel, 2018; Patton, 2002; Trent & Cho, 2020; Willig, 2017). Researchers adopting hermeneutic phenomenology are expected to transcend the description of participants' lived experiences in a study to make an interpretation of what is found within the study context, thereby gaining a clear and detailed understanding of the phenomenon. Description and interpretation are essential in interpretive phenomenology, as the description of the lived experience enables the researcher to make sound interpretations of the lived experiences of study participants. A description of participants' lived experiences of a phenomenon enables the deduction of faithful and credible meanings and interpretations. Rather than bracketing off the subjective perspective of the researcher, interpretive phenomenology understands that the researcher, just like the study participant's life world, cannot be done away with in a study, but rather the prior knowledge and experience of the researcher are valuable in guiding the study (Neubauer et al., 2019). By adopting hermeneutic phenomenology, information science researchers can utilise their experiences and knowledge to provide more nuanced interpretations of participants' lived experiences.

CHARACTERISTICS AND FEATURES OF PHENOMENOLOGY

Phenomenology as a qualitative research design can be best understood by identifying and explaining its features and characteristics, which describe it and serve as the basis for distinguishing it from other qualitative research designs. Figure I outlines the features of phenomenology.

Figure 1. Characteristics of phenomenology (Design by Authors)

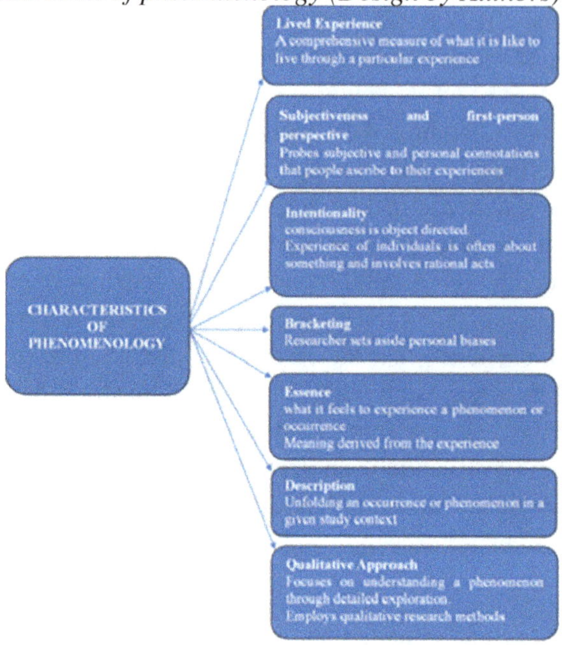

Phenomenology is a qualitative approach, focusing on the descriptive study of participants' lived experiences. It focuses on the essence and intentionality of the experience, capturing the subjectivity of participants. For descriptive phenomenology, bracketing is emphasised, while for hermeneutic phenomenology, reflexivity is emphasised (Alhazmi & Kaufmann, 2022; Badil et al., 2023; Briegel & Müller, 2025; Smith, 2018; Wang et al., 2023).

This research design is criticised for various reasons, including the lengthy time required for data collection and analysis, the lack of generalizability of the findings, the requirement that participants be articulate in explaining their views, the difficulty in dealing with researchers' biases, as well as the subjectivity of the participants. (Annisa, 2024; Oluka, 2025). Phenomenology, however, is an ideal approach for informatics researchers because it provides depth in understanding a phenomenon, offering firsthand accounts of a phenomenon, and allowing for clarity and comprehension of a study's phenomenon (Ayton, 2023; Ayton et al., 2024; Wang et al., 2023).

SAMPLING IN PHENOMENOLOGICAL STUDIES

When conducting research, it is essential to remember that data is collected from participants or subjects who are part of the study population. The process of identifying participants to take part in a study for the purposes of raising the required data to inform your study is called sampling. Sampling involves the process of identifying and selecting participants, objects, or respondents for a study. It involves searching for or selecting participants who offer rich data about the phenomenon under study (Moser & Korstjens, 2018; Polit & Beck, 2017).

In informatics phenomenological studies, the purpose of sampling is to select a suitable sample for the study that is manageable yet information-rich, providing accurate and detailed accounts of the study phenomenon. Sampling enables the researcher to investigate a smaller subset of the target population, making the research process more manageable for data collection. Sampling, therefore, produces a sample which is a set of individuals or elements of a study from which data about the study is obtained (Islam & Aldaihani, 2022; Mcleod, 2023). Prior to drawing or selecting their samples, researchers should clearly identify their target population from which a study population is earmarked by clearly defining its characteristics, before an appropriate sampling strategy and or technique to draw a sample or participants for the study is applied to the study population (Leav Leavy y, 2017; Ritchie et al., 2013). In phenomenology, focus should be put on identifying and selecting only those individuals who possess the experience of the problem, issue or phenomenon under study as participants (Ngulube & Ngulube, 2017).

Sampling Techniques

As an element of qualitative research, phenomenology is guided by qualitative sampling strategies that researchers can explore when selecting study participants. Generally, deliberate sampling is employed in qualitative research, as opposed to the random sampling techniques commonly used in quantitative research approaches. Sampling in qualitative research tasks the researcher to purposefully select study participants who the researcher believes will provide the best data to understand the phenomenon being studied (Creswell, 2021). The most adopted non-random sampling techniques in qualitative studies are convenience sampling, purposive sampling, and snowball sampling (Islam & Aldaihani, 2022; Moser & Korstjens, 2018).

The purposive sampling technique is the most adopted sampling strategy in selecting participants (Badil et al., 2023; Frechette et al., 2020). Purposive sampling enables researchers to identify and select participants who possess rich knowledge or, rather, are informed about the phenomenon under study (Leavy, 2017; Patton, 2002). It enables the researcher to identify specially qualified participants to provide the requisite and vital data about the phenomenon (Frechette et al., 2020; Islam & Aldaihani, 2022; Urcia, 2021). Purposive sampling is based on the premise that identifying and selecting suitable study participants yields suitable data. When researchers clearly identify information-rich participants for their studies in line with the study's objectives, the resultant data from these participants will be credible in answering the study's research questions, as informed by the study's objectives. Generally, phenomenology espouses purposive sampling in which people who have experience with the study phenomenon become study participants (Daruhadi, 2024).

Researchers adopting purposive sampling in phenomenological studies should develop and incorporate explicit criteria that participants must fulfil to be selected for the study (Padilla-Díaz, 2015). Researchers should develop an inclusion and exclusion criterion, the basis on which participants are selected for the study. Because phenomenology focuses on lived experiences, the researcher should clearly define the criteria used to include participants in a study, as well as the exclusion criteria, to identify precisely which individuals cannot be included in the study. An inclusion criterion denotes the characteristics that a researcher deems suitable for a participant to be considered eligible to take part in a study. An exclusion criterion, on the other hand, is the set of factors upon which potential participants are disqualified from taking part in the study (Garg, 2016; Patino & Ferreira, 2018). The purpose of the inclusion and exclusion criteria is to ensure that data is collected from the right participants, thereby achieving the credibility of the study findings. In phenomenological research, the most important factor to consider in purposive sampling is the participants' experience with the phenomenon, as opposed to merely representing the population.

Convenience sampling, on the other hand, involves researchers selecting participants for a study because they are easily accessible, whether as a result of participant availability at a given point in time, geographical proximity and participants' willingness to take part in the study (Etikan et al., 2016; Islam & Aldaihani, 2022). The researchers view the subjects as convenient data sources to inform their understanding of a phenomenon. In contrast to purposive sampling, where expert judgment is required to identify and select study participants, convenience sampling does not require expert judgment (Battaglia, 2008). Convenience sampling is espoused by researchers for its advantages, especially being less time-consuming in selecting samples, taking little effort and not requiring a list of elements of the population, and being less costly (Etikan et al., 2016; Golzar et al., 2022; Jager et al., 2017). Phenomenological researchers need to be aware of the nature of their study elements or participants before adopting convenience sampling. Homogeneous study populations can effectively accommodate convenience sampling to yield rich, detailed, and representative data, as compared to heterogeneous populations, which align well with purposive sampling in phenomenological studies.

A strategy that can be used to get to participants in phenomenology, especially when employing purposive sampling, is snowball sampling. Snowball sampling enables researchers to find participants through other participants. As a technique to find study participants, researchers make contact with a few participants who are linked to other potential subjects who can be contacted to participate in the study. Snowball sampling utilises participants' networks to move from one participant to another. The strategy is most applicable in situations where the researchers know little about the study population elements (Creswell & Poth, 2018; Etikan & Bala, 2017). Purposive and snowball sampling strategies are very common when conducting hermeneutic phenomenology. (Ngulube & Ngulube, 2017).

Sample Sizes in Phenomenology

Although specifying a sample size for a study is critical in research, there is considerable contention about the appropriate sample size required in phenomenology to fully comprehend a phenomenon (Ngulube & Ngulube, 2017; Urcia, 2021). There is no clear position and agreement seen in the literature on sample size for this research approach. The sample sizes in phenomenology lack standardisation, which continuously confuses the understanding of suitable sample sizes for phenomenology (Bartholomew et al., 2021). Different scholars suggest varying sample sizes. Participants in phenomenological studies, ranging from 3 to 4 or even 10 to 15 persons or fewer, who possess experience with the phenomenon under study in a heterogeneous setting, are regarded as suitable (Morse, 1994; Urcia, 2021). According to Smith et al. (2009), sample sizes in phenomenology are better suited in the range of 4 – 10 interviews. In the same vein, Creswell and Creswell (2023) propose a range of 3–10 participants for phenomenological studies, whereas Sarfo et al. (2021) suggest that a range of 6–10 participants is more suitable for this type of study. Similarly, Moser and Korstjens (2018) aver that phenomenological studies should have 10 or fewer interviews. Ngulube and Ngulube (2017) reiterate that fewer participants studied with greater detail and depth form the yardstick for sampling in phenomenology. In the same vein, it is opined that hermeneutic phenomenology is carried out with small samples that are adequate to realise the potential of hermeneutic phenomenology (Bartholomew et al., 2021; Smith, 2011). Some writers argue that sample sizes in phenomenology should be open-ended to allow for participant inclusion in the study until data saturation is achieved, where no new information emerges from the participants (Bowie & Wojnar, 2015; Ngulube & Ngulube, 2017). For studies carried out on equally homogeneous populations, saturation in data can be achieved with a few individual interviews, ranging from 9 to 17, and 4 to 8 group discussions (Hennink & Kaiser, 2022).

There are diverging views in the literature on sample sizes in phenomenology, but, unlike quantitative approaches, rich small samples are considered adequate for phenomenological investigations. Researchers should ensure that the samples or participants identified and selected for the study are rich in information, aligning with the study phenomenon. They should possess the information that can help answer research questions with depth, as required in phenomenology. Selecting sample sizes in phenomenology should be guided by the nature of the study phenomenon and research questions.

DATA COLLECTION IN PHENOMENOLOGY STUDIES

Data in phenomenological studies can be collected using several data-gathering techniques as outlined below.

Interviews

Interviews are the most suitable data collection techniques in phenomenological studies (Dangal & Joshi, 2020; Frechette et al., 2020; Padilla-Díaz, 2015). Interviews are used to collect detailed insights from participants about their lived experiences of the study phenomenon. Interviews have the potential to explore and discover experiences, beliefs, opinions, views and ideas of participants on the study phenomenon by providing detailed accounts and comprehension of the phenomenon (Islam & Aldaihani, 2022). Interviews are verbal exchanges where the interviewer elicits information from another on a given phenomenon (Denzin & Lincoln, 2018). Interviews are suitable in phenomenology because they provide participants with the opportunity to describe in detail the concept or phenomenon as they experience it in their lived context and environment. The researcher is able to dig deeper into the participants' lived experiences of the phenomenon by asking and probing aspects that require clarification from the participants.

The researcher can espouse the different interview types, although the semi-structured and unstructured are the most applicable in phenomenological studies to allow the participants and researcher the opportunity to delve deeper into the lived experiences of participants, in which their reality is captured faithfully (Daruhadi, 2024; Padilla-Díaz, 2015). Like all qualitative research designs, phenomenology requires that the researcher embody the study phenomenon and directs the data gathering process in a way that enables participants to express themselves fully. In phenomenology, the researcher is expected to pay attention and listen attentively to the participants, so as to uncover their lived experiences of the phenomenon through both their spoken responses and nonverbal cues that may prompt further probing. If well applied, probing is critical in provoking rich data in interviews as it offers an opportunity for the researcher to prompt clarification, elaboration and explanation of participants' responses (Robinson, 2023).

The application of data collection triangulation in phenomenological studies, particularly in hermeneutic phenomenology, is essential because each method of data collection complements the others (Frechette et al., 2020). Interview technique is the primary data collection method espoused in qualitative approaches and is often complemented by other qualitative data collection tactics, such as participant observation (Dunwoodie et al., 2023). Interviews, whether structured, semi-structured or unstructured, can be complemented with data from participant observation, focus groups, chat dialogues, diaries, official documents, autobiographies and letters (Ngulube & Ngulube, 2017).

Phenomenological interviews should be well-planned and conducted in an orderly manner to achieve maximum benefit from this data-gathering approach. The who, the where, and the how of the interview are critical for successful interviews (Quinney et al., 2016). For example, in studying the information-seeking behaviour of university students, using interviews and observation techniques can yield detailed responses from a lived experience perspective. The researcher must have the ability to create a conducive environment for the interview and establish rapport with the interviewee. Building rapport is crucial in initiating an interview, as it influences and sets the tone for the participants' experience throughout the study. As such, the researcher should establish a trusting environment that fosters participants' comfort in sharing and disclosing their lived experiences in great depth. Being friendly and respectful to both participants and oneself is key to establishing a trustworthy interview environment. The researcher should be able to pay attention and listen carefully and faithfully to the participants, allowing them time to fully express themselves, so as to elicit detailed descriptions of the phenomenon (Høffding et al., 2022; Islam & Aldaihani, 2022; Tavakol & Sandars, 2025).

Focus Group Discussions (FGD)

Similar to interviews are Focus Group Discussions (FGDs), which are seen as an extended approach to the interview method where a group of participants is interviewed about the phenomenon in the form of a discussion (Gundumogula, 2020). This approach to data collection enables the researcher in a phenomenological study to gather data from more than one individual in the same time interval (Yaakop, 2023). FGDs permit group interaction where discussions on the phenomenon are held, thereby offering a chance for the researcher to record new ideas and notions about the study by intensifying participants' thoughts, clarifying and exploring participants' ideas (Downs et al., 2024). For example, in a study on students' perception of electronic resources use in a university, the researcher can hold an FGD in which participants discuss the topics presented by the researcher as a way to solicit in-depth explanations about the feelings and lived experiences of the students about electronic resources. FGDs enhance individual interviews by allowing for more comprehensive discussions on the topic under investigation (Geampana & Perrotta, 2024).

In a phenomenological study, FGDs help to capture participants' lived experiences by means of sharing insights in groups that are deemed appropriate by the researcher. It is essential for the researcher in phenomenological studies to select focus groups that produce the desired outcomes of the discussions. Focus groups should be made as homogeneous as possible to facilitate appropriate discussions that yield responses that can adequately inform a study. FGDs are ideal as a complementary source of information to individual interviews, as they can illuminate common and divergent aspects about the phenomenon, which may be missed during individual interviews. However, FGDs are prone to various limitations, including managing participants' personalities and dominance, silent and shy participants, difficulties in ensuring confidentiality, and logistical challenges in organising them (Alam et al., 2025; Muijeen et al., 2020).

Just as with individual interviews, to ensure success when using FGDs for phenomenological studies, the researcher must select participants who have experience with the study topic to ensure that they are information-rich. It is essential for the researcher to guide the discussions of the focus group as a moderator, rather than a participant. Using an FGD guide is crucial to directing the flow of discussions and ensuring that the discussion remains on point and is completed within the allocated time. The researcher must create a discussion environment that is conducive for all members, making them feel comfortable and encouraging them to engage in the discussion. All members in the FGD should be given an equal opportunity to share their experiences about the phenomenon, so as to prevent dominant participants from overshadowing others. The researcher should utilise their managerial and interpersonal skills to manage FGD dynamics, such as countering disruptive participants and engaging silent and shy participants. The number of members of the FGD should also be adequate to yield maximum benefit from the discussions. FGDs should have a manageable number of members in phenomenological studies to ensure satisfactory discussions. To ensure that all points are captured well and not lost in the process, the researcher should ensure that the entire discussion is recorded to allow for direct transcription of the recordings (Akyıldız & Ahmed, 2021; Alam et al., 2025; Muijeen et al., 2020; Nyumba et al., 2018; Shoko et al., 2024).

Observation Technique

Another common data collection technique in phenomenology, often adopted in conjunction with interviews, whether individual or focus groups, is the participant observation approach. The observation technique involves collecting data by means of observing events as they happen and unfold (Dangal & Joshi, 2020). Other than asking to solicit information from participants, observation involves capturing and recording occurrences about the phenomenon(Kothari & Garg, 2019). Observation technique in phenomenology should be used in conjunction with other data collection techniques, such as individual and group interviews. Whether Participatory or non-participatory, the observation technique is critical in observing and describing participants' behaviour on a given phenomenon by way of phenomenon watching until study insights are realised. (Badil et al., 2023; Daruhadi, 2024; Islam & Al-daihani, 2022; Kumar & Sharma, 2023).

DATA QUALITY IN PHENOMENOLOGY

Just as applicable to other qualitative research designs, phenomenology requires researchers to be mindful of quality in the research process. Ensuring quality is essential in research if the study is to bring about credible data and knowledge that can support practice (Ngulube, 2019). Qualitative researchers must to come up with a strategy for establishing quality in research, and have plans to safeguard it (Ngulube & Ngulube, 2017). Data quality has attracted several debates in qualitative research in comparison to the quantitative approaches (Forero et al., 2018; Kalu & Bwalya, 2017; Tomaszewski et al., 2020). To ensure quality and rigour in qualitative approaches, researchers should be mindful of the concepts of transferability, dependability, confirmability and credibility (Ahmed, 2024; Guba & Lincoln, 1994; Kalu & Bwalya, 2017; Ngulube & Ngulube, 2017). Researchers must ensure trustworthiness, which in phenomenology warrants that study findings are realistic to the lived experiences of participants and that rigour is ensured throughout the entire process of research (Tavakol & Sandars, 2025).

Credibility

Credibility denotes the degree of accuracy, trustworthiness and believability realised in study findings (Alele & Malau-Aduli, 2023). It conveys the confidence that is placed in the study's findings and whether these findings comprise original data gathered from the participants, embodying an accurate representation of the participants’ original views and perceptions (Korstjens & Moser, 2018). Credibility in phenomenology can be realised through honest and transparent reporting, training, and the researcher's experience (Johnson et al., 2020; Tavakol & Sandars, 2025). Other strategies that can be taken to ensure credibility include triangulation for data collection method, data and the researcher; member checking strategies, peer debriefing, exact and thorough in data recording, excellent records keeping, prolonged field engagements to ensure comprehension of the phenomenon from a lived experience point of view (Creswell, 2014; Nowell et al., 2017; Stahl & King, 2020). For example, in a study on information seeking among postgraduate students in a university, a researcher can apply member checking by contacting the students after data transcription, presenting them with data transcripts for discussion, in which the students can either confirm, correct, or add context to the findings.

Transferability

Transferability encompasses the degree to which findings of a study can be applied to other contexts and settings (Connelly, 2016; Nyirenda et al., 2020; Stahl & King, 2020). Transferability is likened to generalizability in quantitative studies. It is seen as the ability to extrapolate study results to other contexts and settings (Ahmed, 2024). Transferability in phenomenology is ensured by adopting various strategies, including developing a detailed description of the context of the study, documenting the entire research process, including sampling and data collection techniques, and describing the characteristics of participants (Ahmed, 2024; Stahl & King, 2020; Stenfors et al., 2020; Sundler et al., 2019; Tomaszewski et al., 2020).

Dependability

Dependability in phenomenological research signifies constancy, consistency, steadiness and or data stability in contexts or settings that are alike (Alele & Malau-Aduli, 2023; Connelly, 2016; Korstjens & Moser, 2018). Phenomenological researchers must ensure that their study data achieves dependability so as to assure readers of the trustworthiness of their study results. If studies are repeated with similar and comparable settings and environments, the study findings should be consistent with those of the earlier study. To achieve and realise dependability, researchers ought to comprehensively describe the adopted research process, conduct peer debriefing, conduct audit trails, capture and keep the researchers' activities, and ensure data appropriateness (Connelly, 2016; Dangal & Joshi, 2020; Johnson et al., 2020; Kasirye, 2021; Nowell et al., 2017; Nyirenda et al., 2020).

Confirmability

In phenomenology, confirmability refers to the extent to which the study's findings accurately reflect the participants' experiences, rather than the researcher's. It portrays the neutral position of the researcher during the interpretation of findings, as interpretations are based on the lived experience of participants (Alele & Malau-Aduli, 2023; Johnson et al., 2020; Nyirenda et al., 2020). Confirmability ensures that study interpretations are derived from the data gathered, rather than the researcher's preconceptions. Simply stated, confirmability is about objectiveness in data. Interpretation for data should be obtained from the statements of participants, and that fabrications by the researcher are not part of the interpretation (Ahmed, 2024; Korstjens & Moser, 2018; Megheirkouni & Moir, 2023; Nowell et al., 2017; Stahl & King, 2020). To achieve confirmability, all the aspects of credibility, transferability and dependability must be realised (Nowell et al., 2017).

To realise and achieve conformability, phenomenological researchers must capture and keep audit trails, have reflexivity and manage the researcher's personal bias, conduct peer debriefing, conduct member checking and adopt triangulation (Dangal & Joshi, 2020; Nyirenda et al., 2020). Researcher reflexivity is crucial for confirmability, as it can impact all elements of the research process. The data obtained and presented should reflect reality in the same way the participants view it (Ngulube & Ngulube, 2017).

DATA ANALYSIS IN PHENOMENOLOGY

As phenomenology's focus is on exploring participants' lived experiences on a phenomenon, data analysis calls for researchers to pay attention to the views of participants (Tomaszewski et al., 2020). There are several approaches to qualitative data analysis that can be applicable to phenomenology. Ritchie and Spencer promulgated the framework analysis strategy, which takes a five-step sequence, including data familiarisation, identifying and writing themes, indexing and coding, charting and arranging data, and mapping and interpretation of data as developed from participants' narratives (Goldsmith, 2021; Klingberg et al., 2024; Rabiee, 2004).

Thematic analysis is another common strategy adopted by phenomenologists to analyse their data. Thematic analysis can be applied to data, whether inductively or deductively, to produce findings that reflect the participants' lived experiences, making it suitable for both hermeneutic and transcendental or descriptive phenomenology. Thematic analysis is a sequence of stages in data analysis, where each stage builds upon the previous one, resulting in a thorough comprehension of the data and enhancing its replicability and consistency (Naeem et al., 2023). Thematic analysis in phenomenology encompasses stages including data transcription and data familiarisation, identification and selection of keywords, coding, development of themes, interpretation of themes and writing a report (Braun & Clarke, 2023; Islam & Aldaihani, 2022; Naeem et al., 2023; Ngulube & Ngulube, 2017). Specifically, Braun and Clarke illuminate thematic analysis to encompass data familiarisation, code generation, finding themes in the data, reviewing themes, naming and defining themes and finally report writing (Ahmed et al., 2025; Braun & Clarke, 2023; Tavakol & Sandars, 2025).

Moustakas (1994) suggested an approach to data analysis that is regarded by many as adequate for analysing data in phenomenological studies. Moustakas suggests that phenomenological data analysis should entail a systematic process encompassing *epoche*, identifying key statements from participants' data, grouping and categorising the statements into themes, carrying out a structural and textural description of the themes into lived experience descriptions and finally extracting or making an essence of the description (Aguas, 2022; Dangal & Joshi, 2020; Daruhadi, 2024; Moerer-Urdahl & Creswell, 2004; Ngulube & Ngulube, 2017). Phenomenological researchers can choose from a range of data analysis approaches to produce study reports that clearly present the lived experiences of the participants.

Data analysis in phenomenology can be aided by computer-based packages such as NVivo and Alasia. The application of ICTs in data analysis enables researchers to organise and manage data effectively for analysis. A researcher espousing computer-aided analysis can efficiently manage their data and can easily integrate different data types from different sources in the data analysis package (Ngulube, 2023). Although the application of computer-aided data analysis packages can be challenging for novice researchers, it is valuable in achieving robustness in qualitative studies like phenomenology (Dalkin et al., 2021).

ETHICS IN PHENOMENOLOGY

Phenomenology, like any other qualitative research design, demands researchers to espouse ethics in their research process. Ethics in research ensures that studies are conducted responsibly with integrity. Research ethics encompasses professional and moral conduct guidelines in data collection, analysis, and reporting of findings in research (Arafat, 2024; Barrow et al., 2017). Phenomenology studies must be conducted with ethics to ensure study results are credible and trustworthy. Several guidelines for research ethics in phenomenology can be espoused, including:

- Faithfully selecting study participants with a clear explanation of the criteria for inclusion and exclusion

- Clarifying to the participant the purpose of the study
- Ensuring informed consent prior to data collection
- Protecting participants from risks that may arise from the study, such as distress as a result of the effect of a lived experience
- Ensuring the privacy of participants in the study
- Ensuring voluntary participation in the study by the participants
- Faithfully reporting as gathered to portray the lived experience
- Protecting and storing data well
- Avoiding plagiarism in research writing and reporting
- Ensuring bracketing and reflexivity in describing and interpreting participants' lived experience
- Maintaining professionalism throughout the research process
- Acknowledging and professionally managing distress among participants, such as showing empathy, holding or pausing the interview and allowing the participant to dictate continuation or stopping the interview

Research ethics, if adopted in a study, promote truth, error elimination, fairness, mutual respect, trust and accountability (Arafat, 2024; Mirza et al., 2023; Resnik, 2020; Steffen, 2016; Westland et al., 2025; Whitney & Evered, 2022).

FURTHER RESEARCH DIRECTIONS

Phenomenology offers researchers in information science the opportunity to explore the lived experiences of study participants in depth, making it an appropriate study design for understanding a phenomenon from the participants' perspective. In this design, further research can be conducted in exploring sampling and sample sizes in phenomenological research in greater detail, as there are different schools of thought on this aspect. Understanding the application of phenomenology more can also benefit from studies that explore the requirement for bracketing and reflexivity for both transcendental and interpretive phenomenology.

CONCLUSION

Researchers adopting phenomenology need to plan their studies with a phenomenological mindset from the start of the research process until the end i.e. from developing the research topic all through to data analysis for presentation and report writing. Having a mindset tuned to this research design from the beginning enables the researcher to develop research objectives and research questions that can be clearly explored in line with this research design. Researchers need to be clear about the type of phenomenology adopted for their studies to ensure that all guidelines followed, or rather the approaches adopted, are in tandem with the phenomenological type chosen. It is essential that researchers provide and clarify the rationale for adopting phenomenology, making it easier for readers and researchers to understand the study. Data quality is crucial in phenomenology, and researchers must be mindful of the strategies they adopt to ensure the trustworthiness of their data. Phenomenology requires depth in gathering data; therefore, researchers must be familiar with the principles governing the credibility, dependability, transferability, and confirmability of data. Phenomenologists need to ensure that although the data gathered and required for comprehending the lived experiences of participants requires depth, rigour is essential and must be assured throughout the process. The key issues for phenomenological studies that researchers should consider include subjectivity, bracketing, sample selection, data collection approaches adopted, and data analysis approaches employed.

Ultimately, for a field dedicated to streamlining information access and use, phenomenology offers not just a method, but a vital orientation toward understanding the human lived experience at the heart of informatics.

REFERENCES

Aguas, P. P. (2022). Fusing approaches in educational research: Data collection and data analysis in phenomenological research. *The Qualitative Report*, *27*(1), 1–20.

Ahmed, S. K. (2024). The pillars of trustworthiness in qualitative research. *Journal of Medicine, Surgery, and Public Health*, *2*, 100051. DOI: 10.1016/j.glmedi.2024.100051

Ahmed, S. K., Mohammed, R. A., Nashwan, A. J., Ibrahim, R. H., Abdalla, A. Q., Ameen, B. M. M., & Khdhir, R. M. (2025). Using thematic analysis in qualitative research. *Journal of Medicine, Surgery, and Public Health, 6*, 100198. Akyıldız, S. T., & Ahmed, K. H. (2021). An overview of qualitative research and focus group discussion. *International Journal of Academic Research in Education*, *7*(1), 1–15. Retrieved October 10, 2025, from https://dergipark.org.tr/en/pub/ijare/article/866762

Alam, Md. S., Asmawi, A., & Fatema, S. (2025). Focus group discussions (FGDs) for qualitative and mixed-method approaches: Principles, applications, and key considerations. *Integrated Journal for Research in Arts and Humanities*, *5*(4), 51–56. DOI: 10.55544/ijrah.5.4.8

Alele, F., & Malau-Aduli, B. (2023). *An Introduction to research methods for undergraduate health profession students*. James Cook University. Retrieved on 10 October 2025 from https://jcu.pressbooks.pub/intro-res-methods-health/

Alhazmi, A. A., & Kaufmann, A. (2022). Phenomenological qualitative methods applied to the analysis of cross-cultural experience in novel educational social contexts. *Frontiers in Psychology*, *13*, 785134. Retrieved October 10, 2025, from https://www.frontiersin.org/articles/10.3389/fpsyg.2022.785134/full PMID: 35548502

Arafat, M. Y. (2024). Research ethics: Meaning and principles. *International Journal of Academic Multidisciplinary Research*, *8*(1). Retrieved October 10, 2025, from https://papers.ssrn.com/sol3/papers.cfm?abstract_id=4887943

Ayton, D. (2023). *Phenomenology. Qualitative research–a practical guide for health and social care researchers and practitioners*. Retrieved on 10 October 2025 from https://oercollective.caul.edu.au/qualitative-research/chapter/__unknown__-6/

Ayton, D., Tsindos, T., & Berkovic, D. (2024). *Qualitative research: A practical guide for health and social care researchers and practitioners*. Monash University.

Badil, D. M., Aslam, Z., Khan, K., Ashiq, A., & Bibi, U. B. U. (2023). Phenomenology qualitative research inquiry: A review paper. *Pakistan Journal of Health Sciences, 4*(3), 09–13.

Barrow, J. M., Brannan, G. D., & Khandhar, P. B. (2017). *Research ethics*. Retrieved on 10 October 2025 from https://europepmc.org/books/nbk459281

Bartholomew, T. T., Joy, E. E., Kang, E., & Brown, J. (2021). A choir or cacophony? Sample sizes and quality of conveying participants' voices in phenomenological research. *Methodological Innovations, 14*(2), 205979912110400. DOI: 10.1177/20597991211040063

Battaglia, M. (2008). Convenience sampling. *Encyclopedia of Survey Research Methods, 1*, 148–149.

Bowie, B. H., & Wojnar, D. (2015). Using a phenomenology as a research method in community-based research. In *Nursing research using phenomenology: Qualitative designs and methods in nursing* (pp. 145–155). Springer Publishing Company.

Braun, V., & Clarke, V. (2023). Toward good practice in thematic analysis: Avoiding common problems and be(com)ing a *knowing* researcher. *International Journal of Transgender Health, 24*(1), 1–6. DOI: 10.1080/26895269.2022.2129597 PMID: 36713144

Briegel, H. J., & Müller, T. (2025). Phenomenology. In Briegel, H. J., & Müller, T. (Eds.), *Projective simulation in action* (Vol. 507, pp. 333–374). Springer Nature Switzerland., DOI: 10.1007/978-3-031-98119-7_6

Connelly, L. M. (2016). Trustworthiness in qualitative research. *Medsurg Nursing, 25*(6), 435–436. PMID: 30304614

Creswell, J. W. (2014). Qualitative, quantitative and mixed methods approaches. *Sage (Atlanta, Ga.).*

Creswell, J. W. (2021). A concise introduction to mixed methods research. *Sage (Atlanta, Ga.).*

Creswell, J. W., & Creswell, J. D. (2023). *Research design: Qualitative, quantitative, and mixed methods approaches* (6th ed.). Sage.

Creswell, J. W., & Poth, C. N. (2018). Qualitative inquiry and research design choosing among five approaches. *Sage (Atlanta, Ga.).*

Dalkin, S., Forster, N., Hodgson, P., Lhussier, M., & Carr, S. M. (2021). Using computer assisted qualitative data analysis software (CAQDAS; NVivo) to assist in the complex process of realist theory generation, refinement and testing. *International Journal of Social Research Methodology, 24*(1), 123–134. DOI: 10.1080/13645579.2020.1803528

Dangal, M., & Joshi, R. (2020). Hermeneutic phenomenology: Essence in educational research. *Open Journal for Studies in Philosophy*, *4*(1). Advance online publication. DOI: 10.32591/coas.ojsp.0401.03025d

Daruhadi, G. (2024). Phenomenological Method as A Theoretical Basis of Qualitative Methods. *International Journal of Social Health*, *3*(9), 599–613. DOI: 10.58860/ijsh.v3i9.238

Denzin, N. K., & Lincoln, Y. S. (2018). *The Sage handbook of qualitative research*. Sage.

Downs, N., Davidson, J., Haddad, A., & Zisook, S. (2024). Learning from health care counselors' perspectives on health care worker distress: A qualitative analysis. *HCA Healthcare Journal of Medicine*, *5*(3), 12. DOI: 10.36518/2689-0216.1613 PMID: 39015588

Dunwoodie, K., Macaulay, L., & Newman, A. (2023). Qualitative interviewing in the field of work and organisational psychology: Benefits, challenges and guidelines for researchers and reviewers. *Applied Psychology*, *72*(2), 863–889. DOI: 10.1111/apps.12414

Etikan, I., & Bala, K. (2017). Sampling and sampling methods. *Biometrics & Biostatistics International Journal*, *5*(6), 00149. DOI: 10.15406/bbij.2017.05.00149

Etikan, I., Musa, S. A., & Alkassim, R. S. (2016). Comparison of convenience sampling and purposive sampling. *American Journal of Theoretical and Applied Statistics*, *5*(1), 1–4.

Forero, R., Nahidi, S., De Costa, J., Mohsin, M., Fitzgerald, G., Gibson, N., McCarthy, S., & Aboagye-Sarfo, P. (2018). Application of four-dimension criteria to assess rigour of qualitative research in emergency medicine. *BMC Health Services Research*, *18*(1), 1–11. DOI: 10.1186/s12913-018-2915-2 PMID: 29454350

Frechette, J., Bitzas, V., Aubry, M., Kilpatrick, K., & Lavoie-Tremblay, M. (2020). Capturing lived experience: Methodological considerations for interpretive phenomenological inquiry. *International Journal of Qualitative Methods*, *19*, 1609406920907254. DOI: 10.1177/1609406920907254

Gabriel, Y. (2018). Interpretation, reflexivity and imagination in qualitative research. In Ciesielska, M., & Jemielniak, D. (Eds.), *Qualitative methodologies in Organization Studies* (pp. 137–157). Springer International Publishing., DOI: 10.1007/978-3-319-65217-7_8

Gallagher, S. (2022). What is phenomenology? In Gallagher, S. (Ed.), *Phenomenology* (pp. 1–10). Springer International Publishing., DOI: 10.1007/978-3-031-11586-8_1

Garg, R. (2016). Methodology for research I. *Indian Journal of Anaesthesia*, *60*(9), 640–645. Retrieved October 10, 2025, from https://journals.lww.com/ijaweb/fulltext/2016/60090/methodology_for_research_i.6.aspx PMID: 27729690

Geampana, A., & Perrotta, M. (2024). Using interview excerpts to facilitate focus group discussion. *Qualitative Research*. Advance online publication. DOI: 10.1177/14687941241234283 PMID: 40028392

Gill, M. J. (2020). Phenomenology as qualitative methodology. *Qualitative Analysis: Eight Approaches*, 73–94.

Goldsmith, L. J. (2021). Using framework analysis in applied qualitative research. *Qualitative Report*, *26*(6), 2061–2076.

Golzar, J., Noor, S., & Tajik, O. (2022). Convenience sampling. *International Journal of Education & Language Studies*, *1*(2), 72–77. Retrieved October 10, 2025, from https://www.ijels.net/article_162981.html

Guba, E. G., & Lincoln, Y. S. (1994). Competing paradigms in qualitative research. *Handbook of Qualitative Research, 2*(163–194), 105. Retrieved on 10 October 2025 fromchrome-extension://efaidnbmnnnibpcajpcglclefindmkaj/https://ankara.lti.cs.cmu.edu/11780/sites/default/files/10-guba_lincoln_94.pdf

Gundumogula, M. (2020). Importance of focus groups in qualitative research. [IJHSS]. *International Journal of Humanities and Social Science*, *8*(11), 299–302. DOI: 10.24940/theijhss/2020/v8/i11/HS2011-082

Habibullah, K. M., Mohammed, B. U., & Hamza, M. (2023). Conceptual framework in reflexive bracketing techniques in qualitative methodology. *Research Review, 4*(02), 849–857. http://researchreview.in/index.php/rr/article/view/161

Hawamdeh, S., & Madali, N. P. (2023). The emergence of new disciplines and the transformation of the library and information science profession. *Library Trends*, *72*(2), 187–204. https://muse.jhu.edu/pub/1/article/941423/summary

Hennink, M., & Kaiser, B. N. (2022). Sample sizes for saturation in qualitative research: A systematic review of empirical tests. *Social Science & Medicine*, *292*, 114523. DOI: 10.1016/j.socscimed.2021.114523 PMID: 34785096

Høffding, S., Martiny, K., & Roepstorff, A. (2022). Can we trust the phenomenological interview? Metaphysical, epistemological, and methodological objections. *Phenomenology and the Cognitive Sciences*, *21*(1), 33–51. DOI: 10.1007/s11097-021-09744-z

Islam, M. A., & Aldaihani, F. M. F. (2022). Justification for adopting qualitative research method, research approaches, sampling strategy, sample size, interview method, saturation, and data analysis. *Journal of International Business and Management, 5*(1), 01–11. https://doi.org/DOI: 10.37227/JIBM-2021-09-1494

Jager, J., Putnick, D. L., & Bornstein, M. H. (2017). II. More than just convenient: The scientific merits of homogeneous convenience samples. *Monographs of the Society for Research in Child Development*, *82*(2), 13–30. DOI: 10.1111/mono.12296 PMID: 28475254

Järvelin, K., & Vakkari, P. (2022). LIS research across 50 years: Content analysis of journal articles. *The Journal of Documentation*, *78*(7), 65–88. DOI: 10.1108/JD-03-2021-0062

Johnson, J. L., Adkins, D., & Chauvin, S. (2020). A review of the quality indicators of rigor in qualitative research. *American Journal of Pharmaceutical Education*, *84*(1), 7120. DOI: 10.5688/ajpe7120 PMID: 32292186

Kalu, F. A., & Bwalya, J. C. (2017). What makes qualitative research good research? An exploratory analysis of critical elements. *International Journal of Social Science Research*, *5*(2), 43–56. DOI: 10.5296/ijssr.v5i2.10711

Kasirye, F. (2021). *A conceptual paper on ensuring quality in qualitative research.* [Paper presentation], Malaysia. Retrieved on 10 October 2025 from https://www.researchgate.net/profile/Faiswal-Kasirye/publication/350525202_A_Conceptual_Paper_on_Ensuring_Quality_in_Qualitative_Research/links/6064a604299bf13821f866b6/A-Conceptual-Paper-on-Ensuring-Quality-in-Qualitative-Research.pdf

Keshavarz, H. (2020). *Hermeneutic phenomenology in supporting research and information services: Contributions to information science.* Retrieved on 10 October 2025 from https://repository.kisti.re.kr/handle/10580/16183

Klingberg, S., Stalmeijer, R. E., & Varpio, L. (2024). Using framework analysis methods for qualitative research: AMEE Guide No. 164. *Medical Teacher*, *46*(5), 603–610. DOI: 10.1080/0142159X.2023.2259073 PMID: 37734451

Korstjens, I., & Moser, A. (2018). Series: Practical guidance to qualitative research. Part 4: Trustworthiness and publishing. *The European Journal of General Practice*, *24*(1), 120–124. DOI: 10.1080/13814788.2017.1375092 PMID: 29202616

Kothari, C. R., & Garg, G. (2019). *Research methodology: Methods and techniques.* New Age International.

Kumar, A., & Sharma, A. (2023). Observation method: A review study. *Library Philosophy and Practice (e-Journal).* Https://Digitalcommons. Unl. Edu/Libphil-prac. Retrieved on 10 October 2025 from https://www.researchgate.net/profile/Ankit-Kumar-385/publication/383270822_Observation_Method_A_Review_Study/links/66c5a8274b25ef677f72889a/Observation-Method-A-Review-Study.pdf

Leavy, P. (2017). *Research design: Quantitative, qualitative, mixed methods, arts-based, and community-based participatory research approaches.* Guilford Publications.

Mcleod, S. (2023). Sampling methods in research: Types, techniques, and examples. Retrieved on 10 October 2025 from Https://Www. Simplypsychology. Org/Sampling. Html.

Megheirkouni, M., & Moir, J. (2023). Simple but effective criteria: Rethinking excellent qualitative research. *The Qualitative Report*, *28*(3), 848–864. DOI: 10.46743/2160-3715/2023.5845

Mirza, H., Mirza, C., & Bellalem, F. (2023). Ethical considerations in qualitative research: Summary guidelines for novice social science researchers., *11*(1), 441–449. Retrieved on 10 October 2025 from https://asjp.cerist.dz/en/article/220704

Moerer-Urdahl, T., & Creswell, J. W. (2004). Using transcendental phenomenology to explore the "ripple effect" in a leadership mentoring program. *International Journal of Qualitative Methods*, *3*(2), 19–35. DOI: 10.1177/160940690400300202

Morse, J. M. (1994). Designing qualitative research. In Denzin, N. K., & Lincoln, Y. S. (Eds.), *Handbook of qualitative inquiry* (pp. 220–235). Sage.

Moser, A., & Korstjens, I. (2018). Series: Practical guidance to qualitative research. Part 3: Sampling, data collection and analysis. *The European Journal of General Practice*, *24*(1), 9–18. DOI: 10.1080/13814788.2017.1375091 PMID: 29199486

Muijeen, K., Kongvattananon, P., & Somprasert, C. (2020). The key success factors in focus group discussions with the elderly for novice researchers: A review. *Journal of Health Research*, *34*(4), 359–371. https://www.emerald.com/insight/content/doi/10.1108/JHR-05-2019-0114/full/html

Mwadulo, M. W., & Odoyo, C. O. (2020). Phenomenology approach applicability in information systems and technology field. *Computer Science and Information Technology*, *8*(2), 37–42. DOI: 10.13189/csit.2020.080201

Naeem, M., Ozuem, W., Howell, K., & Ranfagni, S. (2023). A step-by-step process of thematic analysis to develop a conceptual model in qualitative research. *International Journal of Qualitative Methods*, *22*, 16094069231205789. DOI: 10.1177/16094069231205789

Ndame, T. (2023). Phenomenological studies. In Okoko, J. M., Tunison, S., & Walker, K. D. (Eds.), *Varieties of qualitative research methods* (pp. 377–382). Springer International Publishing., DOI: 10.1007/978-3-031-04394-9_58

Neubauer, B. E., Witkop, C. T., & Varpio, L. (2019). How phenomenology can help us learn from the experiences of others. *Perspectives on Medical Education*, *8*(2), 90–97. DOI: 10.1007/S40037-019-0509-2 PMID: 30953335

Ngulube, P. (2019). *Handbook of research on connecting research methods for information science research*. IGI Global.

Ngulube, P. (2023). Improving the quality of reporting findings using computer data analysis applications in educational research in context. *Heliyon*, *9*(9). Advance online publication. DOI: 10.1016/j.heliyon.2023.e19683 PMID: 37810136

Ngulube, P., & Ngulube, B. (2017). Application and contribution of hermeneutic and eidetic phenomenology to indigenous knowledge research. In Ngulube, P. (Ed.), *Handbook of research on theoretical perspectives on indigenous knowledge systems in developing countries* (pp. 127–155). IGI Global., DOI: 10.4018/978-1-5225-0833-5.ch006

Nowell, L. S., Norris, J. M., White, D. E., & Moules, N. J. (2017). Thematic analysis: Striving to meet the trustworthiness criteria. *International Journal of Qualitative Methods*, *16*(1), 160940691773384. DOI: 10.1177/1609406917733847

Nyirenda, L., Kumar, M. B., Theobald, S., Sarker, M., Simwinga, M., Kumwenda, M., Johnson, C., Hatzold, K., Corbett, E. L., Sibanda, E., & Taegtmeyer, M. (2020). Using research networks to generate trustworthy qualitative public health research findings from multiple contexts. *BMC Medical Research Methodology*, *20*(1), 13. DOI: 10.1186/s12874-019-0895-5 PMID: 31964333

Nyumba, T. O., Wilson, K., Derrick, C. J., & Mukherjee, N. (2018). The use of focus group discussion methodology: Insights from two decades of application in conservation. *Methods in Ecology and Evolution*, *9*(1), 20–32. DOI: 10.1111/2041-210X.12860

Oluka, A. (2025). Phenomenological research strategy: Descriptive and interpretive approaches. *F1000 Research*, *14*, 725. Retrieved October 10, 2025, from https://f1000research.com/articles/14-725

Padilla-Díaz, M. (2015). Phenomenology in educational qualitative research: Philosophy as science or philosophical science. *International Journal of Educational Excellence*, *1*(2), 101–110. DOI: 10.18562/IJEE.2015.0009

Patino, C. M., & Ferreira, J. C. (2018). Inclusion and exclusion criteria in research studies: Definitions and why they matter. *Jornal Brasileiro de Pneumologia : Publicaçao Oficial da Sociedade Brasileira de Pneumologia e Tisilogia*, *44*, 84–84. DOI: 10.1590/S1806-37562018000000088 PMID: 29791550

Patton, M. Q. (2002). *Qualitative evaluation and research methods*. Sage. Retrieved on 10 October 2025 from https://psycnet.apa.org/record/1990-97369-000

Polit, D. F., & Beck, C. T. (2017). *Nursing research: Generating and assessing evidence for nursing practice* (10th ed.). Wolters Kluwer Health.

Quinney, L., Dwyer, T., & Chapman, Y. (2016). Who, where, and how of interviewing peers: Implications for a phenomenological study. *SAGE Open*, *6*(3), 2158244016659688. DOI: 10.1177/2158244016659688

Rabiee, F. (2004). Focus-group interview and data analysis. *The Proceedings of the Nutrition Society*, *63*(4), 655–660. DOI: 10.1079/pns2004399 PMID: 15831139

Ravn, S. (2023). Integrating qualitative research methodologies and phenomenology—Using dancers' and athletes' experiences for phenomenological analysis. *Phenomenology and the Cognitive Sciences*, *22*(1), 107–127. DOI: 10.1007/s11097-021-09735-0

Resnik, D. B. (2020). *What is ethics in research and why is it important* (Vol. 1). December. https://www.niehs.nih.gov/research/resources/bioethics/whatis/index.cfm

Ritchie, J., Lewis, J., Nicholls, C. M., & Ormston, R. (2013). Qualitative research practice: A guide for social science students and researchers. *Sage (Atlanta, Ga.)*.

Robinson, O. C. (2023). Probing in qualitative research interviews: Theory and practice. *Qualitative Research in Psychology*, *20*(3), 382–397. DOI: 10.1080/14780887.2023.2238625

Sarfo, J. O., Debrah, T., Gbordzoe, N. I., Afful, W. T., & Obeng, P. (2021). Qualitative research designs, sample size and saturation: Is enough always enough. *Journal of Advocacy. Research in Education*, *8*(3), 60–65. Retrieved October 10, 2025, from https://www.academia.edu/download/101557437/jare.2021.3.pdf

Shoko, J., Muboko, N., & Gandiwa, E. (2024). *Focus group discussion method as a tool for eliciting local community views for decision-making: A case study of the Save Valley Conservancy in the Southeast Lowveld of Zimbabwe*. Retrieved on 10 October 2025 from https://www.preprints.org/frontend/manuscript/c7945ac3f5ccefd910faef27424c5943/download_pub

Sloan, A., & Bowe, B. (2014). Phenomenology and hermeneutic phenomenology: The philosophy, the methodologies, and using hermeneutic phenomenology to investigate lecturers' experiences of curriculum design. *Quality & Quantity*, *48*(3), 1291–1303. DOI: 10.1007/s11135-013-9835-3

Smith, D. W. (2018). Phenomenology. In *The Stanford encyclopedia of philosophy*. Metaphysics Research Lab, Stanford University. Retrieved on 10 October 2025 from https://plato.stanford.edu/archives/sum2018/entries/phenomenology

Smith, J. A. (2011). Evaluating the contribution of interpretive phenomenological analysis. *Health Psychology Review*, *5*, 9–27.

Smith, J. A., Flowers, P., & Larkin, M. (2009). Interpretive phenomenological analysis: Theory, method and research. *Sage (Atlanta, Ga.)*.

Squires, V. (2023). Reflexive bracketing. In Okoko, J. M., Tunison, S., & Walker, K. D. (Eds.), *Varieties of qualitative research methods* (pp. 425–430). Springer International Publishing., DOI: 10.1007/978-3-031-04394-9_66

Stahl, N. A., & King, J. R. (2020). Expanding approaches for research: Understanding and using trustworthiness in qualitative research. *Journal of Developmental Education*, *44*(1), 26–28. Retrieved October 10, 2025, from https://files.eric.ed.gov/fulltext/EJ1320570.pdf

Steffen, E. (2016). Ethical considerations in qualitative research. *Analysing Qualitative Data in Psychology*, *2*, 31–44.

Stenfors, T., Kajamaa, A., & Bennett, D. (2020). How to assess the quality of qualitative research. *The Clinical Teacher*, *17*(6), 596–599. DOI: 10.1111/tct.13242 PMID: 32790137

Sundler, A. J., Lindberg, E., Nilsson, C., & Palmér, L. (2019). Qualitative thematic analysis based on descriptive phenomenology. *Nursing Open*, *6*(3), 733–739. DOI: 10.1002/nop2.275 PMID: 31367394

Tavakol, M., & Sandars, J. (2025). Twelve tips for using phenomenology as a qualitative research approach in health professions education. *Medical Teacher*, •••, 1–6. DOI: 10.1080/0142159X.2025.2478871 PMID: 40089895

Thomas, S. P., & Sohn, B. K. (2023). From Uncomfortable squirm to self-discovery: A phenomenological analysis of the bracketing experience. *International Journal of Qualitative Methods*, *22*, 16094069231191635. DOI: 10.1177/16094069231191635

Togia, A., & Malliari, A. (2017). Research methods in library and information science. In Oflazoglu, S. (Ed.), *Qualitative versus quantitative research*. InTech., DOI: 10.5772/intechopen.68749

Tomaszewski, L. E., Zarestky, J., & Gonzalez, E. (2020). Planning qualitative research: Design and decision making for new researchers. *International Journal of Qualitative Methods*, *19*, 160940692096717. DOI: 10.1177/1609406920967174

Trent, A., & Cho, J. (2020). *Interpretation in qualitative research: What, why, how*. Retrieved on 10 October 2025 from https://academic.oup.com/edited-volume/34283/chapter-abstract/290650730

Urcia, I. A. (2021). Comparisons of adaptations in grounded theory and phenomenology: Selecting the specific qualitative research methodology. *International Journal of Qualitative Methods*, *20*, 16094069211045474. DOI: 10.1177/16094069211045474

Wang, X., Chen, L., & Li, Z. (2023). Information phenomenology: An informational interpretation of Husserl's phenomenology. *Computer Sciences & Mathematics Forum*, *8*(1), 86. https://www.mdpi.com/2813-0324/8/1/86

Westland, H., Vervoort, S., Kars, M., & Jaarsma, T. (2025). Interviewing people on sensitive topics: Challenges and strategies. *European Journal of Cardiovascular Nursing*, *24*(3), 488–493. https://academic.oup.com/eurjcn/article-abstract/24/3/488/7761136 PMID: 39295513

Whitney, C., & Evered, J. A. (2022). The qualitative research distress protocol: A participant-centered tool for navigating distress during data collection. *International Journal of Qualitative Methods*, *21*, 16094069221110317. DOI: 10.1177/16094069221110317

Williamson, K., & Johanson, G. (2018). *Research methods: Information, systems, and contexts*. Chandos Publishing.

Willig, C. (2017). *Interpretation in qualitative research. The SAGE Handbook of Qualitative Research in Psychology, 2, 274–288*. Sage.

Yaakop, N. (2023). A content validation of focus group discussions based on need analysis in a physical education training module for primary school teachers. *Retos: Nuevas Tendencias En Educación Física. Deporte y Recreación*, *50*, 1115–1122. DOI: 10.1177/0739456X17723971

Chapter 9
Theoretical and Conceptual Frameworks and Their Role in Shaping Research in Context

Patrick Ngulube
http://orcid.org/0000-0002-7676-3931
University of South Africa, South Africa

ABSTRACT

Theoretical and conceptual frameworks in computing and informatics offer organised methods for analysing and interpreting phenomena. They help researchers understand complex systems and relationships. In intricate, technologically advanced settings, these frameworks guide the development of research questions, influence methodological decisions, and shape the interpretation of results and findings. This chapter examines the differences and connections between theoretical and conceptual frameworks, focusing on how they support innovation while firmly establishing research. Frameworks offer coherence, support methodological rigour, and align research goals with results and findings. By combining theory and practice, researchers can produce work that is both socially meaningful and scientifically robust. By integrating these elements, they can produce research that is not only scientifically rigorous but also relevant to societal needs. This chapter highlights the role of frameworks in enhancing scholarly inquiry and contributing to the evolving digital landscape.

DOI: 10.4018/979-8-3693-4018-9.ch009

INTRODUCTION

Systematic investigation is the lifeblood of rigorous research and academic studies. The disciplines of computing and informatics, which include data science, software engineering, human–computer interaction (HCI), computer science, information systems, and related sciences, have developed into demanding scientific domains that mostly depend on theoretical and conceptual advancement. Research frameworks, including conceptual and theoretical frameworks, provide a structured and coherent approach to the research process (Kivunja, 2018; Luft et al., 2022; Salawu et al., 2023). Collectively, they influence research direction, guide methodology, and aid in result interpretation, all of which contribute to the study's increased rigour and clarity. Additionally, they clearly outline the reasons for conducting the research and connect it to existing knowledge.

Theoretical and conceptual frameworks are identified by critically analyzing the literature, identifying gaps or unresolved issues, and situating the research efforts within the framework (Ngulube et al., 2015; Oyewobi et al., 2024, Paperpal. 2024). This implies that research in many disciplines, including computing and informatics, should be supported intellectually by theoretical and conceptual frameworks (Kivunja, 2018; Salawu, Shamsuddin & Masibo, 2023; Ngulube, 2020; Ngulube et al., 2015; Oyewobi et al., 2024, Paperpal. 2024). Using a theory is a mark of research seriousness and the maturity of a cognitive discipline (Ukwoma & Ngulube, 2021). Without a framework, research can become disorganised and lack a solid theoretical basis (Adom et al., 2018; Ngulube, 2018; Ngulube & Mosha, 2023). According to Corley and Gioia (2011), theory is the currency of the scholarly world and research that is not based on theory "is simply description" (Van House, 1991, p. 87). If the theoretical component of the work is not sufficiently developed, neither scientific nor practical results may be anticipated (Lewin, 1945). Therefore, theories are the cornerstone for generating research problems, hypotheses and meanings (Nachmias & Nachmias, 1996). That implies that theory is both the ultimate destination and the starting point of research (Jeong & Kim, 2005). Moreover, theories offer explanations, predictions, and design principles for technological phenomena, which require a clear conceptual grounding (Gregor, 2006).

While a conceptual framework (CF) maps links between variables and constructs relevant to the study, a theoretical framework (TF) uses well-established theories to explain phenomena (Kivunja, 2018; Luft et al., 2022; Van der Waldt, 2020). Some authors have gone so far as to state that once a theoretical framework is adopted for a research study, it is implied that all the components of the chosen theory will be used, otherwise, using aspects of a selected theory reduces this building block for research to a conceptual framework (Green, 2014; Ngulube, 2020; Ngulube et al., 2015). This issue may seem straightforward, but it is more complex due to the contestations and variations surrounding the CF and TF (Crawford, 2019; Ngulube et al., 2015). Many researchers and students are confused by the differences between the two terms across various academic traditions (Chukwuere, 2021; Crawford, 2019). This chapter contributes to the discourse by adding another layer of explanation. That is achieved through the following objectives:

- Explaining the nature and purpose of theoretical and conceptual frameworks.
- Demonstrating their role in shaping computing and informatics research.
- Providing guidance on developing effective frameworks.
- Exploring challenges and future directions in framework application.

Despite debates surrounding the CF and TF, theory and research methodology are fundamental to research grounded in the scientific method. Research methodology assists in developing and testing a theory. Quantitative research methods are often used to test theories, while qualitative research methods are often used to build and improve them. Mixed methods research can facilitate both theory testing and theory development by combining quantitative and qualitative methodologies within a singular study. Many research methods can help create middle-range theories (Merton, 1967), especially when they aim to link real-world observations with ideas, like combining qualitative insights with quantitative data to improve our understanding of theories.. Middle-range theories are central to empirical studies. Theories unify the various aspects of empirical research. In that regard, theories explain the research variables, develop the research approach and answer the research questions. It is extremely challenging to understand and explain social reality and research variables without the aid of these research building blocks.

Theoretical and conceptual frameworks matter in research because they:

- provide structure and clarity (Anfara & Mertz, 2015; Creswell & Creswell, 2018);
- support coherence and internal consistency in a study (Ravitch & Riggan, 2017);
- justify methodological choices (Creswell & Plano Clark, 2018; Ngulube et al., 2015);
- help identify and define key variables or constructs (Grant & Osanloo, 2014; Miles, et al., 2014);
- shape data collection and analysis (Charmaz, 2014; Corbin & Strauss, 2015);
- enhance the study's credibility, rigour, and scholarly value (Booth et al., 2024; Lincoln & Guba, 1985);
- connect the study to existing knowledge (Marshall & Rossman, 2016; Merriam & Tisdell, 2016; Ngulube et al., 2015);

- support theory building or theory testing (Glaser & Strauss, 1967; Ngulube, 2025; Yin, 2018).

Theories form the basis of developing theoretical and conceptual frameworks that underpin scientific and empirical research. A theoretical framework and a conceptual framework are different, although some schools of thought assume that they are similar (Ngulube, 2020; 2018). A piece of research can be supported by a conceptual framework derived from a theory or set of theories and concepts. On the other hand, another study can be grounded in a theoretical framework. Researchers need to understand the use and application of theory for them to make an explicit distinction between the two research frameworks. That implies that choosing a good theory that may advance knowledge in computing and informatics, for example, is based on a sound understanding of its use and application. Choosing a good theory is also dependent on a thorough review of scholarship in a cognitive discipline. Equally important is the need to distinguish between models, meta-theories and theories when using them in empirical research, as discussed in the next section.

The chapter begins by clarifying concepts and defining theoretical and conceptual frameworks, then analyses their roles in computing research. It then examines frameworks as tools for innovation, outlines steps for developing them, and concludes with emerging trends and ethical considerations. The chapter overview is illustrated in Figure 1. Note that AI in Figure 1 stands for artificial intelligence, and IoT stands for Internet of Things.

Figure 1. Overview of theoretical and conceptual frameworks in computing and informatics research

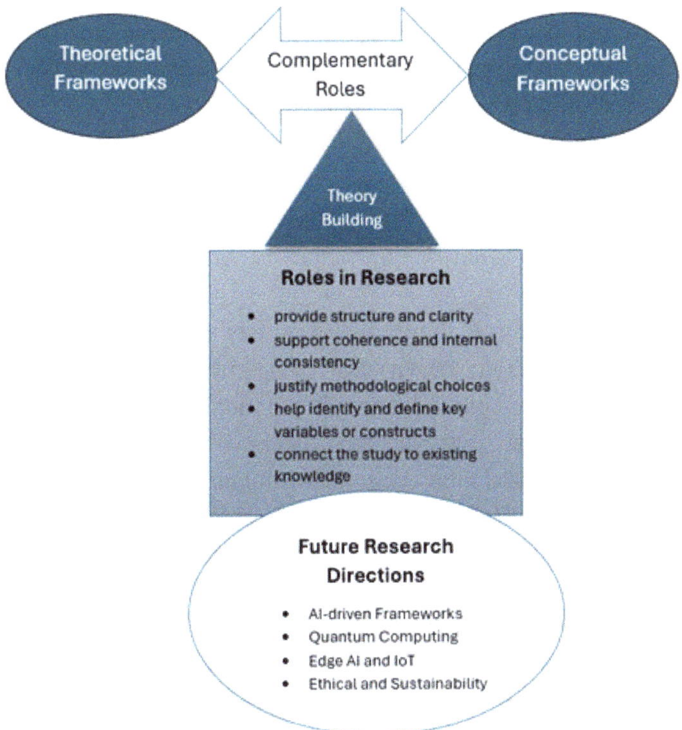

Source: Author's design

UNDERSTANDING THEORY, META-THEORIES AND MODELS

It is human nature to develop theories about reality (Leedy & Ormrod, 2015). Often, human beings explain reality using informal theories. Many informal theories are based on beliefs and subjective impressions (Leedy & Ormrod, 2015). The accuracy of the theories can be ascertained through research. Theories tend to generate further research by generating new research questions and by "suggesting hypotheses about the likely outcomes of particular investigations" (Leedy & Ormrod, 2015, p. 40). Following Matthews (2025), we define a theory as "basically a scholarly way of talking about our understanding in relation to ideas and evidence" (p. 82). This underscores the supremacy of agency in the use of theories.

Theories must be used contextually. In contrast to universal statements, theories are social constructions, and researchers should avoid the trap of universalising and decontextualising them (Ngulube, 2025). Context-specific frameworks are essential in computing and informatics research, as they analyse software and information work within organisational and social contexts, emphasising the importance of contextual design and evaluation (Oates, 2006). Context is important because theories emerge from "historical and cultural processes and human interactions" (Matthews, 2025, p. 87). Prus (1996) offers an excellent perspective on this:

Like all realms of science and knowledge, theory in the social sciences is very much a social product and a social process. [...] ideas about how group life takes place, involves some creativity and initiative on the part of particular thinkers, but is fundamentally rooted in the social, interactive contexts in which these people find themselves (p. 33).

This implies that, like social life, a theory has elements of intersubjectivity, multiperspectivity, reflection, activity-basedness, negotiability, relationality, and processuality (Prus, 1996). This understanding of theory as a social construction will help researchers avoid pitfalls that can lead to theory dropping, theory diversification, theoretical triangulation, and theoretical determinism (Ngulube, 2020; Ukwoma & Ngulube, 2021), thereby improving their ability to deliver high-quality research.

On the other hand, the way researchers interpret data should not be dictated by theories. The usefulness of theories is undermined when theories determine how researchers interpret data. When the understanding of a phenomenon is altered to fit a theory, the result is theoretical determinism (Matthews, 2025). Creativity and innovation in exploring, understanding, and explaining a phenomenon are equally diminished when everything is determined by theory rather than by the data and context. This undermines the trustworthiness, accuracy, utility, and validity of research. Put differently, "theoretical determinism reduces our understanding to ridiculousness, and if we frame research like this, we are not doing good science" (Matthews, 2025, p. 16). A helpful analogy for theoretical determinism is the idea of the tail wagging the dog. This analogy illustrates the determination to make empiricism subservient to theory and not the other way round. Concerns about theoretical determinism led to the development of grounded theory by the Chicago School (Glaser & Strauss, 1967) and to the call by scholars such as Blumer (1969) for "a return to the empirical human world" (p. 34). Theory building and theorisation discussed in the next section is a good principle in research that can lead researchers to critically think about the tools they use to understand and explain the world.

Models and meta-theories need to be conceptually clarified, as they are often misunderstood in discussions of theory and theoretical frameworks (Ngulube, 2020; Ngulube et al., 2015; Ngulube & Mosha, 2023). In academic research, particularly in computing and informatics, understanding the distinction between theories and models is essential for designing coherent studies, making strong contributions, and ensuring conceptual clarity. Theories provide explanations and predictions, while models represent and operationalise concepts. Together, they establish a critical framework for scientific inquiry and knowledge development. A model is a representation of reality (De Vos et al., 2011). Following Whetten's (1989) analogy, a simpler way to differentiate between a model and a theory is to perceive a theory as a story about why and view a model as a visual aid that supports storytellers in highlighting the main components of their explanations. A model is not equivalent to a theory. Thus, a model is a structured representation derived from or associated with a theory but does not usually provide a full causal explanation on its own. In other words, theories provide generalised explanations that apply across multiple contexts (Gregor, 2006; Whetten, 1989), whereas models represent specific components or relationships within a theory, often used to simplify or visualise complex ideas (Blalock, 1969). The purpose and function of theories and models differ (Dubin, 1978; Weber, 2012). Theories answer research questions such as:

- Why does this occur?
- How are these elements related to each other??

On the other hand, models typically answer questions like:

- What components are involved?
- How do they interact within defined boundaries?

Furthermore, theories possess explanatory depth as they describe underlying mechanisms and justify why relationships exist. Models possess representational clarity as they often illustrate constructs, variables, components, pathways, or systems. For example, the Technology Acceptance Theory explains why users adopt systems (Davis, 1989) while the Technology Acceptance Model (TAM) visualises relationships such as perceived usefulness and intention to use. Models are often derived from, tested with, or used to communicate theories because a theory guides inquiry and interpretation while a model structures it (Grand et al., 2024). Ultimately, theories often require models to make them empirically testable (Miles et al., 2014). The argument is that theories provide a conceptual foundation, while models offer analytical or representational tools.

Understanding theory also implies distinguishing it from metatheories. Theories and metatheories play different yet complementary roles in research. Although they are connected, theory and metatheory serve different intellectual functions and operate at different levels of abstraction. The basis for rigorous theorisation, conceptual clarity, and efficient knowledge development is an understanding of their distinctions. Metatheory is defined as "theorizing about theory" or theories about theories (Hamilton, 2017, p. 136). As “theories about theories”, they examine the assumptions, structures, and philosophical foundations upon which theories are built (Ritzer, 1991; Reynolds, 2007). As "theories about theories," they study the philosophical underpinnings, processes, and presumptions that underpin theories (Ritzer, 1991; Reynolds, 2007). While theories describe actual phenomena, metatheories describe how theories ought to be developed, assessed, or comprehended. Therefore, metatheories offer the conceptual framework that defines what constitutes valid knowledge, valid explanation, and legitimate evidence. A theory explains phenomena, while a metatheory describes how theories function and how they should be developed. While theories address empirical questions, metatheories tackle philosophical matters. Moreover, theories focus on specific phenomena (Turner, 1991), while metatheories examine the workings of science rather than the operation of specific phenomena.

Despite its concern with scientific conceptual processes, a metatheory is not a theory, as Henning et al. (2004) and Netshakhuma (2024) suggest. For instance, paradigms as meta-theories do not explain anything, although they offer a lens of understanding how theories are developed. Equally, grounded theory is not a theory in the conventional sense as it is a systematic methodology primarily used in qualitative research that aims to develop a theory from data collected (Ncube & Ngulube, 2025).

THEORY BUILDING AND THEORIZING IN SCHOLARSHIP

The process of developing explanations for observed events, often formalised through models, is known as theory building (Saunders et al., 2015). In order to organise empirical observations and transform them into theories, conceptual and theoretical frameworks are essential in computing and informatics. These disciplines examine the interactions between people and technology in organisational contexts. The research process develops theories (Dubin, 1969; Holton & Lowe, 2007; Lynham, 2002). The process of theorising is a neglected area in many fields (Kawulich, 2009; Schneider, 2006; Ngulube, 2020). This has led to limited theory-building in many fields, including computing and informatics. Consequently, the description of the nature of theory has not advanced very much. Theorising assists in understanding the "systematic reasons for a particular occurrence or non-occurrence" (Sutton & Staw, 1995, p. 378).

Theory building is not the exclusive domain of certain academics and researchers. The work of Glaser and Strauss (1967) was crucial in developing the "liberating thesis that anyone can create their own theory" if it was based on empirical research (Alvesson & Skoldberg, 2000, p. 16). Theory building can be defined as the "process of knowing, perceiving and conceiving relationships, derived from systematic inquiry, which can transcend paradigms if a comprehensive and consistent 'worldview' is developed" (Kettley, 2010, p.10). Different researchers may theorise, organise, conceptualise, and validate their knowledge in different ways. This implies that it is challenging to develop universal theories that explain all contexts. As computing and informatics researchers theorise, they should explain their reality in relation to their context and experiences.

Rather than focusing on finding a distinguishable "theoretical core" as once suggested for the information systems field (Gray, 2003), for instance, computing and informatics scholarship should concentrate on theorising and developing theories to explain their various realities. Theorising should be done with consideration of the historical, social, and moral context of the researcher or observer, as advised by Dilthey (1883/1989). There are pitfalls of theorising without considering "broader moral, social, or historical considerations" (Horkheimer, 1937/2002, pp. 134).

The three steps described by Schneider (2006) offer an instructive framework for theorising and theory building. The process of theorising starts with developing a problem and reading diverse literature that defines the theory (Schneider, 2006; Sutton & Staw, 1995), and drawing on the context of the phenomenon under study and the researcher's experiences (Ngulube, 2020). The theorising process does not generally produce complete theories (Hassan, Lowry & Mathiassen, 2022; Weick, 1995), although it creates a theory to explain the problem.

Theorising is "making certain claims in the form of statements that reflexively apply specific rules of formation to constitute a discourse within a field" (Hassan, Lowry & Mathiassen, 2022, p. 8). Initial theorising can be the building block for further theory development, but the scholar must be transparent about it. Theory development may start "with guesses and speculations and ends with explanations and models" (Weick, 1995, p. 386). Theories can start as concepts that are developed into "propositional argument" (Weick, 1995, p. 386). Theory development should be regarded as a process rather than a product. According to Merton (1967), theorisation results in an approximate theory. Theorising is a continuum rather than a dichotomy (Weick, 1995, p. 386). Indeed, "The process of theorizing consists of activities like abstracting, generalizing, relating, selecting, explaining, synthesizing, and idealizing" (Weick, 1995, p. 389). Computing and informatics scholarship can form different concepts and make claims about its contexts.

Figure 2. Empirical cycle of developing theory

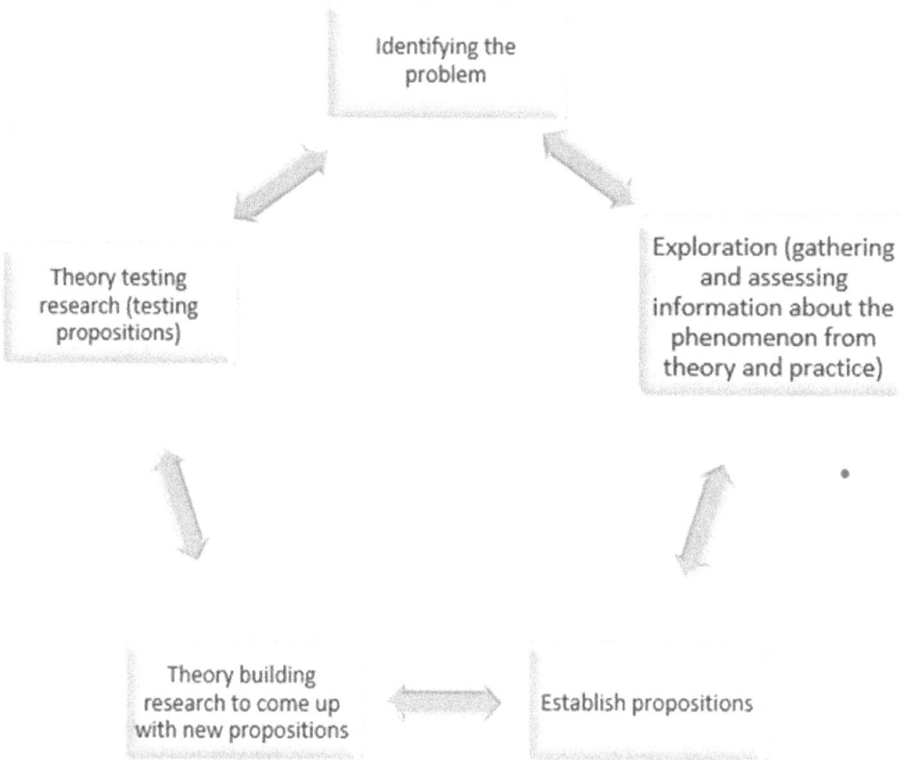

Source: Eisend and Kuss (2019).

Kettley (2010) expanded this cycle (i.e., Figure 2) into 12 steps, in line with the Cambridge School's synthetic and transformative theory-building approach. After problem identification and theory creation, theory evaluation follows. The theory should be tested for robustness, checked for its substantive implications and tested for logical implications against the competing theories (Schneider, 2006). Frameworks help align abstraction, interpretation, and observation, ensuring theory development is methodical rather than haphazard.

UNDERSTANDING THEORETICAL AND CONCEPTUAL FRAMEWORK

Understanding theoretical and conceptual frameworks entails defining them and showing their distinctions and interconnections. It is very common for researchers to confuse the theoretical and conceptual frameworks, or to use them incorrectly (Kivunja, 2018; Luft et al., 2022; Salawu et al., 2023; Ngulube, 2020; Ngulube et al., 2015; Oyewobi et al., 2024). While the terms are often used interchangeably, they are not synonyms. The confusion between these research building blocks stems from their complementary nature and frequent overlap, as illustrated in Figure 3.

A theoretical framework is built on established, well-researched theories that have been tested and validated over time (Anfara & Mertz, 2015; Miles et al., 2024; Whetten, 1989). Its other key features are that it provides explanations, predictions, and causal logic, shapes hypotheses or analytic expectations and positions the study in broader theoretical debates. It serves as a broad, general foundation for a study and is often derived directly from existing literature. You can think of it as utilising a well-recognised theory, such as the Theory of Planned Behaviour, to underpin your entire research.

However, there is some debate over whether, when aspects of a theory are used in research, one can still call it a theoretical framework. Green (2024) and Ngulube (2020) disagree. Their argument is that when aspects of a theory are used, it is appropriate to call it a conceptual framework, since only some of its concepts will inform the study. A good analogy for how theoretical frameworks work can be taken from a tree. If the researcher uses all the components of a tree, they will have a theoretical framework. However, using the branch of a tree or its trunk or bark, for example, cannot be considered as applying all the parts of a tree. In that case, the tree ceases to be the theory as some components are deployed as the theoretical foundation of the study. Therefore, the proper naming of the concepts used should be conceptual rather than theoretical. This is a clear summary of the situation, noting that "a conceptual framework consists of at least three essential dimensions: the researcher's personal interest, relevant topical research (derived from an extensive literature review), and a theory or components of a theory" (Van der Waldt, 2020, p. 3). The parts of a tree in our analogy are the theory or part of the theory, the literature review, and the researcher's experience, which together constitute the conceptual framework, thereby obviating the need to call it a theoretical framework.

This argument begs the question: should research in the human and behavioural sciences be informed by a theoretical or conceptual framework, given that context-relevant theories are rare? A provisional assessment indicates that research in this domain should be grounded in a conceptual framework because there is an element of incorporating some aspects of a theory, the literature, and the personal experience of the researcher. Theories that cover all the constructs in a research project's objectives are rare, making it imperative to look elsewhere to address those not covered by the selected theory. In this way, the researcher will avoid the trap of theoretical determinism discussed in the preceding sections. If a conceptual framework incorporates a theory or some aspect of a theory, is it necessary to have a theoretical and conceptual framework in one study? Instead, are the theoretical foundations of the study supposed to be highlighted as informing the conceptual framework? My answer will be that in the social and behavioural sciences, having one suffices. Diagrams illustrating the theories that underpin the study are unnecessary, as they fail to demonstrate a connection with all the constructs in the research. Diagrams illustrating theories that support a study should be avoided unless all constructs outlined in the study's objectives are included in those theories.

Theoretical frameworks have the following advantages and disadvantages (Kivunja, 2018; Ludviga, 2023; Ngulube et al., 2015; Pretorius, 2026):

- Placing the research within a broader context establishes a foundation for the study.
- By linking the research problem, goals, and hypotheses to established theories or models, the framework ensures that the study meets academic standards.
- By clarifying which established theories guide the research, theoretical frameworks reduce ambiguity for novice researchers and help them define their theoretical perspective early in the research process.
- By focusing on the constructs in the theory, it may restrict the creativity and flexibility of the researcher.
- Constraint risk emanating from overly rigid frameworks may limit creative thinking.
- Confirmation bias arising from the frameworks can lead to seeing only expected patterns.
- Paradigm lock-in coming from dominant frameworks may suppress alternative perspectives.
- Complexity reduction caused by simplification may omit important nuances.
- Applicability boundaries may prevent frameworks developed in one context from being transferable to another.

The best practice is to use frameworks as guides, not straitjackets; remain open to emergent findings.

On the other hand, a conceptual framework is developed specifically for your study. It reflects the researcher's understanding of how key concepts are interconnected (Kulesa et al., 2024; Miles et al., 2014; Ngulube et al., 2015; Ravitch & Riggan, 2017). As explained in the previous paragraphs, this framework can synthesise different theories or ideas from the literature and the researcher's own experience and is often represented by a diagram that illustrates the proposed relationships between variables (Ngulube, 2020; Van der Waldt, 2020). In summary, one can utilise a theoretical framework while simultaneously creating a conceptual framework. In a study, the theoretical foundation should precede the conceptual framework because it helps the reader to see the theories reflected in it.

The literature strongly recommends presenting the conceptual framework graphically (Jacard & Jacoby, 2010; Miles et al., 2014; Ngulube et al., 2015; Ravitch & Riggan, 2017; Van der Waldt, 2020). Conceptual frameworks differ from theoretical frameworks in that they are not limited to a singular established theory; instead, they may combine concepts from various disciplines or sources. They are especially helpful in studies where theory is new, limited, or crosses fields, like research on the Internet of Things (IoT) and cloud computing adoption. In such cases, researchers must be transparent in justifying their use of a conceptual framework by explaining why it was chosen, how it aligns with the research aims, the judgements made in its development, and its strengths and limitations. This transparency enhances the rigour and credibility of the research (Ngulube & Mosha, 2023)."

Equally, conceptual frameworks have advantages and disadvantages as outlined below (Adom et al., 2018; Crawford, 2019; Kulesa et al., 2024; Ludviga, 2023; Master Academia, 2026; Ngulube et al., 2015; Oxbridge Essays, 2026; Pretorius, 2026):

- They provide flexibility and adaptability by enabling researchers to choose and integrate concepts from various theories, tailoring a model to their specific study context. This flexibility renders them valuable in exploratory or interdisciplinary research.
- By creating a visual or narrative map that illustrates the relationships between key variables, constructs, or concepts related to the research problem, a logical structure for the study is provided. This approach clarifies the study's direction and purpose.
- Synthesising literature and concepts, conceptual frameworks reveal gaps in existing research and help formulate research questions.
- Conceptualisation activities assist novice researchers in refining their topics and alleviating early research stress by clarifying which ideas and relationships are significant.
- Conceptual frameworks are defined in various ways. Some authors use the terms interchangeably with theoretical frameworks, which leads to confusion and inconsistent application.
- Conceptual frameworks often incorporate ideas from diverse sources, which can lead to weak theoretical foundations and diminish the academic rigour of the study if not constructed properly.
- Combining multiple theories or constructs can become complicated, making it challenging for novice researchers to defend or explain the framework.

Many novice researchers often confuse the terms “conceptual framework” and “theoretical framework,” using them interchangeably (Kivunja, 2018; Salawu et al., 2023; Ngulube et al., 2015; Oyewobi et al., 2024). However, established scholarship highlights important distinctions and meaningful connections between the two (Kivunja, 2018; Ngulube, 2018; Salawu et al., 2023; Varpio et al., 2020). A possible source of confusion is that these research tools are complementary: theory informs the conceptual map, and the conceptual map translates theory into measurable variables and analytic pathways (Ngulube et al., 2015; Miles et al., 2014). Understanding these differences and relationships is crucial for researchers in selecting the most suitable framework for their projects, ensuring methodological coherence. Figure 3 illustrates the relationship between a conceptual and theoretical framework.

Theoretical framework Conceptual framework

Figure 3. Comparative Illustration of theoretical vs conceptual frameworks

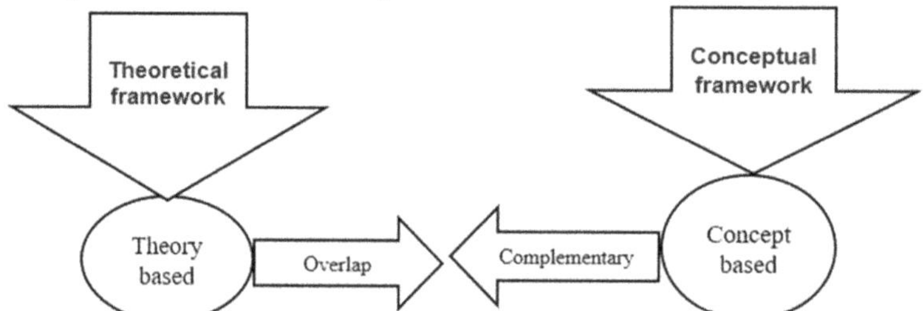

Source	Existing theories: based on a single established theory (or a set of closely related theories) that guides the research	Researcher-developed concepts: researcher's own construction of ideas
Purpose	Explain phenomena: offers a lens for viewing the problem and grounding hypotheses	Illustrate relationships: outlines a roadmap that details the variables and their relationships
Characteristic	Less flexible (rigid): constrained by the assumptions and propositions of the chosen theory	Highly adaptable: combines multiple theories, allowing the inclusion of variables based on the research context, and is adaptable to practical situations and emerging findings
Applicability	Often employed in hypothesis-driven research and studies that aim to test or expand upon a theory	Commonplace in applied research, exploratory studies, and interdisciplinary work
Examples in computing	• Social informatics theory • Complexity theory • Distributed systems theory	• Cloud computing adoption • Artificial intelligence ethics • IoT security

Sources: Crawford (2019); Green (2014); Kivunja (2018); Salawu et al. (2023); Ngulube (2020); Ngulube et al. (2015); Oyewobi et al. (2024; and Paperpal (2024)

Although theoretical and conceptual frameworks differ in origin and structure, they are deeply interconnected in several ways as outlined below (Adom et al., 2018;

Crawford, 2019; Ludviga, 2023; Master Academia, 2026; Ngulube et al., 2015; Oxbridge Essays, 2026; Pretorius, 2026).

- The two frameworks serve as essential guides for research design, data collection, and analytical methods. Each framework offers a vision for the research, allowing the investigator to analyse and interpret data in an organised way.
- Researchers need to actively engage with the existing literature to effectively use either framework. It is common for researchers to indirectly incorporate theoretical concepts into a conceptual framework by drawing from established theories. To ensure that theoretical frameworks remain relevant and applicable in modern contexts, they must also be integrated into the broader body of literature.
- By identifying essential concepts or variables and explaining how they are expected to interact, theoretical frameworks provide a structured approach that aids in the development of research questions and analytical pathways. This shared function enhances the overall coherence of the research design.

Researchers should avoid using two different frameworks simultaneously unless they address distinct aspects of the study (Ngulube et al., 2015; Oxbridge Essays, 2026; Pretorius, 2026).

ROLE OF FRAMEWORKS IN RESEARCH DESIGN

Theoretical and conceptual frameworks play a vital role in organising and guiding scholarly research. They provide an intellectual foundation for studies, shaping methodological choices, clarifying the researcher's perspective, and ensuring that the problem, objectives, and study design are coherent (Adom et al., 2018; Crawford, 2019; Ngulube et al., 2015; Pretorius, 2026). Despite their different origins and structures, both frameworks are essential for establishing rigour, clarity, and scholarly authority in research. In computing, they help prevent inconsistencies across socio-technical layers and clarify the analysis's logic. The framework selected for a study provides a foundation for robust and credible research, whether it is based on existing theories or combines different concepts into a tailored model (Kivunja, 2018; Ngulube, 2018).

The theoretical framework serves as a lens through which researchers analyse phenomena and situate their studies within established theories (Adom et al., 2018; Chakabwata & McKay, 2026; Master Academia, 2026; Ngulube et al., 2015; Pretorius, 2026). It relies on widely recognised and well-tested concepts that have been evaluated over time, enabling researchers to apply broad principles to specific study contexts (William, 2024). This foundation ensures that the conclusions drawn from the study are supported by academic precedents and align with accepted knowledge in the field. By utilising existing theoretical concepts to analyse data and shape the research design, theoretical frameworks guide the exploration of a research problem (Adom et al., 2018; Master Academia, 2026; Ngulube et al., 2015; Pretorius, 2026). Furthermore, they promote analytical clarity by helping researchers identify relevant variables, relationships, and assumptions, as they are rooted in established bodies of knowledge. However, theoretical frameworks may limit flexibility due to their reliance on established theory, particularly in developing sectors where current theories may not adequately reflect the complexity of current concerns (Varpio et al., 2020).

On the other hand, a conceptual framework offers researchers greater flexibility by allowing them to integrate ideas from various theories, empirical data, and contextual insights (Kulesa et al., 2024; Milles et al., 2014; Ravitch & Riggan, 2017). Because conceptual frameworks are constructed rather than simply adopted, researchers can tailor a model to fit the specific circumstances of their study. According to Crawford (2019), conceptual frameworks provide a general direction for the research, making it easier for both the reader and the researcher to understand how the study contributes to scholarly knowledge and how its components relate to one another. Conceptual frameworks are especially beneficial in interdisciplinary or exploratory research, where no single theory can fully explain a phenomenon, as they synthesise diverse ideas (Chakabwata & McKay, 2026; Ngulube & Mosha, 2023). Pretorius (2026) emphasises that by cohesively linking disparate concepts, conceptual frameworks clarify the relationships between key ideas and guide the interpretation of data.

Theoretical and conceptual frameworks are significantly related, despite their differences. Both rely on academic literature, and to achieve the study's goals, the researcher must synthesise the existing body of information. Engaging with literature—whether to identify ideas or extract key concepts—ensures that the research is grounded in current scholarly discussions (Chakabwata & McKay, 2026). This engagement also helps identify gaps that the study aims to address, as highlighted by Adom et al. (2018) and Oxbridge Essays (2026). Furthermore, both frameworks are essential for formulating research questions, hypotheses, and methodological decisions (Chakabwata & McKay, 2026; William, 2024). They establish the study's parameters and guide the analytical framework used to interpret the results by describing the key concepts and relationships vital to the investigation.

FRAMEWORKS IN COMPUTING AND INFORMATICS

Informatics and information systems prioritise socio-technical, contextual, and interpretive frameworks, whereas computing research primarily employs formal, mathematical, and experimental frameworks (Yatsko, 2018). Computing and informatics research is interdisciplinary and rapidly evolving, requiring frameworks that integrate technical performance with human, organisational, and ethical dimensions (Oates, 2006). As multidisciplinary fields that combine technological, human, and organisational aspects, informatics and computing require frameworks that make sense of the connections between ideas, guide methodological choices, and ensure that research is grounded in accepted knowledge. Thus, conceptual and theoretical frameworks dictate the intellectual architecture and significance of research rather than just providing support for it.

Informatics and information systems (IS) foreground context, human and organisational concerns, and plural theoretical sources, leading to socio-technical and interpretive frameworks (Hassan et al., 2022). Theoretical frameworks such as information theory, systems theory, and diffusion of innovations offer robust logics, while conceptual models adapt these logics to address contemporary issues such as data ethics and algorithmic bias (Moschis, 2024; Chukwuere, 2021). As computing and informatics are unique interdisciplinary fields, their rapid technological evolution and socio-technical complexity require flexible yet rigorous frameworks (Oates, 2006). Information theory, systems theory, cognitive load theory, and diffusion of innovations are frequently used to structure computing inquiries (Moschis, 2024). Human-Computer Interaction (HCI) models linking usability and accessibility, data ethics maps relating privacy, fairness, and accountability, and bias frameworks identifying data/model/deployment sources exemplify conceptual applications (Chukwuere, 2021). Conceptual frameworks are essential for elucidating the fundamental concepts and ideas that underpin informatics (Ringelstein & Patel, 2023).

DEVELOPING FRAMEWORKS FOR RESEARCH

Ensuring conceptual clarity, methodological rigour, scholarly legitimacy, and theoretical growth relies on developing robust research frameworks. These frameworks, whether theoretical or conceptual, ensure that research is generative, well-founded, and cohesive—essential elements for producing significant scholarly contributions. Robust frameworks arise from systematic engagement with literature, clear construct definitions, and logical relationships that are refined iteratively and presented visually (Adom et al., 2018; Miles et al., 2014; Tegtmeyer, 2023). Figure 4 illustrates the steps in developing theoretical and conceptual frameworks.

Figure 4. Steps for developing theoretical and conceptual frameworks

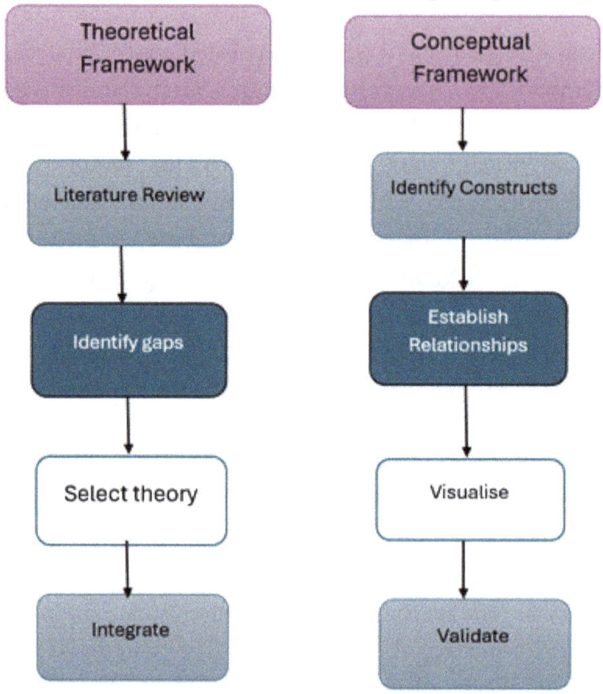

Steps of Building a Theoretical Framework

A crucial first step in organising academic research is creating a theoretical framework, which serves to ground the investigation in accepted wisdom and direct the interpretation of results. Usually, the procedure consists of several methodical steps that guarantee theoretical rigour, conceptual clarity, and methodological alignment (Crawford, 2019; Kivunja, 2018). The first step is to identify, from the keywords, the key variables or concepts relevant to the research problem (Adom et al., 2018). This requires an in-depth engagement with existing literature to understand how scholars have defined, operationalised, and related these concepts. Kivunja (2018) states that developing a theoretical framework begins by understanding existing theories related to the topic, which helps the researcher identify the key variables central to the phenomenon under investigation.

The researcher must then review and select suitable existing theories. According to Master Academia (2026), a theoretical framework is developed by identifying which established theories offer the most coherent explanations for the research context. This selection process involves comparing different theories, evaluating their underlying assumptions, and determining which theory aligns best with the study's objectives (Ringelstein & Patel, 2023).

The third step is to establish relationships among the chosen concepts or variables. According to Pretorius (2024), a theoretical framework should outline the key concepts and explain the relationships that will guide data analysis. Clearly articulating these connections helps transform abstract theory into a structured model that supports empirical research. Once relationships are defined, the researcher must construct a narrative or diagrammatic representation of the framework. This communicates the logical flow of ideas and provides a blueprint for how the theory informs the research design. Kivunja (2018) stresses that a good theoretical framework also demonstrates how the researcher interprets the theory in relation to the study. Finally, the framework must be justified and contextualised within the broader literature (Adom et al., 2018; Kivunja, 2018). This step ensures that the chosen theory is appropriate, relevant, and aligned with established academic discourse. In summary, developing a theoretical framework involves identifying key concepts, selecting appropriate theories, clarifying their relationships, constructing a coherent model, and grounding the framework in existing research. All these steps are essential for maintaining theoretical integrity and methodological coherence.

Steps in Developing a Conceptual Framework

Developing a conceptual framework is a systematic process that enables researchers to map the major concepts, relationships, and assumptions that guide their study (Crawford, 2019; Miles et al., 2014). It offers the intellectual framework through which the researcher expresses the study's contribution to current knowledge. As illustrated in Figure 4, the steps involved in developing such a framework ensure that the resulting structure is coherent, evidence-based, and methodologically sound (Crawford, 2019; Oxbridge Essays, 2026). The first step is to review relevant literature to identify key concepts, theories, and variables related to the research problem. Ngulube (2020) highlight that conceptual frameworks often draw from multiple theories or knowledge domains, especially in interdisciplinary fields such as information science. This literature review serves as the foundation for identifying patterns, gaps, and conceptual relationships (Miles et al., 2014; Ravitch & Riggan, 2017).

The next step for the researcher is to select and define the core concepts or constructs that will form the study's framework. Crawford (2019) underscores that conceptual frameworks can be broad, drawing on existing theories or literature reviews, or narrow, focusing on specific variables aligned with the study's objectives. It is crucial to provide clear, operational definitions to ensure consistency and eliminate ambiguity in the concepts (Oxbridge Essays, 2026). This step is followed by establishing relationships among the identified concepts or constructs. This requires determining how variables or constructs interact, support each other, or contribute to explaining the phenomenon of interest. Articulating these relationships strengthens both the conceptual clarity and analytical direction of the study (Miles et al., 2014; Ravitch & Riggan, 2017). The next step is to construct a visual or narrative model that represents these relationships (Miles et al., 2014; Ngulube, 2018; Ravitch & Riggan, 2017). This may take the form of diagrams, flowcharts, or a logically structured narrative (Miles et al., 2014). However, narrative conceptual frameworks are not encouraged (Ngulube & Mosha, 2023). As Crawford (2019) and Oxbridge Essays (2026) state, the purpose of graphics is to provide a clear orientation that helps both the researcher and the reader understand the study's structure. Finally, the conceptual framework must be justified and refined to ensure it aligns with the research questions, methodology, and existing knowledge. This justification illustrates the framework's relevance, coherence, and its contribution to knowledge. Validation can also be achieved through expert reviews or pilot studies.

Challenges and Best Practices

Creating and implementing frameworks in computing and informatics involves unique challenges; following best practices helps reduce these risks (Kivunja, 2018; Chukwuere, 2021). Table 1 outlines the common pitfalls and best practices in building theoretical conceptual frameworks.

Table 1. Challenges and best practices in framework development

Challenges	Best Practices
Ambiguity in framework selection	Begin with a clear problem statement and objectives (Kivunja, 2018).
Overcomplexity of constructs	Focus on essential constructs and relationships (Miles et al., 2014; Pretorius, 2026).
Misalignment with objectives	Ground frameworks in literature; ensure contextual relevance (Ngulube, 2018; Ravitch & Riggan, 2017).
Rapid technological change	Continuously improve frameworks as new insights are gained. (Oates, 2006).
Interdisciplinary conflicts	Visualize relationships to aid collaboration (Crawford, 2019; Miles et al., 2014).

FRAMEWORKS AS TOOLS FOR INNOVATION

Frameworks are not just structural tools for research; they also act as catalysts for innovation. As tools of innovation, frameworks drive new research directions by inspiring novel approaches and bridging theory and practice through translating abstract concepts into practical solutions. By clarifying key concepts and explaining their relationships, frameworks promote the generation of new hypotheses, thereby encouraging innovative research directions (Creswell & Creswell, 2018). With this methodical approach, creativity is guaranteed to arise from a cumulative, cohesive evolution of information rather than at random. For example, conceptual frameworks for AI ethics help operationalise abstract principles like fairness and transparency into measurable constructs. Interdisciplinary knowledge is integrated by combining computing theories with social science concepts to address issues such as digital inclusion and cybersecurity governance (Djatmiko et al., 2025).

Frameworks also foster innovation by synthesising insights across disciplines, allowing researchers to transform abstract theoretical principles into actionable models. According to Eisenhardt (1989), theoretical frameworks serve as scaffolding that enables researchers to develop new conceptual contributions from empirical observations. In this way, frameworks provide pathways for extending existing theories or generating entirely new ones, especially when empirical findings challenge conventional assumptions.

Theoretical frameworks play a crucial role in connecting theory to practice. They assist practitioners in interpreting complex problems and guide decision-making by providing evidence-based explanations and predictive models (Van de Ven, 2007). This relationship ensures that innovations are rooted in rigorous research while also addressing real-world needs. For instance, innovation theories such as Rogers' Diffusion of Innovation offer practical insights into how new ideas spread and how their adoption can be accelerated (Rogers, 2003). These frameworks enable practitioners to refine their strategies, enhance processes, and promote organisational learning. Furthermore, conceptual models for AI ethics translate abstract principles into actionable guidelines and in Big Data analytics, frameworks have enabled scalable architectures by integrating theoretical principles of distributed computing with practical design patterns (Kandiero et al., 2024).

There are several benefits for innovation offered by the conceptual and theoretical frameworks. Including (Kandiero et al., 2024; Ngulube et al., 2015):

- encouraging creativity by visualizing unexplored relationships;
- supporting interdisciplinary research by integrating diverse theoretical perspectives; and

- facilitating practical implementation by translating abstract ideas into actionable models.

FUTURE RESEARCH DIRECTIONS

Future research directions for theoretical and conceptual frameworks in computing and informatics include AI-driven, quantum, edge AI/IoT, ethical and sustainability, Interdisciplinary and data-centric. Table 2 summarises future research directions for computing and informatics frameworks (Kumar & Dhir, 2021; National Academies of Sciences, Engineering, and Medicine, 2022; Omar & Mwakondo, 2024; Tiwari & Darbari, 2025; Tiwari et al., 2024).

Table 2. Future research directions for frameworks in computing and informatics

Trend	Description
AI-driven frameworks	• Automating framework generation using machine learning and real-time data. • Dynamic adaptation of conceptual models based on predictive analytics.
Quantum computing frameworks	• Frameworks for quantum algorithms, error correction, and hybrid systems. • Addressing scalability and security challenges in quantum environments
Edge AI and IoT frameworks	• Models for distributed intelligence, latency optimization, and device interoperability. • Integrating privacy and ethical considerations into IoT frameworks.
Ethical & sustainability frameworks	• Embedding fairness, accountability, and linking computing efficiency with sustainability. • Linking computing efficiency with environmental sustainability principles
Interdisciplinary frameworks	• Combining computing theories with social sciences, law, and psychology. • Frameworks for digital inclusion, cybersecurity governance, and human-centred computing
Data-Centric Adaptive frameworks	• Prioritizing data governance, privacy, and adaptive evolution with regulations. • Adaptive frameworks that evolve with regulatory and technological changes

CONCLUSION

For computer and informatics research to be cohesive, rigorous, and influential, theoretical and conceptual frameworks are essential. They clarify problems, harmonise techniques, and put discoveries in context by linking theory and practice. Conceptual and theoretical frameworks are essential to rigorous theorisation, methodical theory-building, and reliable knowledge production in computing and informatics research; they are not optional academic extras. Theoretical frameworks guarantee coherence, validity, and consistency with current knowledge by firmly establishing studies within preexisting ideas. Conversely, conceptual frameworks offer adaptability and inventiveness, allowing researchers to investigate new fields where formal theories might not be as useful and to visualise the relationships between constructs.

Throughout this chapter, we have demonstrated:

- The importance of frameworks in shaping research questions, guiding methodology, and ensuring systematic inquiry.
- The complementary roles of theoretical and conceptual frameworks in computing research.
- Frameworks as tools for innovation, bridging theory and practice, and driving interdisciplinary research.
- Future directions, including AI-driven frameworks, quantum computing paradigms, edge AI/IoT models, and ethical sustainability considerations.

Finally, as computing technologies continue to evolve, frameworks will be essential for organising research, promoting innovation, and tackling ethical and societal challenges. Researchers should view frameworks not just as guiding tools but also as dynamic instruments that can adapt to technological advancements and the demands of different disciplines.

REFERENCES

Adom, D., Hussein, E. K., & Agyem, J. A. (2018). Theoretical and conceptual framework: Mandatory ingredients of a quality research. *International Journal of Scientific Research, 7*(1), 438–441.

Alvesson, M., & Skoldberg, K. (2000). *Reflexive methodology: New vistas for qualitative research.* SAGE.

Anfara, V. A., & Mertz, N. T. (Eds.). (2015). *Theoretical frameworks in qualitative research* (2nd ed.). SAGE.

Blalock, H. M. (1969). *Theory construction: From verbal to mathematical formulations.* Prentice-Hall.

Blumer, H. (1969). *Symbolic interaction: Perspectives and method.* University of California Press.

Booth, W. C., Colomb, G. G., & Williams, J. M. (2024). *The craft of research* (4th ed.). University of Chicago Press.

Chakabwata, W., & McKay, V. (2026). Applying theoretical and conceptual frameworks to doctoral research in education. In Throne, R. (Ed.), *Navigating theory, methodology, and researcher development in doctoral research* (pp. 1–28). IGI Global., DOI: 10.4018/979-8-3373-5676-1.ch001

Charmaz, K. (2014). *Constructing grounded theory* (2nd ed.). SAGE.

Chukwuere, J. (2021). Theoretical and conceptual framework: A critical part of information systems research process and writing. *Review of International Geographical Education, 11*(9), 2678–2683. DOI: 10.48047/rigeo.11.09.234

Corbin, J., & Strauss, A. (2015). *Basics of qualitative research: Techniques and procedures for developing grounded theory* (4th ed.). SAGE.

Corley, G. K., & Gioia, A. D. (2011). Building theory about theory building: What constitutes a theoretical contribution? *Academy of Management Review, 36*(1), 12–32.

Crawford, L. M. (2019). Conceptual and theoretical frameworks in research. In Burkholder, G. J., Cox, K. A., Crawford, L. M., & Hitchcock, J. H. (Eds.), *Research design and methods: An applied guide for the scholar-practitioner* (pp. 35–48). SAGE.

Creswell, J. W., & Creswell, J. D. (2018). *Research design: Qualitative, quantitative, and mixed methods approaches* (5th ed.). SAGE.

Creswell, J. W., & Plano Clark, V. L. (2018). *Designing and conducting mixed methods research* (3rd ed.). SAGE.

De Vos, A. S., Strydom, H., Fouché, C. B., & Delport, C. S. L. (2011). *Research at grass roots: For the social sciences and human services profession* (4th ed.). Van Schaik.

Dilthey, W. (1989). *Selected works: Introduction to the human sciences* (Vol. 1). Princeton University Press. (Original work published 1883)

Djatmiko, G. H., Sinaga, O., & Pawirosumarto, S. (2025). Digital transformation and social inclusion in public services: A qualitative analysis of e-government adoption for marginalized communities in sustainable governance. *Sustainability, 17*(7), 2908. DOI: 10.3390/su17072908

Dubin, R. (1969). *Theory building*. 2nd ed. Free Press, Collier-Macmillan

Eisenhardt, K. M. (1989). Building theories from case study research. *Academy of Management Review*, *14*(4), 532–550.

Glaser, B. G., & Strauss, A. L. (1967). *The discovery of grounded theory: Strategies for qualitative research*. Aldine.

Grand, J. A., Braun, M. T., & Kuljanin, G. (2024). Hello world! Building computational models to represent social and organizational theory. *Organizational Research Methods*, *28*(3), 487–539. DOI: 10.1177/10944281241261913

Grant, C., & Osanloo, A. (2014). Understanding, selecting, and integrating a theoretical framework in dissertation research. *Administrative Issues Journal*, 4(2), 12–26. https://doi.org/DOI: 10.5929/2014.4.2.9

Gray, P. (2003). Introduction to the debate on the core of the information systems field. *Communications of the AIS*, 12(1), Art. 42.

Green, H. E. (2014). Use of theoretical and conceptual frameworks in qualitative research. *Nurse Researcher*, *21*(6), 34–38. DOI: 10.7748/nr.21.6.34.e1252 PMID: 25059086

Gregor, S. (2006). The nature of theory in information systems. *Management Information Systems Quarterly*, *30*(3), 611–642. DOI: 10.2307/25148742

Hamilton, , S. (2017). A genealogy of metatheory in IR: How 'ontology' emerged from the inter-paradigm debate. *International Theory, 9*(1), 136–170. https://doi.org/10.1017/S1752971916000257

Hassan, N. R., Lowry, P. B., & Mathiassen, L. (2022). Useful products in information systems theorizing: A discursive formation perspective. *Journal of the Association for Information Systems, 23*(2), 418–446. DOI: 10.17705/1jais.00730

Henning, E., Van Rensburg, W., & Smit, B. (2004). *Finding your way in qualitative research*. Van Schaik Publishers.

Holton, E. F., & Lowe, J. S. (2007). Toward a general research process for using Dublin's theory building model. *Human Resource Development Review, 6*(3), 297–320.

Horkheimer, M. (2002). *Critical theory: Selected essays*. Continuum. (Original work published 1937).

Jacard, J., & Jacoby, J. (2010). *Theory construction and model-building skills: A practical guide for social scientists*. The Guilford Press.

Jeong, D. Y., & Kim, S. J. (2005). Knowledge structure of library and information science. *Library & Information Science Research, 27*, 51–72.

Kandiero, A., Bigirimana, S., & Chizwina, S. (2024). *Theoretical and Conceptual Frameworks in ICT Research*. IGI Global (case studies on technology adoption and innovation).

Kawulich, B. (2009). The role of theory in research. In Garner, M., Wagner, C., & Kawulich, B. (Eds.), *Teaching research methods in the social sciences* (pp. 37–47). Routledge.

Kettley, N. (2010). *Theory building in educational research*. Continuum.

Kivunja, C. (2018). Distinguishing between theory, theoretical framework, and conceptual framework: A systematic review of lessons from the field. *International Journal of Higher Education, 7*(6), 44–53. DOI: 10.5430/ijhe.v7n6p44

Kulesa, J., Induru, S., Hubbard, E., & Bhansali, P. (2024). The conceptual framework: A practical guide. *Hospital Pediatrics, 14*(11), e503–e508. DOI: 10.1542/hpeds.2024-007794 PMID: 39363881

Kumar, D., & Dhir, S. (Eds.). (2021). *Trends in future informatics and emerging technologies*. Bentham Books.

Leedy, P. D., & Ormrod, J. E. (2015). *Practical research: Planning and design* (11th ed.). Pearson Education Limited.

Lewin, K. (1945). The Research Centre for Group Dynamics at Massachusetts Institute of Technology. *Sociometry, 8*(2), 126–136.

Lincoln, Y. S., & Guba, E. G. (1985). *Naturalistic inquiry*. SAGE.

Ludviga, I. (2023). Theoretical and conceptual frameworks and models: What are they, when, and how to apply them in teaching research methodology to master's and PhD students?

Luft, , J. AJeong, , SIdsardi, , RGardner, , G. (2022). Literature reviews, theoretical frameworks, and conceptual frameworks: An introduction for new biology education researchers. *CBE Life Sciences Education, 21*(3). DOI: 10.1187/cbe.21-05-0134

Lynham, S. A. (2002). Theory building: Dubin's method. *Advances in Developing Human Resources, 4*(3), 242–276.

Marshall, C., & Rossman, G. B. (2016). *Designing qualitative research* (6th ed.). SAGE.

Master Academia. (2026). Theoretical vs. conceptual frameworks: Simple definitions and an overview of key differences. Retrieved on 13 February 2026 from https://master-academia.com/theoretical-vs-conceptual-frameworks-definitions-differences/

Matthews, C. R. (2025). *Doing good social science: Lessons from immersion, understanding social life and exploring the in-between*. Routledge.

Merriam, S. B., & Tisdell, E. J. (2016). *Qualitative research: A guide to design and implementation* (4th ed.). Jossey-Bass.

Merton, R. K. (1967). *On theoretical sociology*. Free Press.

Miles, M. B., Huberman, A. M., & Saldaña, J. (2014). *Qualitative data analysis: A methods sourcebook* (3rd ed.). SAGE.

Moschis, G. P. (2024). Conceptual framework and theory development. In *Academic research in business and the social sciences: A guidebook for early career researchers* (pp. 83–99). Springer., DOI: 10.1007/978-3-031-56548-9_5

Nachmias, C., & Nachmias, D. F. (1996). *Research methods in the social science* (5th ed.). Arnold.

National Academies of Sciences, Engineering, and Medicine. (2022). *Fostering responsible computing research: Foundations and practices*. The National Academies Press.

Ncube, M. M., & Ngulube, P. (2025). Mapping theoretical approaches: A scoping review of data analytics in higher education. *Discover Education, 4*, 201. DOI: 10.1007/s44217-025-00647-8

Netshakhuma, N. S. (2024). Application of interpretivism theoretical framework on the management of university records: Comparison of selected universities. In A. Kandiero, S. Bigirimana & S. Chizwina. (2024). *Theoretical and conceptual frameworks in ICT research.* (pp. 29–49). IGI Global.

Ngulube, P. (2018). Overcoming the difficulties associated with using conceptual and theoretical frameworks in heritage studies. In Ngulube, P. (Ed.), *Handbook of research on heritage management and preservation* (pp. 1–23). IGI Global.

Ngulube, P. (2020). Theory and theorising in information science scholarship. In Ngulube, P. (Ed.), *Handbook of research on connecting research methods for information science research* (pp. 18–39). IGI Global.

Ngulube, P. (2025). Indigenous theories and theorisation in African scholarship in the quest for epistemic freedom. In Ngulube, P. (Ed.), *Harnessing Indigenous epistemologies for sustainable progress* (pp. 139–172). IGI Global., DOI: 10.4018/979-8-3693-8447-3.ch008

Ngulube, P., Mathipa, E. R., & Gumbo, M. T. (2015). Theoretical and conceptual framework in the social sciences. In Mathipa, E. R., & Gumbo, M. T. (Eds.), *Addressing research challenges: making headway in developing researchers* (pp. 43–66). Mosala-MASEDI Publishers & Booksellers., DOI: 10.13140/RG.2.1.3210.7680

Ngulube, P., & Mosha, N. F. V. (2023). Transparency in the application of theoretical frameworks to the advancement of knowledge in selected library and information science journals: A systematic review. *African Journal of Library Archives and Information Science, 33*(2), 129–146.

Oates, B. J. (2006). Researching information systems and computing. *Sage (Atlanta, Ga.).*

Omar, A. S., & Mwakondo, F. (2024). Evolution of cloud computing: Trends, issues, and future directions: A systematic literature review. *International Journal of Computer Science Trends and Technology, 12*(3), 102–111.

Oxbridge Essays. (2026). Literature review, theoretical framework, and conceptual framework. Retrieved on 13 February 2026 from https://www.oxbridgeessays.com/blog/literature-review-theoretical-framework-conceptual-framework/

Oyewobi, L. O., Okanlawon, T. T., Medayese, S. O., Ogunbode, E. B., & Jimoh, R. A. (2024). Is pursuing a PhD without theoretical and conceptual framework a journey without roadmap? *Environmental Technology and Science Journal, 15*(2), 138–149. DOI: 10.4314/etsj.v15i2.15

Paperpal. (2024). Conceptual framework vs theoretical framework – Difference and examples. *Paperpal Blog*. Retrieved on 3 November 2025 from https://paperpal.com/blog/researcher/conceptual-framework-vs-theoretical-framework-difference-and-examples

Pretorius, L. (2026). Theoretical and conceptual frameworks in research. Retrieved on 13 February 2026 from https://www.lynettepretorius.com/the_scholars_way_blog/theoretical-and-conceptual-frameworks-in-research/

Prus, R. (1996). *Symbolic interaction and ethnographic research: Intersubjectivity and the study of human experience*. State University of New York Press.

Ravitch, S. M., & Riggan, M. (2017). *Reason & rigor: How conceptual frameworks guide research* (2nd ed.). SAGE.

Reynolds, P. D. (2007). *A primer in theory construction: An A&B classics edition*. Routledge.

Ringelstein, D., & Patel, K. (2023). A conceptual framework for informatics. [files.eric.ed.gov]. *E-Journal of Business Education & Scholarship of Teaching*, *17*(1), 9–14. https://files.eric.ed.gov/fulltext/EJ1376059.pdf

Ritzer, G. (1991). *Metatheorizing in sociology*. Lexington Books.

Salawu, R. O., Shamsuddin, A.-O. S., & Masibo, S. (2023). Theoretical and conceptual frameworks in research: Conceptual clarification. *European Chemical Bulletin*, *12*(12), 2103–2117.

Saunders, M. N. K., Gray, D. E., Tosey, P., & Sadler-Smith, E. (2015). Concepts and theory building. In Anderson, L., Gold, J., Stewart, J., & Thorpe, R. (Eds.), *A guide to professional doctorates in business and management* (pp. 35–56). Sage., DOI: 10.4135/9781473921412.n3

Schneider, M. A. (2006). *The theory primer: A sociological guide*. Rowman and Littlefield.

Sutton, R. I., & Staw, B. M. (1995). What theory is not. *Administrative Science Quarterly*, *40*(3), 371–384. DOI: 10.2307/2393788

Tegtmeyer, J. (2023). Building a dissertation conceptual and theoretical framework: A recent doctoral graduate narrates behind the curtain development. *Perspectives on Urban Education*, 20(1). Retrieved on 14 February 2026 from https://urbanedjournal.gse.upenn.edu/archive/volume-20-issue-1-fall-2022

Tiwari, A., & Darbari, M. (2025). *Emerging trends in computer science and its application*. CRC Press., DOI: 10.1201/9781003606635

Tiwari, A. B., Rai, D. A., Singh, A. M., & Jaiswal, K. J. (2024). Ethical and security considerations in the era of quantum computing: A framework for responsible implementation strategies. *International Journal of Research Publication and Reviews, 5*(5), 7266–7271. DOI: 10.55248/gengpi.5.0524.1295

Turner, J. H. (1991). *The structure of sociological theory* (5th ed.). Wadsworth.

Ukwoma, S. C., & Ngulube, P. (2021). (in press). To borrow or not to borrow is the question? Theory borrowing in library information science postgraduate research in Nigeria and South Africa. *The International Information & Library Review, 53*(1). Advance online publication. DOI: 10.1080/10572317.2020.1790261

Van de Ven, A. H. (2007). *Engaged scholarship: A guide for organizational and social research.* Oxford University Press., DOI: 10.1093/oso/9780199226290.001.0001

Van der Waldt, G. (2020). Constructing conceptual frameworks in social science research. *The Journal for Transdisciplinary Research in Southern Africa, 16*(1), 758. DOI: 10.4102/td.v16i1.758

Van House. N. A. (1991). Assessing the quantity, quality, and impact of LIS research. In C. R. McClure and P. Hernon (Eds.), *Library and information science research: Perspectives and strategies for improvement* (pp. 85-100). Ablex.

Varpio, L., Paradis, E., Uijtdehaage, S., & Young, M. (2020). The distinctions between theory, theoretical framework, and conceptual framework. *Academic Medicine, 95*(7), 989–994. DOI: 10.1097/ACM.0000000000003075 PMID: 31725464

Weber, R. (2012). Evaluating and developing theories in the information systems discipline. *Journal of the Association for Information Systems, 13*(1), 1–30. DOI: 10.17705/1jais.00284

Weick, K. E. (1995). What theory is not, theorizing is. *Administrative Science Quarterly, 40*, 385–390.

Whetten, D. A. (1989). What constitutes a theoretical contribution? *Academy of Management Review, 14*(4), 490–495.

William, F. K. A. (2024). Unveiling the significance and influence of theoretical and conceptual frameworks in social science research. *International Journal of Research Publications, 142*(1), 137–142. DOI: 10.47119/ijrp1001421220246059

Yatsko, V. A. (2018). Informatics, Information Science, and Computer Science. *Scientific and Technical Information Processing, 45*, 235–240. DOI: 10.3103/S0147688218040081

Yin, R. K. (2018). *Case study research and applications: Design and methods* (6th ed.). SAGE.

ADDITIONAL READING

Bradford, A. (2017). What is a scientific theory. Retrieved on 12 September 2020 from https://www.livescience.com/21491-what-is-a-scientific-theory-definition-of-theory.html#

Breslin, D., & Gatrell, C. (2020). Theorizing through literature reviews: The miner-prospector continuum. Organizational Research Methods, 26(1), 139–167. https://doi.org/ (Original work published 2023)DOI: 10.1177/1094428120943288

Chen, Z., & Chen, Y. (2024). Computing grounded theory: A quantitative method to develop theories. *The Journal of Chinese Sociology*, *11*, 17. Advance online publication. DOI: 10.1186/s40711-024-00218-8

Eisenhardt, K. M. (1989). Building theories from case study research. *Academy of Management Review*, *14*(4), 532–550.

Poschin-Young, M., Haines-Young, R., Görg, C., Heink, U., Jax, K., & Schleyer, C. (2018). Understanding the role of conceptual frameworks: Reading the ecosystem service cascade. *Ecosystem Services*, *29*, 428–440. PMID: 31008045

Sale, J. E., & Carlin, L. (2025). The reliance on conceptual frameworks in qualitative research: A way forward. *BMC Res Methodol*, *25*, 36. DOI: 10.1186/s12874-025-02461-0 PMID: 39948487

Sandberg, J., & Alvesson, M. (2021). Meanings of theory: Clarifying theory through typification. *Journal of Management Studies*, *58*(2), 487–516. DOI: 10.1111/joms.12587

Tamene, E. H. (2016). Theorizing conceptual framework. *International Journal of Social Sciences and Humanities Invention*, *10*, 7623–7649. DOI: 10.18535/ijsshi/v10i01.08

Tiwari, A. B., Rai, D. A., Singh, A. M., & Jaiswal, K. J. (2024). Ethical and security considerations in the era of quantum computing: A framework for responsible implementation strategies. *International Journal of Research Publication and Reviews*, *5*(5), 7266–7271. DOI: 10.55248/gengpi.5.0524.1295

Truex, D., Holmström, J., & Keil, M. (2006). Theorizing in information systems research: A reflexive analysis of the adaptation of theory in information systems research. *Journal of the Association for Information Systems*, *7*(12), 797–821. DOI: 10.17705/1jais.00109

Wallis, S. E. (2010). Toward a science of metatheory. *Integral Review*, *6*(3), 73–120.

KEY TERMS AND DEFINITIONS

AI ethics: A set of guidelines for AI systems that guarantee equity, openness, and responsibility.

Big data analytics: Analysing big, complicated data sets to find trends, connections, and information to help with decision-making.

Cloud computing: An internet-based framework for delivering computing services that includes networking, computation, and storage.

Conceptual framework: A visual or descriptive model showing how different variables and concepts in a study are related.

Edge computing: A distributed computing approach that minimises latency by placing processing and data storage closer to data sources.

Human-computer interaction (HCI): Analysing human-computer interaction and design concepts that improve usability and user satisfaction.

Internet of Things (IoT): A network of physically connected devices with sensors and software that can exchange data online.

Theoretical framework: A structure that helps shape research questions and methods, based on established theories.

Chapter 10
Testing of Hypotheses: A Practical Approach for Computing and Informatics

R. Krishna Kumari
http://orcid.org/0000-0002-1802-628X
Department of Mathematics, College of Engineering and Technology, SRM Institute of Science and Technology, Kattankulathur, India

ABSTRACT

Hypothesis testing plays a crucial role in validating research claims in the domains of computing and informatics. It provides a systematic framework for data-driven decision-making. This chapter provides a comprehensive and accessible guide to hypothesis testing, specifically tailored for researchers and practitioners in computing and informatics. The chapter begins by introducing the fundamental concepts, including null and alternative hypotheses, significance levels, p-values, and types of statistical errors. It then presents both parametric and non-parametric tests commonly used in empirical computing research, such as t-tests, ANOVA, and the Chi-square test. Real-world case studies—ranging from A/B testing in web applications to algorithm benchmarking—are used to contextualise each technique. The chapter also highlights the integration of hypothesis testing in machine learning model evaluation and provides hands-on guidance using Python and R. By connecting theory with practice, this chapter equips researchers to apply hypothesis testing effectively in computing projects.

INTRODUCTION

Hypothesis testing has long served as a foundational pillar in computing, data science, and informatics, offering a structured framework for drawing inferences

DOI: 10.4018/979-8-3693-4018-9.ch010

from data, validating system behaviours, and comparing algorithmic performance. As computational systems grow in scale and complexity, the need for reliable statistical reasoning becomes even more critical. Classical statistical principles—originating from the pioneering work of Neyman (1942) on formulating null and alternative hypotheses—have evolved into sophisticated tools tailored for modern computational research (Anderson, Burnham, and Thompson, 2000). These classical formulations provide the basis for evaluating uncertainties, quantifying evidence, and deriving objective conclusions about digital processes, algorithms, and models. Over the years, the philosophy and interpretation of hypothesis testing have been further examined by Mizrahi (2020) and Keller and Egbert (2020), who highlighted the differences between ordinary and scientific usages of the term, underscoring the need for clarity and methodological rigour across disciplines.

The fundamental building blocks—significance levels, p-values, Type I and Type II errors, and statistical power—form the core analytical machinery behind hypothesis testing. McKenzie (2004) emphasised that these elements support judgment and decision-making across computational problem settings, from evaluating algorithmic robustness to quantifying evidence in simulation-driven analyses. With the rise of data-driven computational models, these principles have found renewed importance in machine learning model validation (Raschka, 2020), where performance comparisons, generalization studies, and error estimation rely heavily on sound statistical inference. Hypothesis testing also plays an instrumental role in regression modelling, as highlighted in applied computational research by Al-Khaiat, Noori, and Cengiz (2022), and in simulation-based studies involving complex systems (Scerbo, Calhoun, and Hui, 2019).

Information theory has further enriched the landscape of hypothesis testing by providing mathematical frameworks for studying uncertainty, data transmission, and error behaviour. Foundational contributions by Cover (1969) and Blahut (2003) demonstrate how hypothesis testing can be used to evaluate information bounds and perform error analysis in constrained communication environments. Ahlswede and Csiszár (2003) extended these ideas by describing decision-making under communication constraints, and Haroutunian, Haroutunian, and Harutyunyan (2008) provided theoretical insights on reliability and statistical distinguishability in noisy systems. Similar contributions by Tuncel (2005) and Zeitouni and Gutman (2002) explored error exponents and large deviation principles, establishing statistical guarantees even for rare-event regimes. This theoretical foundation is complemented by research on hypothesis testing for arbitrarily varying sources by Fangwei and Shiyi (1996), and earlier work by Hogg (1961) on the resolution of competing statistical hypotheses. Collectively, these contributions demonstrate the deep connections between hypothesis testing and information-processing tasks central to modern computing. In applied computing domains, researchers often confront data dependencies—serial, temporal, or structural—that violate the classical assumption of independent observations. Sobhiyeh and Naraghi-Pour (2016) examined hypothesis testing under dependent measurements, particularly relevant for signal processing applications where noise, correlation, and structured variability must be accounted for. Nonparametric and permutation-based approaches, such as those detailed by Good (2013), have become increasingly important for handling such data irregularities, especially when dealing with small sample sizes or violations of distributional assumptions.

The practical implementation of hypothesis testing has been significantly influenced by statistical software ecosystems. Tools such as SPSS, R, and Python have become indispensable among computing researchers. Haslwanter (2016) discussed the broad applicability of these tools for statistical computations, while McCormick and Salcedo (2017) provided practical guidance for users performing statistical tests in applied settings. In the context of applied linguistics and digital humanities, Larson-Hall and Mizumoto (2019) illustrated how hypothesis testing supports empirical validation across language-related computational workflows. Computational researchers have increasingly turned to Python-based environments for reproducible and scalable statistical analysis. Wang, Manolas, and Xanthidis (2022) highlighted how Python facilitates robust statistical evaluation and integrative workflows, and Al-Khaiat et al. (2022) compared Python with SPSS and Excel to emphasise its versatility for regression and inferential tasks. Hypothesis testing also plays a pivotal role in modern artificial intelligence and machine learning pipelines. Gonçalves and Silva (2023) examined how statistical tests inform model selection, helping researchers avoid arbitrary or biased decisions when evaluating models. Wu et al. (2022) proposed score-based testing frameworks for unnormalized models, demonstrating how emerging ML methods demand new forms of statistical validation. Raschka (2020) provided comprehensive instructional materials emphasising sound statistical practice when applying machine learning algorithms, particularly in evaluating classifier performance, hyperparameter tuning, and cross-validation outcomes.

The literature has also addressed methodological limitations and philosophical debates surrounding the use of hypothesis testing. Sedgwick (2013, 2014) highlighted challenges such as underpowered studies, misuse of p-values, and misinterpretations of statistical significance. These concerns echo those raised by Anderson et al. (2000), who proposed alternatives to classical null-hypothesis significance testing in ecological and computational contexts. Recognising these limitations, researchers have explored broader testing frameworks such as Bayesian model evaluation (Raftery et al., 1995), composite hypothesis testing (Feder & Merhav, 2002), and universal hypothesis testing methods (Zeitouni & Gutman, 2002). These approaches help address uncertainties associated with high-dimensional models, limited data environments, and complex system structures. Advanced computational applications further demonstrate the relevance of hypothesis testing across diverse domains. Xiao, Wei, and Tai (2012) applied hypothesis testing to multi-source signal estimation, showing its utility in communication systems and sensor networks. Carterette and Smucker (2007) explored hypothesis testing in information retrieval systems where relevance judgments are incomplete—a common challenge in evaluating search algorithms. Beyond traditional computing, Newman (2008) discussed hypothesis testing in network science applications, while Mizrahi (2020) explored its epistemological grounding within computational scientific practice.

Overall, hypothesis testing has progressed from a classical statistical concept to a dynamic, interdisciplinary framework central to computing, machine learning, information theory, and digital system evaluation. With theoretical foundations strengthened by decades of research and practical applications supported by advanced software tools, hypothesis testing remains indispensable for validating knowledge claims, benchmarking AI systems, guiding decision-making, and ensuring scientific rigour in an increasingly data-centric technological landscape.

Objectives of the Chapter

- To introduce the key concepts of hypothesis testing in the context of computing and informatics
- To demonstrate the application of both parametric and non-parametric tests with real-world examples
- To illustrate how hypothesis testing supports model evaluation and system design in machine learning and software engineering
- To provide practical exposure to hypothesis testing using statistical tools like Python, R, and SPSS
- To emphasise best practices and common pitfalls in hypothesis testing for computing researchers

FUNDAMENTALS OF HYPOTHESIS TESTING

The hypothesis testing workflow begins with a precise formulation of the null hypothesis (H_0) and the alternative hypothesis (H_1), grounded in the central research question. Once the hypotheses are defined, an appropriate test statistic is selected based on the type of data, its distributional characteristics, and the study design. The researcher then specifies a significance level, often $\alpha = 0.05$, although stricter thresholds such as $\alpha = 0.01$ may be adopted in high-stakes domains (e.g., cybersecurity, medical diagnostics). Using sample data, a test statistic—such as a z, t, or χ^2 value—is computed, from which the corresponding p-value is derived. This p-value is compared against the chosen α to determine whether the evidence is strong enough to reject H_0. Throughout the entire process, it is critical to examine test assumptions, evaluate effect sizes, and interpret findings within the practical and computational context of the research domain, rather than relying solely on statistical significance.

Null vs. Alternative Hypotheses

In hypothesis testing (see Fig. 1), we begin with two competing statements: the null hypothesis (H_0) and the alternative hypothesis (H_1 or Ha). The null hypothesis represents the default or status quo assumption about a population parameter. For example, H_0 might state that the mean performance of two algorithms is equal. The alternative hypothesis, on the other hand, suggests that there is a significant effect or difference.

Mathematically, for a population mean μ, the hypotheses could be:

- Two-tailed test: H_0: $\mu = \mu_0$ vs. H_1: $\mu \neq \mu_0$
- Left-tailed test: H_0: $\mu \geq \mu_0$ vs. H_1: $\mu < \mu_0$
- Right-tailed test: H_0: $\mu \leq \mu_0$ vs. H_1: $\mu > \mu_0$

The choice between a one-tailed or two-tailed test depends on the research question and the direction of the expected effect. Clearly defining these hypotheses is critical to choosing the correct statistical test.

Significance Levels, p-values, and Statistical Errors

Once the hypotheses are defined, we choose a significance level (α), which is the probability of rejecting the null hypothesis when it is actually true. Common values are 0.05, 0.01, or 0.10.

The p-value is the probability, under the null hypothesis, of observing a result at least as extreme as the one obtained. If the p-value $\leq \alpha$, we reject the null hypothesis in favour of the alternative.

Two types of errors can occur:

- Type I Error (α): Rejecting H_0 when it is true.
- Type II Error (β): Failing to reject H_0 when H_1 is true.

The power of a test, defined as 1 - β, is the probability of correctly rejecting a false null hypothesis. Balancing these errors is essential in test design.

Test Selection and Distributional Assumptions

Selecting an appropriate statistical test is a critical stage in hypothesis testing and depends primarily on the measurement scale of the variables, the number of groups or conditions being compared, and the distributional characteristics of the data. The validity of any inferential conclusion rests on whether the chosen test aligns with these underlying properties. Commonly used tests include:

- t-test (for comparing means when population variance is unknown)
- z-test (for comparing means with known variance)
- ANOVA (for comparing more than two group means)
- Chi-square test (for categorical data and goodness-of-fit)
- Mann-Whitney U test or Kruskal-Wallis test (non-parametric equivalents)

These tests assume different underlying distributions (e.g., normality for t-tests). Diagnostic tools, such as Q-Q plots, Shapiro-Wilk tests, and histograms, can help assess normality before selecting a test. Additionally, assumptions about independence, homogeneity of variance, and scale of measurement play key roles in determining the validity and reliability of test outcomes.

In parametric testing frameworks, the procedures typically rely on assumptions such as normality of the population distribution, independence of observations, and homogeneity of variance. These assumptions ensure that the sampling distributions of test statistics follow known theoretical forms, allowing for the accurate estimation of probabilities and p-values. When these conditions are met, parametric methods generally provide more statistical power because they utilise the full information contained in the data. In contrast, non-parametric methods relax many of these stringent assumptions. Instead of relying on specific distributional forms, they operate on the ranks or ordinal structure of the data. Such tests are particularly useful when dealing with skewed distributions, ordinal scales, outliers that cannot be reasonably removed, or small sample sizes for which normality cannot be assured. However, the reduction in distributional assumptions typically results in reduced power when parametric assumptions are met, underscoring the need for thoughtful test selection.

The decision process for choosing between parametric and non-parametric procedures is therefore anchored in an assessment of distributional assumptions. Tools such as normality assessments, variance checks, and independence verification serve as preliminary diagnostics that guide the selection of tests. Failure to consider these assumptions can lead to inflated error rates, misleading conclusions, and reduced generalizability of results. Thus, test selection is not merely a procedural choice but an integral component of methodological rigour. Ensuring alignment between the data characteristics and the inferential technique strengthens the credibility of hypothesis testing and enhances the interpretability of statistical evidence in computing, informatics, and interdisciplinary research contexts.

Figure 1. Testing of Hypothesis Workflow (Designed by Author)

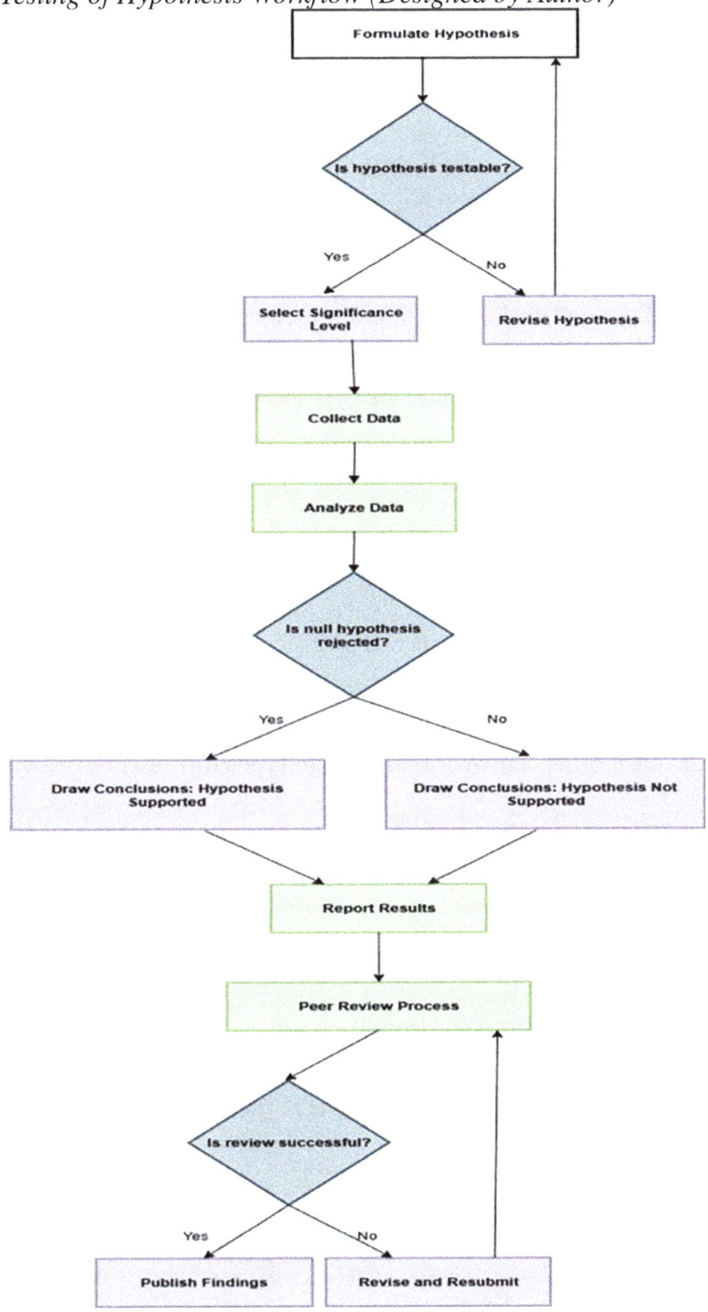

STATISTICAL TESTS IN COMPUTING: PARAMETRIC AND NON-PARAMETRIC APPROACHES

Statistical testing is a foundational aspect of empirical research, particularly in computing and informatics, where the evidence-based validation of results is crucial. Whether comparing the efficiency of algorithms, evaluating system performance, or interpreting user feedback, researchers must apply appropriate statistical tests to draw valid and reliable conclusions.

Broadly, statistical tests can be categorised into parametric and non-parametric methods. The selection between them hinges on the characteristics of the dataset, the scale of measurement, and the assumptions that can be reasonably met. Understanding these tests, along with their assumptions and use cases, is crucial for anyone engaged in empirical computing research.

Parametric Tests and Their Applications

Parametric tests (see Table 1) are based on assumptions about the distribution of the data—typically that it follows a normal distribution. These tests also require interval or ratio scale data and homogeneity of variance. They are more statistically powerful when assumptions are satisfied.

t-Test

The t-test is used to determine whether the means of two groups are statistically different from each other. There are three main types of t-tests:

- Independent Samples t-test: Compares the means of two independent groups.
- Paired Samples t-test: Compares the means from the same group at two different times.
- One-sample t-test: Compares the sample mean to a known value.

Formula (Independent Samples t-test):

$$t = \frac{\overline{X}_1 - \overline{X}_2}{\sqrt{\frac{s_1^2}{n_1} + \frac{s_2^2}{n_2}}}$$

Use Case: Comparing the execution times of two sorting algorithms in Python.

ANOVA (Analysis of Variance)

Used to compare the means across three or more groups by examining between-group and within-group variance.

Formula (F-ratio):

$$F = \frac{\text{Between-group variability}}{\text{Within-group variability}} = \frac{MS_{between}}{MS_{within}}$$

Use Case: Evaluating task completion times across different UI designs.

z-Test

Used when the population standard deviation is known, and the sample size is large (n ≥ 30).

Formula:

$$z = \frac{\overline{X} - \mu}{\sigma / \sqrt{n}}$$

Use Case: Assessing whether server response time meets SLA benchmark.

Table 1. Comparative Overview of Parametric Tests

Test	Purpose	Computing Use Case	Assumptions
t-Test	Compare means between two groups	Comparing latency of two network protocols	Normal distribution, equal variances
ANOVA	Compare means among ≥ 3 groups	Evaluating user performance across UI layouts	Homogeneity of variances
z-Test	Compare sample mean to population mean	Verifying query response time against SLA benchmark	Known population variance

Non-Parametric Tests and Their Role

Non-parametric tests (see Table 2) are used when the assumptions of parametric tests cannot be met—such as with ordinal, categorical, or non-normally distributed data. They are robust and useful in usability studies, HCI, and small-sample research.

Chi-Square Test

Used to assess relationships between categorical variables or evaluate fit to expected frequencies.

Formula (Goodness of Fit):

$$\chi^2 = \sum \frac{(O_i - E_i)^2}{E_i}$$

Use Case: Assessing if user preference differs between menu layouts.

Mann–Whitney U Test

Used to compare two independent groups when data is ordinal or non-normally distributed.

Formula:

$$U = n_1 n_2 + \frac{n_1(n_1 + 1)}{2} - R_1$$

Use Case: Comparing user ratings of two mobile apps on an ordinal scale.

Wilcoxon Signed-Rank Test

Used for comparing two related samples. It's based on signed ranks of the differences between pairs.

Test Statistic: Sum of positive and negative signed ranks from paired differences

Use Case: Evaluating system vulnerability scores before and after patching.

Table 2. Comparative Overview of Non-Parametric Tests

Test	Purpose	Computing Use Case	Data Type
Chi-Square Test	Compare categorical distributions	User behaviour analysis in menu design	Categorical
Mann–Whitney U Test	Compare two independent ordinal groups	Comparing usability ratings of two apps	Ordinal/Non-normal
Wilcoxon Signed-Rank Test	Compare two related samples	System performance pre- and post-update	Ordinal/Pairs

In computing research and applications, selecting the correct statistical test is essential for obtaining valid and reliable results. Parametric tests are appropriate when assumptions such as normality and equal variance are met, offering greater statistical power. Non-parametric tests, while less powerful, are more robust and applicable when data is ordinal, categorical, or does not follow a normal distribution. Both classes of tests are indispensable in computing disciplines—from evaluating algorithms and interfaces to measuring performance and user experience. The choice of test must align with the nature of the data and research objectives to ensure credible and actionable insights.

APPLICATIONS IN MACHINE LEARNING AND DATA SCIENCE

In computing and informatics, hypothesis testing plays a vital role in validating models, comparing performance metrics, and ensuring statistically sound conclusions. Whether comparing classifiers or evaluating model performance using cross-validation, statistical testing forms the backbone of reliable machine learning experimentation.

Comparing Classifier Performance

Machine learning practitioners often face the need to compare the performance of multiple classifiers to determine whether one model significantly outperforms another. Common metrics include accuracy, precision, recall, F1-score, and area under the ROC curve (AUC). However, small differences in these metrics may arise due to chance, especially when datasets are limited or imbalanced.

Table 3. Popular Statistical Tests for Classifier Comparison

Test	Use Case	Assumptions
Paired t-test	Comparing two classifiers on the same dataset using cross-validation	Normality, paired samples
Wilcoxon Signed-Rank Test	Non-parametric alternative to the paired t-test	No assumption of normal distribution
McNemar's Test	Comparing classification results for paired binary outcomes	Suitable for 2×2 contingency tables
ANOVA	Comparing three or more classifiers using accuracy or F1-scores	Normality, homogeneity of variance

Example:

Suppose a researcher compares two spam detection algorithms (Naive Bayes and Random Forest) using 10-fold cross-validation. The mean accuracy values across the folds can be used to perform a paired t-test.

Significance Testing in Cross-Validation

Cross-validation is a fundamental component of modern machine learning pipelines, as it provides an empirical estimate of model performance and reduces the risk of overfitting. Among its variants, *k*-fold cross-validation is the most widely adopted. However, a critical challenge arises from the fact that the performance scores obtained across folds are not statistically independent. Each fold shares a large portion of the data with other folds, leading to correlated estimates. Since many classical significance tests, such as the traditional paired *t*-test, assume independence of observations, directly applying them to cross-validation scores can yield misleading or overly optimistic conclusions.

To address these issues, significance testing in the context of cross-validation requires methods that explicitly account for dependency structures and repeated sampling. Two widely used approaches are nested cross-validation and the 5×2 cross-validated paired *t*-test.

Nested Cross-Validation

Nested cross-validation provides an unbiased estimate of model generalization and is especially important when hyperparameter tuning is part of the workflow. It consists of two loops: an inner loop that performs model selection or hyperparameter optimization, and an outer loop that evaluates the chosen model on unseen data. The performance scores collected from the outer loop represent independent assessments of generalization capability because hyperparameter decisions have been isolated to the inner loop.

Once the outer-loop performance values are obtained, statistical hypothesis testing can be applied to compare models. The independence of these outer-loop scores makes them suitable for significance tests designed for multiple repeated measures. This structure prevents information leakage and ensures that the comparison of models reflects true generalization differences rather than artifacts of the tuning process.

5×2 Cross-Validated Paired t-Test (Dietterich's Test)

Dietterich proposed the 5×2 cross-validated paired *t*-test as a more reliable alternative to the traditional paired *t*-test for comparing two classifiers. This method uses five replications of two-fold cross-validation. In each replication, the dataset is randomly split into two equal halves; the models are trained on one half and tested on the other, and then the roles of the halves are swapped. This process generates ten paired performance differences across the replications.

A key advantage of this design is that it reduces the overlap of training data across runs, thereby mitigating the dependency problem inherent in *k*-fold cross-validation. The variance of the performance differences is estimated across these ten values, providing a more stable and conservative evaluation of whether one model statistically outperforms another. Dietterich demonstrated that this method maintains a more appropriate Type I error rate compared to naïve testing approaches, which tend to falsely detect differences between classifiers due to correlated folds.

Overall, significance testing in cross-validation is essential to ensure that model comparisons are statistically justified and not driven by random fluctuation in the data splits. Techniques such as nested cross-validation and the 5×2 cross-validated paired *t*-test offer robust solutions tailored to the unique dependency structure of machine learning evaluation.

REAL-WORLD CASE STUDIES

In the field of computing and informatics, hypothesis testing is extensively applied in real-world scenarios to support empirical findings with statistical rigour. Two significant and practical applications include A/B testing in Human-Computer Interaction (HCI) and performance assessment of computer networks and systems.

A/B Testing in Human-Computer Interaction (HCI)

A/B testing, a form of controlled experimentation, is widely used in HCI to compare two versions of a user interface, system feature, or interaction mechanism. The core idea is to determine whether a modification in the system (e.g., new layout, colour scheme, navigation button position) leads to a significant improvement in user performance or satisfaction.

For instance, assume two different versions of a login page are designed (Version A and Version B). Researchers hypothesize that Version B, with a cleaner design and larger call-to-action buttons, will result in a higher login success rate within 30 seconds. In this context:

- Null Hypothesis (H_0): There is no difference in login success rates between Version A and Version B.
- Alternative Hypothesis (H_1): Version B leads to a significantly higher login success rate than Version A.

Participants are randomly assigned to both version, and data such as task completion time, error rate, and subjective usability scores (like SUS or NASA-TLX) are collected. Depending on the nature of the data:

- A t-test or ANOVA may be used for normally distributed continuous variables.
- A Mann-Whitney U test or Chi-square test is used for ordinal or categorical data.

This helps determine whether observed improvements are statistically significant or due to random variation. Beyond web interfaces, A/B testing is applied to software menus, mobile app designs, gesture recognition algorithms, and adaptive user interfaces in real-time systems.

Network and System Performance Assessment

Hypothesis testing is also vital in evaluating the efficiency and stability of computing systems and network infrastructures. Whether testing a new routing algorithm, comparing virtualization platforms, or assessing the impact of a caching strategy, statistical testing provides an objective basis for decision-making.

Example: Evaluating a New Routing Protocol

Consider a scenario where a researcher introduces a new routing protocol (Protocol X) and aims to compare it with an existing protocol (Protocol Y) in terms of end-to-end delay and packet loss. The hypotheses could be:

- H_0: There is no significant difference in the average delay between Protocol X and Protocol Y.
- H_1: Protocol X significantly reduces the average delay compared to Protocol Y.

A set of controlled experiments is conducted in a network simulation environment (e.g., NS2, NS3, or OMNeT++) or real testbed, where metrics are recorded over multiple runs. Statistical tests such as:

- A paired t-test (if the same network conditions are used for both protocols),
- A Wilcoxon signed-rank test (if normality cannot be assumed), or
- ANOVA (if more than two protocols are compared)

can be used to determine the significance of the performance difference. Similarly, hypothesis testing is also applied in:

- Cloud computing: Comparing VM performance under different resource allocation policies.
- Database systems: Evaluating query optimization techniques.
- IoT networks: Assessing communication latency or energy efficiency with different protocols.
- Cybersecurity: Determining the effectiveness of intrusion detection systems using precision-recall or false positive rates.

Through such case studies, hypothesis testing not only validates proposed improvements but also contributes to the reproducibility and robustness of research in computing and informatics.

TOOLS AND TECHNIQUES

Hypothesis testing in computing and informatics often requires the integration of statistical tools with computational platforms. A range of software tools and programming environments are used to perform hypothesis testing, ranging from general-purpose programming languages like Python and R to specialized statistical tools like SPSS. This section provides a detailed overview of these tools and their role in facilitating hypothesis testing in both academic research and practical computing applications.

Using Python for Hypothesis Testing

Python has become the preferred programming language in data science, AI, and informatics due to its open-source nature, ease of learning, and vast ecosystem of libraries. When it comes to hypothesis testing, Python offers robust support through two principal libraries: SciPy and statsmodels.

SciPy Library

SciPy's scipy.stats module includes a comprehensive set of statistical functions. These include:

- Independent t-tests (ttest_ind) for comparing means between two groups.
- Paired t-tests (ttest_rel) for comparing means from the same group at different times.
- ANOVA (f_oneway) for comparing means across multiple groups.
- Chi-square tests (chisquare, chi2_contingency) for categorical data analysis.
- Non-parametric tests such as mannwhitneyu, wilcoxon, and kruskal.

These functions provide both the test statistics and p-values, enabling researchers to make informed decisions about their null hypotheses. For example, testing whether two different sorting algorithms have significantly different execution times becomes straightforward using SciPy.

statsmodels Library

While SciPy is focused on individual statistical tests, statsmodels is ideal for building and evaluating entire statistical models. It supports:

- Linear regression models, with hypothesis testing on regression coefficients.
- Generalized linear models (GLMs) for binary or count data.
- Time-series models, including ARIMA and SARIMAX.
- ANOVA tables, confidence intervals, and diagnostic tests.

Statsmodels is particularly valuable when hypothesis testing is part of a broader statistical modeling process, such as determining the significance of features in predictive modeling or evaluating the impact of interventions in A/B testing.

Jupyter-based Examples

Jupyter Notebooks provide an interactive, flexible, and reproducible environment for hypothesis testing. They allow researchers to write live Python code, annotate it with rich text (Markdown), and visualize results—all in one document.

Example: T-test to Compare Algorithm Runtimes

```
python
CopyEdit
import scipy.stats as stats
runtime_algo1 = [2.1, 2.4, 2.3, 2.2, 2.5]
runtime_algo2 = [2.8, 2.7, 2.9, 3.0, 2.6]
t_stat, p_value = stats.ttest_ind(runtime_algo1, runtime_
algo2)
print("T-statistic:", t_stat)
print("P-value:", p_value)
```

This code performs a two-sample independent t-test to compare the average execution times of two algorithms. If the p-value is less than a chosen significance level (e.g., 0.05), we reject the null hypothesis that the two algorithms have the same average runtime.

Jupyter's capability to interleave code, output, charts, and explanations makes it an excellent tool for educational demonstrations, prototyping, and even publishing computational research. Users can also integrate libraries like Matplotlib and Seaborn to visualize data distributions before and after testing.

Using R for Hypothesis Testing

R is a language built specifically for statistical computing and is widely used in research communities across informatics, bioinformatics, and social sciences. It has built-in support for almost all forms of hypothesis testing, including parametric and non-parametric methods.

Built-in Testing Functions:

- t-test: t.test() for one-sample, independent two-sample, and paired tests.
- ANOVA: aov() for comparing more than two means.
- Chi-square tests: chisq.test() for independence and goodness-of-fit.
- Non-parametric tests: wilcox.test() (Wilcoxon), kruskal.test() (Kruskal-Wallis).

Advantages of R:

- Highly customizable: Tests can be performed with numerous options for variance assumptions, confidence levels, and alternative hypotheses.
- Extensive libraries: Packages like car, ggpubr, infer, and MASS enhance hypothesis testing capabilities.
- Graphical strength: R's plotting libraries (e.g., ggplot2) allow visualization of statistical distributions, confidence intervals, and outliers—crucial for interpreting test results.

R is often the first choice in academia for its mathematical rigor and reproducibility. Coupled with RMarkdown or R Notebooks, users can generate comprehensive reports with embedded code and results.

SPSS and GUI-Based Tools

SPSS (Statistical Package for the Social Sciences) is a commercial tool that caters to users who may not have a programming background. It is particularly useful in applied computing scenarios where rapid and user-friendly statistical analysis is required.

Key Features:

- User interface: Drag-and-drop functionality for performing t-tests, ANOVA, correlation tests, and more.
- Extensive output: Automatically generates tables and charts for easy interpretation.
- Syntax editor: For advanced users, SPSS offers a command syntax mode to script repeatable analyses.
- Wide test coverage:

  a) *Parametric*: t-tests, ANOVA, ANCOVA
  b) *Non-parametric*: Mann-Whitney, Wilcoxon, Kruskal-Wallis, Friedman
  c) *Descriptive statistics*: Mean, median, standard deviation, skewness, kurtosis

Relevance in Informatics:

In fields like human-computer interaction (HCI), user surveys and usability testing often use SPSS for analyzing Likert-scale data, task completion times, and user satisfaction metrics. SPSS is also widely adopted in healthcare informatics and education technology research.

Moreover, GUI-based tools like Jamovi and PSPP (an open-source alternative to SPSS) are gaining popularity for providing similar statistical functionality in a more accessible format.

From powerful programming environments like Python and R to GUI-based applications such as SPSS, a variety of tools are available to carry out hypothesis testing effectively in the field of computing and informatics. Python excels in integration with machine learning workflows, R is preferred for high-level statistical rigor, and SPSS remains a staple in applied research for its ease of use. Jupyter Notebooks enhance reproducibility and interactivity, making them ideal for teaching and experimentation. Together, these tools empower researchers and practitioners to validate assumptions, test interventions, and derive actionable insights from data.

COMMON PITFALLS AND BEST PRACTICES

In the fast-evolving domains of computing, informatics, and data science, statistical hypothesis testing plays a foundational role in drawing inferences from data. However, incorrect application or interpretation of statistical tests can lead to flawed conclusions, suboptimal models, or even invalid research claims. This section provides a detailed analysis of common pitfalls encountered in practice and outlines best practices to ensure robust, interpretable, and reproducible results.

Misinterpretation of Results

Overreliance on P-Values

One of the most widespread misinterpretations in statistical inference is the overdependence on p-values without proper context. In computing and informatics research:

- A small p-value (e.g., $p < 0.05$) is often misunderstood as proof of a strong effect. In reality, it simply indicates that the observed data is unlikely under the null hypothesis.
- Conversely, a large p-value ($p > 0.05$) is not evidence of the null hypothesis being true; it may suggest insufficient data or poor study design.

Implication in Informatics: For example, in an A/B testing study of user interface changes in human-computer interaction (HCI), a small p-value may not imply a meaningful improvement unless accompanied by practical significance metrics, such as effect size or user satisfaction scores.

Confusion between Correlation and Causation

In data science workflows, particularly in machine learning pipelines, discovering a statistically significant relationship between two variables is often misinterpreted as indicating causality.

- Tools like Python's statsmodels.OLS() may indicate a strong correlation, but this does not imply one feature causes the other.
- Spurious correlations often arise in high-dimensional data, especially in time series or behavioural data without causal modelling frameworks.

Example: A spike in app downloads may correlate with social media mentions, but attributing the cause solely to one source without controlling for confounders is a misinterpretation.

Ignoring Assumptions of Statistical Tests

Most parametric tests (e.g., t-tests, ANOVA) come with assumptions such as:

- Normality of data distribution
- Homogeneity of variances
- Independence of observations

Violating these assumptions without verification (e.g., via Shapiro-Wilk test, Levene's test) can invalidate the results. In computing, this is particularly dangerous when processing non-normal log files, server performance data, or user behavioural logs, where assumptions are often not met.

Data Snooping and Multiple Comparisons

In computing and machine learning tasks, repeated hypothesis testing (e.g., testing several features or hyperparameter settings) increases the risk of Type I error (false positives).

- Conducting multiple t-tests on dozens of variables (e.g., feature selection) without adjustment leads to p-hacking.

- Failing to apply Bonferroni correction, Holm's method, or False Discovery Rate (FDR) control results in misleading conclusions.

Emphasis on Effect Size and Reproducibility

Effect Size: Moving Beyond P-Values

Effect size quantifies the magnitude of a result, providing more actionable insights than p-values alone. Examples include:

- Cohen's d for mean differences
- Eta squared (η^2) in ANOVA
- Odds ratio or relative risk in classification models

In informatics applications, such as evaluating the performance of a new caching algorithm, a statistically significant speed improvement ($p < 0.05$) may not be practically meaningful unless the effect size is large enough to justify the implementation costs.

Confidence Intervals and Practical Significance

Confidence intervals (CIs) provide a range of values within which the true effect likely lies. Reporting CIs:

- Encourages transparency and understanding of uncertainty
- Prevents binary thinking associated with fixed thresholds like 0.05
- Offers deeper insight into real-world decision making, such as resource allocation for IT infrastructure upgrades

Reproducibility and Transparency in Data Science

Reproducibility is a cornerstone of scientific computing and model development. Best practices include:

- Open source code (e.g., via GitHub, GitLab)
- Data provenance (documenting data sources, preprocessing)
- Version control and use of Jupyter notebooks to trace experiments
- Automated logging of model runs, random seeds, and statistical settings using platforms like MLFlow, Weights & Biases, or TensorBoard

Scenario: A data scientist comparing multiple classifiers using cross-validation should publish not only the performance metrics but also the exact split strategy, random seed, and preprocessing steps to ensure replicability by peers.

Registered Hypotheses and Pre-Specified Analyses

- In fields like HCI, cybersecurity, or computational biology, pre-registration of analysis plans improves credibility.
- This helps avoid confirmation bias and post hoc rationalization of results.

Tools like Open Science Framework (OSF) and Jupyter Notebooks with embedded Markdown facilitate transparent reporting.

CONCLUSION AND KEY TAKEAWAYS

Hypothesis testing remains one of the foundational elements of statistical reasoning in computing and informatics. From traditional parametric methods, such as t-tests and ANOVA, to non-parametric approaches like the Mann-Whitney U test and the chi-square test, the field has witnessed the versatile application of statistical tests in various computing contexts. These include machine learning performance evaluation, A/B testing in HCI environments, and systems analysis for network optimisation. A crucial takeaway is the alignment of the choice of hypothesis test with the nature of the data and the assumptions that can be reasonably made. Parametric tests are powerful when their assumptions are met—such as normality and homogeneity of variances—but they may yield misleading results in their absence. Non-parametric tests serve as robust alternatives, especially when ordinal data, small samples, or distributional assumptions cannot be confirmed.

Statistical testing in computing research is not limited to theoretical domains—it has practical implications in model selection, system tuning, and user experience evaluation. For example, when comparing multiple classifiers using cross-validation, statistical significance tests such as the Wilcoxon signed-rank test or the paired t-test can provide deeper insight into the performance differences beyond simple accuracy metrics. Another key point is the increasing emphasis on effect size and reproducibility. Reporting p-values without considering the magnitude of the observed effect can lead to overemphasis on statistical significance without understanding the practical implications. Modern research in informatics is progressively recognising this by incorporating effect size indicators and confidence intervals in reporting standards.

Best practices also emphasise transparency in hypothesis formulation, pre-registration of tests, proper control of multiple comparisons, and ensuring data quality prior to testing. Researchers are encouraged to document their assumptions, choice of test, and software implementation details to foster reproducibility. Bayesian hypothesis testing is emerging as a valuable complement to frequentist approaches, providing a probabilistic framework that enables researchers to quantify the strength of evidence and update their beliefs as more data become available. This aligns well with the dynamic and iterative nature of computing research, particularly in areas like adaptive systems, learning algorithms, and real-time analytics.

Finally, the integration of hypothesis testing frameworks into interactive environments, such as Jupyter notebooks, along with the use of tools like Python's scipy.stats or R's stats and car packages, democratises access to rigorous statistical analysis, enabling both students and professionals to validate their assumptions and make data-driven decisions.

In summary, effective hypothesis testing in computing encompasses not only statistical correctness but also interpretability, contextual understanding, and reproducibility. The evolving landscape encourages researchers to blend traditional methods with emerging practices to enhance the validity, impact, and ethical rigour of their research.

REFERENCES

Ahlswede, R., & Csiszár, I. (2003). Hypothesis testing with communication constraints. *IEEE Transactions on Information Theory*, *32*(4), 533–542.

Al-Khaiat, S. S. J., Noori, M. Z., & Cengiz, M. A. (2022). Application of the regression analysis in Python, SPSS and Microsoft Excel programs. *Journal of Current Research on Educational Studies*, *12*(2), 27–46.

Anderson, D. R., Burnham, K. P., & Thompson, W. L. (2000). Null hypothesis testing: Problems, prevalence, and an alternative. *The Journal of Wildlife Management*, •••, 912–923.

Blahut, R. (2003). Hypothesis testing and information theory. *IEEE Transactions on Information Theory*, *20*(4), 405–417.

Carterette, B., & Smucker, M. D. (2007, November). Hypothesis testing with incomplete relevance judgments. In *Proceedings of the Sixteenth ACM Conference on Information and Knowledge Management* (pp. 643–652). ACM.

Cover, T. M. (1969). Hypothesis testing with finite statistics. *Annals of Mathematical Statistics*, *40*(3), 828–835.

Fangwei, F., & Shiyi, S. (1996). Hypothesis testing for arbitrarily varying source. *Shuxue Xuebao*, *12*(1), 33–39.

Feder, M., & Merhav, N. (2002). Universal composite hypothesis testing: A competitive minimax approach. *IEEE Transactions on Information Theory*, *48*(6), 1504–1517.

Gonçalves, M. C., & Silva, R. (2023, September). The effect of statistical hypothesis testing on machine learning model selection. In *Brazilian Conference on Intelligent Systems* (pp. 415–427). Springer Nature.

Good, P. (2013). *Permutation tests: A practical guide to resampling methods for testing hypotheses* (3rd ed.). Springer Science and Business Media.

Haroutunian, E. A., Haroutunian, M. E., & Harutyunyan, A. N. (2008). Reliability criteria in information theory and in statistical hypothesis testing. *Foundations and Trends in Communications and Information Theory*, *4*(2–3), 97–263.

Haslwanter, T. (2016). *An introduction to statistics with Python: With applications in the life sciences*. Springer International Publishing.

Hogg, R. V. (1961). On the resolution of statistical hypotheses. *Journal of the American Statistical Association*, *56*(296), 978–989.

Keller, D., & Egbert, J. (2020). Hypothesis testing ordinary meaning. *Brooklyn Law Review*, *86*, 489.

Larson-Hall, J., & Mizumoto, A. (2019). Using statistical analysis software (R, SPSS). In *The Routledge handbook of research methods in applied linguistics* (pp. 385–397). Routledge.

McCormick, K., & Salcedo, J. (2017). *SPSS statistics for data analysis and visualization*. John Wiley and Sons.

McKenzie, C. R. (2004). Hypothesis testing and evaluation. In Koehler, D. J., & Harvey, N. (Eds.), *Blackwell handbook of judgment and decision making* (pp. 200–219). Blackwell Publishing.

Mizrahi, M. (2020). Hypothesis testing in scientific practice: An empirical study. *International Studies in the Philosophy of Science*, *33*(1), 1–21.

Newman, M. C. (2008). "What exactly are you inferring?" A closer look at hypothesis testing. *Environmental Toxicology and Chemistry*, *27*(5), 1013–1019. PMID: 18419185

Neyman, J. (1942). Basic ideas and some recent results of the theory of testing statistical hypotheses. *Journal of the Royal Statistical Society*, *105*(4), 292–327.

Python Software Foundation. (2024). Python official documentation. Retrieved on 10 October 2025 from https://docs.python.org/

Raftery, A. E., Gilks, W. R., Richardson, S., & Spiegelhalter, D. (1995). Hypothesis testing and model. In W. R. Gilks, S. Richardson, & D. Spiegelhalter (Eds.), *Markov Chain Monte Carlo in practice* (pp. 165–187). Chapman and Hall.

Raschka, S. (2020). [*Machine learning lecture notes*. University of Wisconsin–Madison.]. *Stat*, •••, 451.

Scerbo, M. W., Calhoun, A. W., & Hui, J. (2019). Research and hypothesis testing: Moving from theory to experiment. In *Healthcare simulation research: A practical guide* (pp. 161–167). Springer International Publishing.

SciPy Community. (2024). SciPy user guide and tutorials. Retrieved on 10 October 2025 from https://docs.scipy.org/

Sedgwick, P. (2013). Sample size: How many participants are needed in a trial? *BMJ (Clinical Research Ed.)*, •••, 346. PMID: 23418280

Sedgwick, P. (2014). Pitfalls of statistical hypothesis testing: Multiple testing. *BMJ (Clinical Research Ed.)*, *349*, g6288. PMID: 25172184

Sobhiyeh, S., & Naraghi-Pour, M. (2016). Hypothesis testing with dependent observations. *IEEE Transactions on Signal Processing*, *65*(5), 1183–1195.

Statsmodels Developers. (2024). Statsmodels documentation and examples. Retrieved on 10 October 2025 from https://www.statsmodels.org/

Tuncel, E. (2005). On error exponents in hypothesis testing. *IEEE Transactions on Information Theory*, *51*(8), 2945–2950.

Wang, H. I., Manolas, C., & Xanthidis, D. (2022). Statistical analysis with Python. In Hamada, T. M. (Ed.), *Handbook of computer programming with Python* (pp. 373–408). Chapman and Hall/CRC.

Wu, S., Diao, E., Elkhalil, K., Ding, J., & Tarokh, V. (2022). Score-based hypothesis testing for unnormalized models. *IEEE Access : Practical Innovations, Open Solutions*, *10*, 71936–71950.

Xiao, M., Wei, P., & Tai, H. M. (2012). Estimation of the number of sources based on hypothesis testing. *Journal of Communications and Networks (Seoul)*, *14*(5), 481–486.

Zeitouni, O., & Gutman, M. (2002). On universal hypothesis testing via large deviations. *IEEE Transactions on Information Theory*, *37*(2), 285–290.

Compilation of References

Abatan, S. M., & Olayemi, M. S. (2014). The role of statistical software in data analysis. *International Journal of Applied Research and Studies*, *3*(8), 1–15.

Abbasi, A., Parsons, J., Pant, G., Liu Sheng, O. R., & Sarker, S. (2024). Pathways for design research on artificial intelligence. *Information Systems Research*, *35*(2), 441–459. DOI: 10.1287/isre.2024.editorial.v35.n2

Abdul-Jabbar, S. S., & Farhan, A. K. (2022). Data analytics and techniques: A review. *ARO-The Scientific Journal of Koya University*, *10*(2), 142–164. DOI: 10.14500/aro.10975

Adeleye, J. O. (2017). *Pragmatism and its implications on the teaching and learning in Nigerian schools*. Retrieved on 10 October 2025 from https://www.isres.org/pragmatism-and-its-implications-on-teaching-and-learning-in-nigerian-schools-51-s.html

Adhikari, R., & Timsina, T. P. (2024). An Educational study focused on the application of mixed method approach as a research method. *OCEM Journal of Management Technology & Social Sciences*, *3*(1), 94–109. DOI: 10.3126/ocemjmtss.v3i1.62229

Adom, D., Hussein, E. K., & Agyem, J. A. (2018). Theoretical and conceptual framework: Mandatory ingredients of a quality research. *International Journal of Scientific Research*, *7*(1), 438–441.

Afifuddin, M., & Seo, W. (2025). Predicting technological convergence through patent network analysis. *Journal of Innovation and Technology*, *18*(3), 112–130. DOI: 10.1016/j.jit.2025.01.003

Afradi, K., & Nourian, F. (2022). Understanding ICT's impacts on urban spaces: A qualitative content analysis of literature. *GeoJournal*, *87*(2), 701–731. DOI: 10.1007/s10708-020-10277-2

Agrawal, M., Peterson, J. C., & Griffiths, T. L. (2020). Scaling up psychology via scientific regret minimization. *Proceedings of the National Academy of Sciences of the United States of America, 117*(16), 8825–8835. DOI: 10.1073/pnas.1915841117 PMID: 32241896

Aguas, P. P. (2022). Fusing approaches in educational research: Data collection and data analysis in phenomenological research. *The Qualitative Report, 27*(1), 1–20.

Agudamu, Li, Y., Mi, N., & Pan, X. (2025). A computational grounded theory–based analysis of research on China's old-age social welfare system. *Frontiers in Public Health, 13*, 1556302. Advance online publication. DOI: 10.3389/fpubh.2025.1556302

Ahlswede, R., & Csiszár, I. (2003). Hypothesis testing with communication constraints. *IEEE Transactions on Information Theory, 32*(4), 533–542.

Ahmed, S. K. (2024). The pillars of trustworthiness in qualitative research. *Journal of Medicine, Surgery, and Public Health, 2*, 100051. DOI: 10.1016/j.glmedi.2024.100051

Ahmed, S. K., Mohammed, R. A., Nashwan, A. J., Ibrahim, R. H., Abdalla, A. Q., Ameen, B. M. M., & Khdhir, R. M. (2025). Using thematic analysis in qualitative research. *Journal of Medicine, Surgery, and Public Health, 6*, 100198. Akyıldız, S. T., & Ahmed, K. H. (2021). An overview of qualitative research and focus group discussion. *International Journal of Academic Research in Education, 7*(1), 1–15. Retrieved October 10, 2025, from https://dergipark.org.tr/en/pub/ijare/article/866762

Åkerblad, L., Seppänen-Järvelä, R., & Haapakoski, K. (2020). Integrative strategies in mixed methods research. *Journal of Mixed Methods Research, 15*(2), 152–170. DOI: 10.1177/1558689820957125

Akhai, S. (2024). A review on optimizations in µ-edm machining of the biomedical material Ti6Al4V using the Taguchi method: Recent advances since 2020. *Latest Trends in Engineering and Technology*, 395-402.

Alam, Md. S., Asmawi, A., & Fatema, S. (2025). Focus group discussions (FGDs) for qualitative and mixed-method approaches: Principles, applications, and key considerations. *Integrated Journal for Research in Arts and Humanities, 5*(4), 51–56. DOI: 10.55544/ijrah.5.4.8

Alele, F., & Malau-Aduli, B. (2023). *An Introduction to research methods for undergraduate health profession students*. James Cook University. Retrieved on 10 October 2025 from https://jcu.pressbooks.pub/intro-res-methods-health/

Alhazmi, A. A., & Kaufmann, A. (2022). Phenomenological qualitative methods applied to the analysis of cross-cultural experience in novel educational social contexts. *Frontiers in Psychology, 13*, 785134. Retrieved October 10, 2025, from https://www.frontiersin.org/articles/10.3389/fpsyg.2022.785134/full PMID: 35548502

Al-Khaiat, S. S. J., Noori, M. Z., & Cengiz, M. A. (2022). Application of the regression analysis in Python, SPSS and Microsoft Excel programs. *Journal of Current Research on Educational Studies, 12*(2), 27–46.

Alkhalil, A., Abdallah, M. A. E., Alogali, A., & Aljaloud, A. (2021). Applying big data analytics in higher education. *International Journal of Information and Communication Technology Education, 17*(3), 29–51. DOI: 10.4018/ijicte.20210701.oa3

Almalki, S. (2016). Integrating quantitative and qualitative data in mixed methods research: Challenges and benefits. *Journal of Education and Learning, 5*(3), 288–296. DOI: 10.5539/jel.v5n3p288

Alonso-Garcia, J., Pablo-Martí, F., & Nunez-Barriopedro, E. (2021). Omnichannel Management in a B2B context: Concept, research agenda and bibliometric review. *International Journal of Industrial Engineering and Management, 12*(1), 37.

Alvesson, M., & Skoldberg, K. (2000). *Reflexive methodology: New vistas for qualitative research.* SAGE.

Alzahrani, A. S., Tsai, Y.-S., Aljohani, N., Whitelock-Wainwright, E., & Gasevic, D. (2023). Do teaching staff trust stakeholders and tools in learning analytics? A mixed methods study. *International Journal of Educational Technology in Higher Education, 20*(1), 13. Advance online publication. DOI: 10.1186/s41239-023-00387-1 PMID: 36874535

American Psychological Association. (2024). Journal article reporting standards for mixed methods research (JARS). Retrieved on 21 November from https://apastyle.apa.org/jars/mixed-methods

Andersen, N. (2021). Mapping the expatriate literature: A bibliometric review of the field from 1998 to 2017 and identification of current research fronts. *International Journal of Human Resource Management, 32*(22), 4687–4724. DOI: 10.1080/09585192.2019.1661267

Anderson, D. R., Burnham, K. P., & Thompson, W. L. (2000). Null hypothesis testing: Problems, prevalence, and an alternative. *The Journal of Wildlife Management*, •••, 912–923.

Anderson, V. (2017). Criteria for evaluating qualitative research. *Human Resource Development Quarterly, 28*(2), 125–133.

Anfara, V. A., & Mertz, N. T. (Eds.). (2015). *Theoretical frameworks in qualitative research* (2nd ed.). SAGE.

Arafat, M. Y. (2024). Research ethics: Meaning and principles. *International Journal of Academic Multidisciplinary Research, 8*(1). Retrieved October 10, 2025, from https://papers.ssrn.com/sol3/papers.cfm?abstract_id=4887943

Aria, M., & Cuccurullo, C. (2017). Bibliometrix: An R-tool for comprehensive science mapping analysis. *Journal of Informetrics, 11*(4), 959–975.

Arnold, M. E., & Pistilli, M. D. (2012). Course signals at Purdue: Using learning analytics to increase student success. In *Proceedings of the 2nd International Conference on Learning Analytics and Knowledge* (pp. 267-270). ACM. https://doi.org/ DOI: 10.1145/2330601.2330666

Atkinson, P. A., & Coffey, A. (1997). Analysing documentary realities. *Qualitative Research: Theory, Method and Practice,* 56–75.

Atkins, S., Launiala, A., Kagaha, A., & Smith, H. (2012). Including mixed methods research in systematic reviews: Examples from qualitative syntheses in TB and malaria control. *BMC Medical Research Methodology, 12*(1). Advance online publication. DOI: 10.1186/1471-2288-12-62 PMID: 22545681

Atuase, D., & Maluleka, J. (2023). Are libraries up to the task? Bridging the gap between libraries and distance-learning students in Ghana. *Journal of Library & Information Services in Distance Learning, 17*(3-4), 131–153. DOI: 10.1080/1533290X.2023.2239221

Ayaz, M., Khan, S., & Rahman, A. (2025). Multilayer dynamic data forwarding in underwater sensor networks. *IEEE Transactions on Mobile Computing, 24*(2), 45–67. DOI: 10.1109/TMC.2025.1234567

Ayton, D. (2023). *Phenomenology. Qualitative research–a practical guide for health and social care researchers and practitioners*. Retrieved on 10 October 2025 from https://oercollective.caul.edu.au/qualitative-research/chapter/__unknown__-6/

Ayton, D., Tsindos, T., & Berkovic, D. (2024). *Qualitative research: A practical guide for health and social care researchers and practitioners*. Monash University.

Babbie, E., & Mouton, J. (2001). *The practice of social research: An introduction to reading research*. Oxford University Press.

Badil, D. M., Aslam, Z., Khan, K., Ashiq, A., & Bibi, U. B. U. (2023). Phenomenology qualitative research inquiry: A review paper. *Pakistan Journal of Health Sciences, 4*(3), 09–13.

Bans-Akutey, A., & Tiimub, B. M. (2021). Triangulation in research. *Academia Letters. Article, 3392*, 1–6. DOI: 10.20935/AL3392

Barbour, R. S., & Barbour, M. K. (2003). Evaluating and synthesizing qualitative research: The need to develop a distinctive approach. *Journal of Evaluation in Clinical Practice, 9*(2), 179–186. DOI: 10.1046/j.1365-2753.2003.00371.x PMID: 12787181

Barrow, J. M., Brannan, G. D., & Khandhar, P. B. (2017). *Research ethics*. Retrieved on 10 October 2025 from https://europepmc.org/books/nbk459281

Bartholomew, T. T., Joy, E. E., Kang, E., & Brown, J. (2021). A choir or cacophony? Sample sizes and quality of conveying participants' voices in phenomenological research. *Methodological Innovations, 14*(2), 205979912110400. DOI: 10.1177/20597991211040063

Bartlett, C., Marshall, M., & Marshall, A. (2012). Two-eyed seeing and other lessons learned within a co-learning journey of bringing together indigenous and mainstream knowledges and ways of knowing. *Journal of Environmental Studies and Sciences, 2*(4), 331–340. DOI: 10.1007/s13412-012-0086-8

Battaglia, M. (2008). Convenience sampling. *Encyclopedia of Survey Research Methods, 1*, 148–149.

Bazeley, P. (2004). Issues in mixing qualitative and quantitative approaches to research. In Buber, R., Gadner, J., & Richards, L. (Eds.), *Applying qualitative methods to marketing management research* (pp. 141–156). Palgrave Macmillan.

Bazeley, P. (2009). Editorial: Integrating data analyses in mixed methods research. *Journal of Mixed Methods Research, 3*(3), 203–207. DOI: 10.1177/1558689809334443

Bazeley, P. (2018). *Integrating analyses in mixed methods research*. SAGE Publications.

Bellaj, M., Bendahmane, A., Boudra, S., & Lamarti, M. S. (2024). Big data analytics in higher education: A new adaptive learning analytics model integrating traditional approaches. [IJIM]. *International Journal of Interactive Mobile Technologies, 18*(06), 24–39. DOI: 10.3991/ijim.v18i06.46289

Bhattacherjee, A. (2012). *Social Science Research: Principles, Methods, and Practices*. Retrieved on 10 October 2025 from https://digitalcommons.usf.edu/oa_textbooks/3

Billups, F. (2014). The quest for rigor in qualitative studies; strategies for institutional research. *The NERA Researcher, 52*, 10–12.

Blackstone, A. (2019). *Social research: Qualitative and quantitative methods*. Flat World.

Blahut, R. (2003). Hypothesis testing and information theory. *IEEE Transactions on Information Theory*, *20*(4), 405–417.

Blalock, H. M. (1969). *Theory construction: From verbal to mathematical formulations*. Prentice-Hall.

Blaxter, L., Hughes, C., & Tight, M. (2006). *How to research* (3rd ed.). Open University Press.

Blumer, H. (1969). *Symbolic interaction: Perspectives and method*. University of California Press.

Boell, S. K., & Cecez-Kecmanovic, D. (2014). A hermeneutic approach for conducting literature reviews and literature searches. *Communications of the Association for Information Systems*, *34*(1), 12.

Booth, W. C., Colomb, G. G., & Williams, J. M. (2024). *The craft of research* (4th ed.). University of Chicago Press.

Bordens, K. S., & Abbott, B. B. (2018). *Research design and methods: A process approach* (10th ed. ed.). McGraw-Hill.

Bowen, G. (2009). Document Analysis as a Qualitative Research Method. *Qualitative Research Journal*, *9*(2), 27–40.

Bowie, B. H., & Wojnar, D. (2015). Using a phenomenology as a research method in community-based research. In *Nursing research using phenomenology: Qualitative designs and methods in nursing* (pp. 145–155). Springer Publishing Company.

Brannen, J. (2005). Working qualitatively and quantitatively. In Seale, C., Gobo, G., Gubrium, J. F., & Silverman, D. (Eds.), *Qualitative Research Practice*. Sage.

Braun, V., Clarke, V., Hayfield, N., & Terry, G. (2019). Thematic analysis. *In P. Liamputtong (Ed.), Handbook of research methods in health social sciences* (pp. 843-860). Springer. California State Department of Education. 10 quick ways to analyze children's books for racism and sexism. Retrieved on 16 October 2025 fromhttps://cmascanada.ca/wp-content/uploads/2011/11/article-10-ways-to-analyze-childrens-books-for-sexism-and racism.pdf

Braun, V., & Clarke, V. (2013). *Successful qualitative research: A practical guide for beginners*. Sage Publishers.

Braun, V., & Clarke, V. (2023). Toward good practice in thematic analysis: Avoiding common problems and be(com)ing a *knowing* researcher. *International Journal of Transgender Health, 24*(1), 1–6. DOI: 10.1080/26895269.2022.2129597 PMID: 36713144

Breiman, L. (2001). Statistical modeling: The two cultures (with comments and a rejoinder by the author). *Statistical Science, 16*(3), 199–231.

Breslin, D., & Gatrell, C. (2023). Theorizing through literature reviews: The miner-prospector continuum. *Organizational Research Methods, 26*(1), 139–167.

Briegel, H. J., & Müller, T. (2025). Phenomenology. In Briegel, H. J., & Müller, T. (Eds.), *Projective simulation in action* (Vol. 507, pp. 333–374). Springer Nature Switzerland., DOI: 10.1007/978-3-031-98119-7_6

Bryman, A. (2006). Integrating quantitative and qualitative research: How is it done? *Qualitative Research, 6*(1), 97–113.

Bryman, A. (2007). Barriers to integrating quantitative and qualitative research. *Journal of Mixed Methods Research, 1*(1), 8–22. DOI: 10.1177/2345678906290531

Bryman, A. (2016). *Social research methods* (5th ed.). Oxford University Press.

Burrell, J. (2016). How the machine "thinks": Understanding opacity in machine learning algorithms. *Big Data & Society, 3*(1), 1–12. DOI: 10.1177/2053951715622512

Buunaaisie, C., Manyara, A. M., Annett, H., Bird, E. L., Bray, I., Ige, J., Jones, M., Orme, J., Pilkington, P., & Evans, D. (2018). Employability and career experiences of international graduates of MSc public health: A mixed methods study. *Public Health, 160*, 62–69. DOI: 10.1016/j.puhe.2018.03.032 PMID: 29751223

Calicdan, T. N. P., & Gementiza-Cubio, L. (2025). Student engagement in learning algebra: A mixed method study. *EPRA International Journal of Multidisciplinary Research (IJMR)*, 858–872. https://doi.org/.DOI: 10.36713/epra19894

Cameron, R., & Miller, P. (2007). Mixed methods research: Phoenix of the paradigm wars. In R. Chapman (Ed.), *Managing our intellectual and social capital: Proceedings of the 21st ANZAM Conference, Sydney, 4–7 December 2007.*

Cameron, R., Dwyer, T., Richardson, S., Ahmed, E., & Sukumaran, A. (2013). Lessons from the field: Applying the Good Reporting of a Mixed Methods Study (GRAMMS) framework. *Electronic Journal of Business Research Methods, 11*(2), 55–66.

Cardno, C. (2018). Policy Document Analysis: A practical educational leadership tool and a qualitative research method. *Educational Administration: Theory & Practice*, *24*(4), 623–640.

Carterette, B., & Smucker, M. D. (2007, November). Hypothesis testing with incomplete relevance judgments. In *Proceedings of the Sixteenth ACM Conference on Information and Knowledge Management* (pp. 643–652). ACM.

Caruth, G. D. (2013). Demystifying mixed methods research design: A review of the literature. *Mevlana International Journal of Education*, *3*(2), 112–122.

Caspari-Sadeghi, S. (2023). Learning assessment in the age of big data: Learning analytics in higher education. *Cogent Education*, *10*(1), 2162697. Advance online publication. DOI: 10.1080/2331186X.2022.2162697

Chakabwata, W., & McKay, V. (2026). Applying theoretical and conceptual frameworks to doctoral research in education. In Throne, R. (Ed.), *Navigating theory, methodology, and researcher development in doctoral research* (pp. 1–28). IGI Global., DOI: 10.4018/979-8-3373-5676-1.ch001

Chaparro-Amaya, E. J., Restrepo-Calle, F., & Ramírez-Echeverry, J. J. (2023). Discovering insights in learning analytics through a mixed-methods framework: Application to computer programming education. *Journal of Information Technology Education*, *22*, 339–372. DOI: 10.28945/5182

Charmaz, K. (2014). *Constructing grounded theory* (2nd ed.). SAGE.

Charmaz, K. (2014). *Constructing grounded theory: A practical guide through qualitative analysis* (2nd ed.). Sage.

Chaturvedi, K. (2016). Sampling methods. Retrieved on 10 October 2025 from https://www.pitt.edu/~super7/43011-44001/43911.pdf

Chen, Z., & Chen, Y. (2024). Computing grounded theory: A quantitative method to develop theories. *The Journal of Chinese Sociology*, *11*, 17. Advance online publication. DOI: 10.1186/s40711-024-00218-8

Chilisa, B., & Tsheko, G. N. (2014). Mixed methods in indigenous research: Building relationships for sustainable intervention outcomes. *Journal of Mixed Methods Research*, *8*(3), 222–233. DOI: 10.1177/1558689814527878

Chukwuere, J. (2021). Theoretical and conceptual framework: A critical part of information systems research process and writing. *Review of International Geographical Education*, *11*(9), 2678–2683. DOI: 10.48047/rigeo.11.09.234

Cicero, M. T., Sutton, E. W., & Rackham, H. (1977). *De Oratore; De Fato; Paradoxa Stoicorum; De Partitione Oratoria*. Harvard University Press.

CIPESA—Collaboration on International ICT Policy for East and Southern Africa. (2024, March 11). *Towards ethical AI regulation in Africa*. https://cipesa.org/2024/03/towards-ethical-ai-regulation-in-africa/

Clark, J.-A., & Tuffley, D. (2019). Learning analytics implementations in universities: Towards a model of success, using multiple case studies. In *ASCILITE 2019 Conference Proceedings: Personalised Learning. Diverse Goals. One Heart* (pp. 167-177). ASCILITE. https://doi.org/DOI: 10.14742/apubs.2019.15

Cleland, J. A. (2017). The qualitative orientation in medical education research. *Korean Journal of Medical Education*, (2), 61–71. PMID: 28597869

Collins, K. M. T. (2017). Sampling decisions in educational research. In Wyse, D., Smith, E., Suter, L., & Selwyn, N. (Eds.), *British Educational Research Association (BERA/SAGE handbook of educational research)* (Vol. 1, pp. 280–292). Sage.

Collins, K. M. T. (2017). Sampling designs in educational research. In Wyse, D., Smith, E., Suter, L., & Selwyn, N. (Eds.), *SAGE Handbook of Educational Research* (pp. 280–292). Sage.

Collins, K. M. T., Onwuegbuzie, A. J., & Jiao, Q. G. (2007). A mixed methods investigation of mixed methods sampling designs in social and health science research. *Journal of Mixed Methods Research*, *1*(3), 267–294. DOI: 10.1177/1558689807299526

Connelly, L. M. (2016). Trustworthiness in qualitative research. *Medsurg Nursing*, *25*(6), 435–436. PMID: 30304614

Coontz, P. (2008). The responsible conduct of social research. In Yang, K., & Miller, G. J. (Eds.), *Handbook of research methods in public administration*. Taylor and Francis Group.

Coppock, A. (2025). *Human-in-the-loop AI for qualitative research: Challenges and opportunities* [Working paper]. Harvard Kennedy School.

Corbin, J., & Strauss, A. (2015). *Basics of qualitative research: Techniques and procedures for developing grounded theory* (4th ed.). SAGE.

Corley, G. K., & Gioia, A. D. (2011). Building theory about theory building: What constitutes a theoretical contribution? *Academy of Management Review*, *36*(1), 12–32.

Corrigan, J. A., & Onwuegbuzie, A. J. (2020). Toward a meta-framework for conducting mixed methods representation analyses to optimize meta-inferences. *The Qualitative Report*. Advance online publication. DOI: 10.46743/2160-3715/2020.3579

Cover, T. M. (1969). Hypothesis testing with finite statistics. *Annals of Mathematical Statistics*, *40*(3), 828–835.

Crawford, L. M. (2019). Conceptual and theoretical frameworks in research. In Burkholder, G. J., Cox, K. A., Crawford, L. M., & Hitchcock, J. H. (Eds.), *Research design and methods: An applied guide for the scholar-practitioner* (pp. 35–48). SAGE.

Creedon, R. (2016). *A mixed method study to explore the impact of nurse prescribing in clinical practice* (Unpublished doctoral dissertation). University College Cork, Cork, Ireland. Retrieved on 10 October 2025 from https://cora.ucc.ie/server/api/core/bitstreams/d8c1d35a-4e3b-4978-8518-95f7334c1828/content

Creswell, J. W. (1994). Research designs: Qualitative and quantitative approaches. *Sage (Atlanta, Ga.).*

Creswell, J. W. (2009). *Research design: Qualitative, quantitative and mixed methods approaches* (3rd ed.). Sage.

Creswell, J. W. (2014). Qualitative, quantitative and mixed methods approaches. *Sage (Atlanta, Ga.).*

Creswell, J. W. (2021). A concise introduction to mixed methods research. *Sage (Atlanta, Ga.).*

Creswell, J. W., & Creswell, J. D. (2018). Research design: Qualitative, quantitative, and mixed methods approaches. *Sage (Atlanta, Ga.).*

Creswell, J. W., & Creswell, J. D. (2022). *Research design: Qualitative, quantitative, and mixed methods approaches* (6th ed.). SAGE Publications.

Creswell, J. W., & Plano Clark, V. L. (2007). Designing and conducting mixed methods research. *Sage (Atlanta, Ga.).*

Creswell, J. W., & Plano Clark, V. L. (2018). *Designing and conducting mixed methods research* (3rd ed.). SAGE Publications., DOI: 10.1177/1937586719832223

Creswell, J. W., & Poth, C. N. (2018). Qualitative inquiry and research design choosing among five approaches. *Sage (Atlanta, Ga.).*

Creswell, J. W., & Poth, C. N. (2024). *Qualitative inquiry and research design: Choosing among five approaches* (5th ed.). SAGE Publications.

Critical Appraisal Skills Programme. (2025). *CASP checklists.* Retrieved on 15 November 2025 from https://casp-uk.net/casp-tools-checklists/

da Silva, G., & de Souza Santos, R. (2023, October). Comparing mobile testing tools using documentary analysis. In *2023 ACM/IEEE International Symposium on Empirical Software Engineering and Measurement (ESEM)* (pp. 1-6). IEEE.'

Dalkin, S., Forster, N., Hodgson, P., Lhussier, M., & Carr, S. M. (2021). Using computer assisted qualitative data analysis software (CAQDAS; NVivo) to assist in the complex process of realist theory generation, refinement and testing. *International Journal of Social Research Methodology*, *24*(1), 123–134. DOI: 10.1080/13645579.2020.1803528

Dangal, M., & Joshi, R. (2020). Hermeneutic phenomenology: Essence in educational research. *Open Journal for Studies in Philosophy*, *4*(1). Advance online publication. DOI: 10.32591/coas.ojsp.0401.03025d

Daruhadi, G. (2024). Phenomenological Method as A Theoretical Basis of Qualitative Methods. *International Journal of Social Health*, *3*(9), 599–613. DOI: 10.58860/ijsh.v3i9.238

Dastin, J. (2018, October 10). Amazon scraps secret AI recruiting tool that showed bias against women. *Reuters*. Retrieved on 19 October 2025 from https://www.reuters.com/article/us-amazon-com-jobs-ai-idUSKCN1MK08G

Dawadi, S., Shrestha, S., & Giri, R. A. (2021). Mixed methods research: A discussion on its types, challenges, and criticisms. *Journal of Practical Studies in Education*, *2*(2), 25–36.

Dawadi, S., Shrestha, S., & Giri, R. A. (2021). Mixed-methods research: A discussion on its types, challenges, and criticisms. *Journal of Practical Studies in Education*, *2*(2), 25–36. https://doi.org/DOI:10.46809/jpse.v2i2.20

De Vos, A. S., Strydom, H., Fouché, C. B., & Delport, C. S. L. (2011). *Research at grass roots: For the social sciences and human services profession* (4th ed.). Van Schaik.

de Waal, C. (2001). *On Peirce*. Wadsworth.

Denzin, N. K. (2017). *The research act: A theoretical introduction to sociological methods*. Routledge.

Dewasiri, N. J., Weerakoon, Y. K. B., & Azeez, A. A. (2018). Mixed methods in finance research: The rationale and research designs. *International Journal of Qualitative Methods*, *17*, 1–13.

Dilthey, W. (1989). *Selected works: Introduction to the human sciences* (Vol. 1). Princeton University Press. (Original work published 1883)

Djatmiko, G. H., Sinaga, O., & Pawirosumarto, S. (2025). Digital transformation and social inclusion in public services: A qualitative analysis of e-government adoption for marginalized communities in sustainable governance. *Sustainability, 17*(7), 2908. DOI: 10.3390/su17072908

Donthu, N., Kumar, S., Mukherjee, D., Pandey, N., & Lim, W. M. (2021). How to conduct a bibliometric analysis: An overview and guidelines. *Journal of Business Research, 133*, 285–296.

Dovetail Editorial Team. (2023). Mixed methods research guide with examples. https://dovetail.com/research/mixed-methods-research/

Downs, N., Davidson, J., Haddad, A., & Zisook, S. (2024). Learning from health care counselors' perspectives on health care worker distress: A qualitative analysis. *HCA Healthcare Journal of Medicine, 5*(3), 12. DOI: 10.36518/2689-0216.1613 PMID: 39015588

Dubin, R. (1969). *Theory building*. 2nd ed. Free Press, Collier-Macmillan

Dunne, B., Pettigrew, J., & Robinson, K. (2016). Using historical documentary methods to explore the history of occupational therapy. *British Journal of Occupational Therapy, 79*(6), 376–384.

Dunwoodie, K., Macaulay, L., & Newman, A. (2023). Qualitative interviewing in the field of work and organisational psychology: Benefits, challenges and guidelines for researchers and reviewers. *Applied Psychology, 72*(2), 863–889. DOI: 10.1111/apps.12414

Dyulicheva, Y. Yu. (2024). Application of learning analytics in higher education: Datasets, methods and tools. *Vysshee Obrazovanie v Rossii Higher Education in Russia, 33*(5), 86–111. https://doi.org/10.31992/0869-3617-2024-33-5-86-111.

Eisenhardt, K. M. (1989). Building theories from case study research. *Academy of Management Review, 14*(4), 532–550.

Eke, D. (2024). Ethics and governance of neurotechnology in Africa: Lessons from AI. *JMIR Neurotechnology, 3*, e56665. Advance online publication. DOI: 10.2196/56665

Ellard-Gray, A., Jeffrey, N. K., Choubak, M., & Crann, S. E. (2015). Finding the hidden participant: Solutions for recruiting hidden, hard-to-reach, and vulnerable populations. *International Journal of Qualitative Methods, 14*(5), 1–10.

Etikan, I., & Bala, K. (2017). Sampling and sampling methods. *Biometrics & Biostatistics International Journal, 5*(6), 00149.

Etikan, I., Musa, S. A., & Alkassim, R. S. (2016). Comparison of convenience sampling and purposive sampling. *American Journal of Theoretical and Applied Statistics*, *5*(1), 1–4.

Fàbregues, S., & Paré, M. H. (2018). Appraising the quality of mixed methods research in nursing: A qualitative case study of nurse researchers' views. *Nursing Inquiry*, *25*(4). Advance online publication. DOI: 10.1111/nin.12247 PMID: 29927008

Fàbregues, S., Younas, A., Escalante-Barrios, E. L., Molina-Azorin, J. F., & Vázquez-Miraz, P. (2024). Toward a framework for appraising the quality of integration in mixed methods research. *Journal of Mixed Methods Research*, *18*(3), 270–280. DOI: 10.1177/15586898241257555

Fahimnia, B., Sarkis, J., & Davarzani, H. (2015). Green supply chain management: A review and bibliometric analysis. *International Journal of Production Economics*, *162*, 101–114.

Fan, Y., Zhang, L., & Li, H. (2023). Bridging the methodological gap: Computational and qualitative research in social media studies. *Proceedings of the ACM on Human-Computer Interaction*, *7*(CSCW1), 1–25. https://doi.org/DOI: 10.1145/3576232

Fan, D., Breslin, D., Callahan, J. L., & Iszatt-White, M. (2022). Advancing literature review methodology through rigour, generativity, scope and transparency. *International Journal of Management Reviews*, *24*(2), 171–180.

Fangwei, F., & Shiyi, S. (1996). Hypothesis testing for arbitrarily varying source. *Shuxue Xuebao*, *12*(1), 33–39.

Fan, Y., Matcha, W., Ahmad Uzir, N., Wang, Q., & Gašević, D. (2021). Learning analytics to reveal links between learning design and self-regulated learning. *International Journal of Artificial Intelligence in Education*, *31*(4), 980–1021. DOI: 10.1007/s40593-021-00249-z

Feder, M., & Merhav, N. (2002). Universal composite hypothesis testing: A competitive minimax approach. *IEEE Transactions on Information Theory*, *48*(6), 1504–1517.

Felderer, M., & Travassos, G. H. (2020). The evolution of empirical methods in software engineering. In Felderer, M., & Travassos, G. H. (Eds.), *Contemporary empirical methods in software engineering* (pp. 1–24). Springer International Publishing., DOI: 10.1007/978-3-030-32489-6_1

Fetters, M. D. (2020). The mixed methods research workbook: Activities for designing, implementing, and publishing projects. *Sage (Atlanta, Ga.)*.

Fetters, M. D., Curry, L. A., & Creswell, J. W. (2013). Achieving integration in mixed methods designs: Principles and practices. *Health Services Research, 48*(6 Pt 2), 2134–2156. DOI: 10.1111/1475-6773.12117 PMID: 24279835

Fitzgerald, T. (2012). Documents and documentary analysis. *Research Methods in Educational Leadership and Management, 3*, 296–308.

Flick, U. (2018). *An introduction to qualitative research* (6th ed.). SAGE Publications.

Forero, R., Nahidi, S., De Costa, J., Mohsin, M., Fitzgerald, G., Gibson, N., McCarthy, S., & Aboagye-Sarfo, P. (2018). Application of four-dimension criteria to assess rigour of qualitative research in emergency medicine. *BMC Health Services Research, 18*(1), 1–11. DOI: 10.1186/s12913-018-2915-2 PMID: 29454350

Frechette, J., Bitzas, V., Aubry, M., Kilpatrick, K., & Lavoie-Tremblay, M. (2020). Capturing lived experience: Methodological considerations for interpretive phenomenological inquiry. *International Journal of Qualitative Methods, 19*, 1609406920907254. DOI: 10.1177/1609406920907254

Friese, S. (2025). AI-assisted conversation analysis (CAAI): A five-step methodology for qualitative research. *Qualitative Methods Forum.*

Gabriel, Y. (2018). Interpretation, reflexivity and imagination in qualitative research. In Ciesielska, M., & Jemielniak, D. (Eds.), *Qualitative methodologies in Organization Studies* (pp. 137–157). Springer International Publishing., DOI: 10.1007/978-3-319-65217-7_8

Gallagher, S. (2022). What is phenomenology? In Gallagher, S. (Ed.), *Phenomenology* (pp. 1–10). Springer International Publishing., DOI: 10.1007/978-3-031-11586-8_1

Garcia, G. A., & Ramirez, J. J. (2021). Proposing a methodological borderland: Combining Chicana feminist theory with transformative mixed methods research. *Journal of Mixed Methods Research, 15*(2), 240–260. DOI: 10.1177/1558689820954023

Garg, R. (2016). Methodology for research I. *Indian Journal of Anaesthesia, 60*(9), 640–645. Retrieved October 10, 2025, from https://journals.lww.com/ijaweb/fulltext/2016/60090/methodology_for_research_i.6.aspx PMID: 27729690

Gautam, A. (2024). Reconfiguring participatory design to resist AI realism. *arXiv.* Retrieved on 19 October 2025 from https://arxiv.org/abs/2406.03245

Geampana, A., & Perrotta, M. (2024). Using interview excerpts to facilitate focus group discussion. *Qualitative Research.* Advance online publication. DOI: 10.1177/14687941241234283 PMID: 40028392

Giddings, L., & Grant, B. (2009). From rigour to trustworthiness: Validating mixed methods. In Andrew, S., & Halcomb, E. J. (Eds.), *Mixed methods research for nursing and health science*. Blackwell Publishing Ltd.

Gill, M. J. (2020). Phenomenology as qualitative methodology. *Qualitative Analysis: Eight Approaches*, 73–94.

Glaser, B. G., & Strauss, A. L. (1967). *The discovery of grounded theory: Strategies for qualitative research*. Aldine.

Goldsmith, L. J. (2021). Using framework analysis in applied qualitative research. *Qualitative Report*, *26*(6), 2061–2076.

Goldstein, D. M. (2014). *Qualitative research in dangerous places: Becoming an "ethnographer" of violence and personal safety*. Social Science Research Council.

Golzar, J., Noor, S., & Tajik, O. (2022). Convenience sampling. *International Journal of Education & Language Studies*, *1*(2), 72–77. Retrieved October 10, 2025, from https://www.ijels.net/article_162981.html

Gonçalves, M. C., & Silva, R. (2023, September). The effect of statistical hypothesis testing on machine learning model selection. In *Brazilian Conference on Intelligent Systems* (pp. 415–427). Springer Nature.

Good, P. (2013). *Permutation tests: A practical guide to resampling methods for testing hypotheses* (3rd ed.). Springer Science and Business Media.

Goundar, S. (2012). *Research methodology and research method*. Retrieved on 10 October 2025 from https://www.researchgate.net/publication/333015026_Chapter_3_Research_Methodology_and_Research_Method

Grand, J. A., Braun, M. T., & Kuljanin, G. (2024). Hello world! Building computational models to represent social and organizational theory. *Organizational Research Methods*, *28*(3), 487–539. DOI: 10.1177/10944281241261913

Granikova, V., Hong, Q. N., Crist, E. C., & Pluye, P. (2020). Mixed methods research in library and information science: A methodological review. *Library & Information Science Research*, *42*(1), 1–10. DOI: 10.1016/j.lisr.2020.101003

Grant, C., & Osanloo, A. (2014). Understanding, selecting, and integrating a theoretical framework in dissertation research. *Administrative Issues Journal*, 4(2), 12–26. https://doi.org/DOI: 10.5929/2014.4.2.9

Gray, P. (2003). Introduction to the debate on the core of the information systems field. *Communications of the AIS*, 12(1), Art. 42.

Greene, J. C. (2007). *Mixed methods in social inquiry*. Jossey-Bass.

Greene, J. C., & Caracelli, V. J. (Eds.). (1997). *Advances in mixed-method evaluation: The challenges and benefits of integrating diverse paradigms: New directions for evaluation*. Jossey-Bass.

Greene, J. C., Caracelli, V. J., & Graham, W. F. (1989). Toward a conceptual framework for mixed method evaluation designs. *Educational Evaluation and Policy Analysis, 11*, 255–274.

Greene, J. C., Caracelli, V. J., & Graham, W. F. (1989). Toward a conceptual framework for mixed-method evaluation designs. *Educational Evaluation and Policy Analysis, 11*(3), 255–274. DOI: 10.3102/01623737011003255

Green, H. E. (2014). Use of theoretical and conceptual frameworks in qualitative research. *Nurse Researcher, 21*(6), 34–38. DOI: 10.7748/nr.21.6.34.e1252 PMID: 25059086

Greenhalgh, T., Thorne, S., & Malterud, K. (2018). Time to challenge the spurious hierarchy of systematic over narrative reviews? *European Journal of Clinical Investigation, 48*(6), e12931. PMID: 29578574

Green, J. L., & Chian, M. M. (2018). Triangulation. In Frey, B. (Ed.), *The Sage encyclopedia of educational research, measurement, and evaluation* (pp. 1718–1720). Sage.

Gregor, S. (2006). The nature of theory in information systems. *Management Information Systems Quarterly, 30*(3), 611–642. DOI: 10.2307/25148742

Gregor, S., & Hevner, A. (2013). Positioning and presenting design science research for maximum impact. *Management Information Systems Quarterly, 37*(2), 337–355.

Gross, J. M. (2018). Document analysis. *The SAGE encyclopedia of educational research, measurement, and evaluation,* 1, 545-548.

Guba, E. G., & Lincoln, Y. S. (1994). Competing paradigms in qualitative research. *Handbook of Qualitative Research, 2*(163–194), 105. Retrieved on 10 October 2025 fromchrome-extension://efaidnbmnnnibpcajpcglclefindmkaj/https://ankara.lti.cs .cmu.edu/11780/sites/default/files/10-guba_lincoln_94.pdf

Guetterman, T. C. (2020). Qualitative, quantitative, and mixed methods research sampling strategies. https://doi.org/DOI: 10.1093/obo/9780199756810-0241

Guetterman, T. C., Fetters, M. D., & Creswell, J. W. (2015). Integrating quantitative and qualitative results in health science mixed methods research through joint displays. *Annals of Family Medicine, 13*(6), 554–561. DOI: 10.1370/afm.1865 PMID: 26553895

Gunasekare, U. L. T. P. (2015). Mixed research method as the third research paradigm: A literature review. *International Journal of Scientific Research*, *4*(8), 361–367.

Gundumogula, M. (2020). Importance of focus groups in qualitative research. [IJHSS]. *International Journal of Humanities and Social Science*, *8*(11), 299–302. DOI: 10.24940/theijhss/2020/v8/i11/HS2011-082

Habibullah, K. M., Mohammed, B. U., & Hamza, M. (2023). Conceptual framework in reflexive bracketing techniques in qualitative methodology. *Research Review, 4*(02), 849–857. http://researchreview.in/index.php/rr/article/view/161

Hamilton, , S. (2017). A genealogy of metatheory in IR: How 'ontology' emerged from the inter-paradigm debate. *International Theory, 9*(1), 136–170. https://doi .org/10.1017/S1752971916000257

Hands, A. S. (2022). Integrating quantitative and qualitative data in mixed methods research: An illustration. *Canadian Journal of Information and Library Science*, *45*(1), 1–20. DOI: 10.5206/cjilsrcsib.v45i1.10645

Hangel, N. F., Schönweitz, F., McLennan, S., & Fiske, A. (2022). Solidarity behavior and its limits: A qualitative study about German and Swiss residents' behaviors towards public health measures during COVID-19 lockdown in April 2020. *SSM. Qualitative Research in Health*, *2*(2), 100051. PMID: 35975169

Haroutunian, E. A., Haroutunian, M. E., & Harutyunyan, A. N. (2008). Reliability criteria in information theory and in statistical hypothesis testing. *Foundations and Trends in Communications and Information Theory*, *4*(2–3), 97–263.

Harris, J. (2021). Mixed methods research in developing country contexts: Lessons from field research in six countries across Africa and the Caribbean. *Journal of Mixed Methods Research*. Advance online publication. DOI: 10.1177/15586898211032825

Harvey, L. (2012). Phenomenological approach to validity. *Quality Research International*. Retrieved on 10 October 2025 from https://www.qualityresearchinternational.com/methodology/RRW1pt8pt3Phenvalidity.php

Hashimov, A. (2025). Max Weber's Bureaucratic Theory: Its Strengths and Weaknesses, and Adaptation to the Modern Era - A Theoretical and Empirical Analysis. *Acta Globalis Humanitatis et Linguarum*, *2*(4), 128–166.

Haslwanter, T. (2016). *An introduction to statistics with Python: With applications in the life sciences*. Springer International Publishing.

Hassan, N. R., Lowry, P. B., & Mathiassen, L. (2022). Useful products in information systems theorizing: A discursive formation perspective. *Journal of the Association for Information Systems*, *23*(2), 418–446. DOI: 10.17705/1jais.00730

Hassna, G. (2022). Big data and analytics to transform higher education: A value chain perspective. *Perspectives Policy and Practice in Higher Education*, *27*(4), 160–168. DOI: 10.1080/13603108.2022.2099475

Hawamdeh, S., & Madali, N. P. (2023). The emergence of new disciplines and the transformation of the library and information science profession. *Library Trends*, *72*(2), 187–204. https://muse.jhu.edu/pub/1/article/941423/summary

Hecker, J., & Kalpokas, N. (2024). The ultimate guide to qualitative research - Part 2: Handling qualitative data. Retrieved on 10 October 2025 from https://atlas.ti.com/guides/qualitative-research-guide-part-2/qualitative-data-analysis

Heneghan, N. R., Jagodzinska, J., Tyros, I., Johnson, W., Nazareth, M., Yeung, E., Sadi, J., Gillis, H., & Rushton, A. (2021). Telehealth e-mentoring in postgraduate musculoskeletal physiotherapy education: A mixed methods case study. *Musculoskeletal Science & Practice*, *56*, 102448. Advance online publication. DOI: 10.1016/j.msksp.2021.102448 PMID: 34416558

Henning, E., Van Rensburg, W., & Smit, B. (2004). *Finding your way in qualitative research.* Van Schaik Publishers.

Hennink, M., & Kaiser, B. N. (2022). Sample sizes for saturation in qualitative research: A systematic review of empirical tests. *Social Science & Medicine*, *292*, 114523. DOI: 10.1016/j.socscimed.2021.114523 PMID: 34785096

Henrys, P. A., Mondain-Monval, T. O., & Jarvis, S. G. (2024). Adaptive sampling in ecology: Key challenges and future opportunities. *Methods in Ecology and Evolution*, *15*(9), 1483–1496. https://doi.org/https://doi.org/10.1111/2041-210X.14393

Hevner, A. R., March, S. T., Park, J., & Ram, S. (2004). Design science in information systems research. *Management Information Systems Quarterly*, *28*(1), 75–105.

Hilliger, I., Ortiz-Rojas, M., Pesántez-Cabrera, P., Scheihing, E., Tsai, Y.-S., Muñoz-Merino, P. J., Broos, T., Whitelock-Wainwright, A., Gašević, D., & Pérez-Sanagustín, M. (2020). Towards learning analytics adoption: A mixed methods study of data-related practices and policies in Latin American universities. *British Journal of Educational Technology*, *51*(4), 915–937. DOI: 10.1111/bjet.12933

Hirose, M., & Creswell, J. W. (2023). Applying core quality criteria of mixed methods research to an empirical study. *Journal of Mixed Methods Research*, *17*(1), 12–28. DOI: 10.1177/15586898221086346

Høffding, S., Martiny, K., & Roepstorff, A. (2022). Can we trust the phenomenological interview? Metaphysical, epistemological, and methodological objections. *Phenomenology and the Cognitive Sciences*, *21*(1), 33–51. DOI: 10.1007/s11097-021-09744-z

Hofman, J. M., Watts, D. J., Athey, S., Garip, F., Griffiths, T. L., & Kleinberg, J.. (2021). Integrating explanation and prediction in computational social science. *Nature*, *595*, 181–188. PMID: 34194044

Hogg, R. V. (1961). On the resolution of statistical hypotheses. *Journal of the American Statistical Association*, *56*(296), 978–989.

Holton, E. F., & Lowe, J. S. (2007). Toward a general research process for using Dublin's theory building model. *Human Resource Development Review*, *6*(3), 297–320.

Hoover, R. S., & Koerber, A. L. (2011). Using NVivo to answer the challenges of qualitative research in professional communication: Benefits and best practices tutorial. *Transactions on Professional Communication*, *54*(1), 68–82. DOI: 10.1109/TPC.2009.2036896

Horkheimer, M. (2002). *Critical theory: Selected essays.* Continuum. (Original work published 1937).

Hosseini-Ashlaghi, M., Gholamhosseinzadeh, M., Zameni, F., & Siamian, H. (2023). Identifying and ranking critical success factors for cloud computing deployment at Mazandaran University of Medical Sciences. *Educational Research in Medical Sciences, 12*(2).

Howlett, M. (2022). Looking at the 'field' through a Zoom lens: Methodological reflections on conducting online research during a global pandemic. *Qualitative Research*, *22*(3), 387–402. PMID: 35663097

Hunter, A. D., & Brewer, J. (2015). Designing multimethod research. In Hesse-Biber, S. N., & Johnson, R. B. (Eds.), *The Oxford Handbook of Multimethod and Mixed Methods Research Inquiry* (pp. 185–205). Oxford University Press.

Ifenthaler, D., & Schumacher, C. (2016). Student perceptions of privacy principles for learning analytics. *Educational Technology Research and Development*, *64*(5), 923–938. DOI: 10.1007/s11423-016-9477-y

Islam, M. A., & Aldaihani, F. M. F. (2022). Justification for adopting qualitative research method, research approaches, sampling strategy, sample size, interview method, saturation, and data analysis. *Journal of International Business and Management, 5*(1), 01–11. https://doi.org/DOI: 10.37227/JIBM-2021-09-1494

ISO, I. (2019). I. 21839, "*Iso/IEC/IEEE International Standard–Systems and Software Engineering–System of Systems (SOS) Considerations in Life Cycle Stages of a System,*" (pp. 1-40). ISO/IEC/IEEE 21839: 2019, Kabanda, G. (2021). Cybersecurity risk management plan for a blockchain application model. *Trans Eng Comput Sci*, *2*(1), 221.

Ivankova, N. V., Creswell, J. W., & Stick, S. L. (2006). Using mixed-methods sequential explanatory design: From Theory to practice. *Field Methods*, *18*(1), 3–20. DOI: 10.1177/1525822x05282260

Ivankova, N., & Plano Clark, V. (2018). Teaching mixed methods research: Using a socio-ecological framework as a pedagogical approach for addressing the complexity of the field. *International Journal of Social Research Methodology*, *21*(4), 409–424. https://doi.org/DOI:10.1080/13645579.2018.1427604

Izgar, G., & Aktürk, A. O. (2018). A mixed method research on peer assessment. [IJERE]. *International Journal of Evaluation and Research in Education*, *7*(2), 118. DOI: 10.11591/ijere.v7i2.12770

Jacard, J., & Jacoby, J. (2010). *Theory construction and model-building skills: A practical guide for social scientists*. The Guilford Press.

Jackson-Gordon, R., & Clark, V. L. P. (2023). Using a joint display for building integration in a sequential study: Informing data collection for a participatory second phase. *Journal of Mixed Methods Research*, *18*(2), 137–146. DOI: 10.1177/15586898231179848

Jager, J., Putnick, D. L., & Bornstein, M. H. (2017). II. More than just convenient: The scientific merits of homogeneous convenience samples. *Monographs of the Society for Research in Child Development*, *82*(2), 13–30. DOI: 10.1111/mono.12296 PMID: 28475254

James, T. G., DeJonckheere, M., & Guetterman, T. C. (2022). Integrating transformative considerations and quantitative results through a participant selection joint display in explanatory sequential mixed methods studies. *Journal of Mixed Methods Research*, *18*(1), 14–30. DOI: 10.1177/15586898221149470 PMID: 38344136

Janice, P. S., Mouton, J., & Louw, W. L. (2018). *The SAGE handbook of qualitative data analysis* (2nd ed.). SAGE Publications.

Järvelin, K., & Vakkari, P. (2022). LIS research across 50 years: Content analysis of journal articles. *The Journal of Documentation*, *78*(7), 65–88. DOI: 10.1108/JD-03-2021-0062

Jeong, D. Y., & Kim, S. J. (2005). Knowledge structure of library and information science. *Library & Information Science Research*, *27*, 51–72.

Jivet, I., Scheffel, M., Schmitz, M., Robbers, S., Specht, M., & Drachsler, H. (2020). From students with love: An empirical study on learner goals, self-regulated learning and sense-making of learning analytics in higher education. *The Internet and Higher Education*, *47*, 100758. Advance online publication. DOI: 10.1016/j.iheduc.2020.100758

Johnson, J. L., Adkins, D., & Chauvin, S. (2020). A review of the quality indicators of rigor in qualitative research. *American Journal of Pharmaceutical Education*, *84*(1), 7120. DOI: 10.5688/ajpe7120 PMID: 32292186

Johnson, R. B., & Christensen, L. B. (2017). *Educational research: Quantitative, qualitative, and mixed approaches* (6th ed.). Sage.

Johnson, R. B., & Onwuegbuzie, A. J. (2004). Mixed methods research: A research paradigm whose time has come. *Educational Researcher*, *33*(7), 14–26. DOI: 10.3102/0013189X033007014

Kadushin, C., Hecht, S., Sasson, T., & Saxe, L. (2008). Triangulation and mixed methods designs: Practicing what we preach in the evaluation of an Israel experience educational program. *Field Methods*, *20*(1), 46–65. DOI: 10.1177/1525822x07307426

Kahn, P., Everington, L., Kelm, K., Reid, I., & Watkins, F. (2017). Understanding student engagement in online learning environments: The role of reflexivity. *Educational Technology Research and Development*, *65*(1), 203–218. DOI: 10.1007/s11423-016-9484-z

Kalpokas, N., & Hecker, J. (2023). The guide to mixed methods research. ATLAS.ti Research Hub. Retrieved on 20 October from https://atlasti.com/guides/the-guide-to-mixed-methods-research

Kalu, F. A., & Bwalya, J. C. (2017). What makes qualitative research good research? An exploratory analysis of critical elements. *International Journal of Social Science Research*, *5*(2), 43–56. DOI: 10.5296/ijssr.v5i2.10711

Kandiero, A., Bigirimana, S., & Chizwina, S. (2024). *Theoretical and Conceptual Frameworks in ICT Research*. IGI Global (case studies on technology adoption and innovation).

Karatzas, D., ... (2016, April). Human-document interaction systems-A new frontier for document image analysis. *In 2016 12th IAPR Workshop on Document Analysis Systems (DAS)* (pp. 369-374). IEEE.

Karppinen, K., & Moe, H. (2019). Texts as data I: Document analysis. In *The Palgrave handbook of methods for media policy research* (pp. 249–262). Springer International Publishing.

Kasirye, F. (2021). *A conceptual paper on ensuring quality in qualitative research.* [Paper presentation], Malaysia. Retrieved on 10 October 2025 from https://www.researchgate.net/profile/Faiswal-Kasirye/publication/350525202_A_Conceptual_Paper_on_Ensuring_Quality_in_Qualitative_Research/links/6064a604299bf13821f866b6/A-Conceptual-Paper-on-Ensuring-Quality-in-Qualitative-Research.pdf

Katchmarchi, A. B., Taliaferro, A. R., & Kipfer, H. (2018). A Document Analysis of Drowning Prevention Education Resources in the United States. *International Journal of Injury Control and Safety Promotion, 25*(3). PMID: 28701104

Kawulich, B. (2009). The role of theory in research. In Garner, M., Wagner, C., & Kawulich, B. (Eds.), *Teaching research methods in the social sciences* (pp. 37–47). Routledge.

Keller, D., & Egbert, J. (2020). Hypothesis testing ordinary meaning. *Brooklyn Law Review, 86*, 489.

Kelle, U., Kühberger, C., & Bernhard, R. (2019). How to use mixed-methods and triangulation designs: An introduction to history education research. *History Education Research Journal, 16*(1). Advance online publication. DOI: 10.18546/herj.16.1.02

Kemper, E. A., Stringfield, S., & Teddlie, C. (2003). Mixed methods sampling strategies in social science research. In Tashakkori, A., & Teddlie, C. (Eds.), *Sage handbook of mixed methods in social and behavioural research* (pp. 1–20). Sage.

Keshavarz, H. (2020). *Hermeneutic phenomenology in supporting research and information services: Contributions to information science.* Retrieved on 10 October 2025 from https://repository.kisti.re.kr/handle/10580/16183

Kettles, A., Creswell, J. W., & Zhang, W. (2011). Mixed methods research in mental health nursing. *Journal of Psychiatric and Mental Health Nursing, 18*(6), 535–542. DOI: 10.1111/j.1365-2850.2011.01701.x PMID: 21749560

Kettley, N. (2010). *Theory building in educational research.* Continuum.

Khalid, M., Akanmu, A., Murzi, H., Lee, S. W., Awolusi, I., Manesh, D., & Okonkwo, C. (2023). Industry perception of the knowledge and skills required to implement sensor data analytics in construction. *Journal of Civil Engineering Education, 150*(1), 04023011. Advance online publication. DOI: 10.1061/JCEECD.EIENG-1902

Kijewski, S., Ronchi, E., & Vayena, E. (2025). The rise of checkbox AI ethics: A review. *AI and Ethics*, *5*(3), 1931–1940. DOI: 10.1007/s43681-025-00358-9

Kitchin, R. (2014). *The data revolution: Big data, open data, data infrastructures and their consequences*. SAGE Publications.

Kitto, S. C., Chesters, J., & Grbich, C. (2008). Quality in qualitative research. *The Medical Journal of Australia*, *188*(4), 243–246. PMID: 18279135

Kivunja, C. (2018). Distinguishing between theory, theoretical framework, and conceptual framework: A systematic review of lessons from the field. *International Journal of Higher Education*, *7*(6), 44–53. DOI: 10.5430/ijhe.v7n6p44

Klingberg, S., Stalmeijer, R. E., & Varpio, L. (2024). Using framework analysis methods for qualitative research: AMEE Guide No. 164. *Medical Teacher*, *46*(5), 603–610. DOI: 10.1080/0142159X.2023.2259073 PMID: 37734451

Korstjens, I., & Moser, A. (2018). Series: Practical guidance to qualitative research. Part 4: Trustworthiness and publishing. *The European Journal of General Practice*, *24*(1), 120–124. DOI: 10.1080/13814788.2017.1375092 PMID: 29202616

Kothari, C. R., & Garg, G. (2019). *Research methodology: Methods and techniques*. New Age International.

Kridel, C. (2015). The biographical and documentary milieu. In He, M. F., Schultz, B. D., & Schubert, W. H. (Eds.), *The Sage guide to curriculum in education* (pp. 311–318). Sage Publishers.

Kulesa, J., Induru, S., Hubbard, E., & Bhansali, P. (2024). The conceptual framework: A practical guide. *Hospital Pediatrics*, *14*(11), e503–e508. DOI: 10.1542/hpeds.2024-007794 PMID: 39363881

Kumar, A., & Sharma, A. (2023). Observation method: A review study. *Library Philosophy and Practice (e-Journal).* Https://Digitalcommons. Unl. Edu/Libphilprac. Retrieved on 10 October 2025 from https://www.researchgate.net/profile/Ankit-Kumar-385/publication/383270822_Observation_Method_A_Review_Study/links/66c5a8274b25ef677f72889a/Observation-Method-A-Review-Study.pdf

Kumar, D., & Dhir, S. (Eds.). (2021). *Trends in future informatics and emerging technologies*. Bentham Books.

Larson-Hall, J., & Mizumoto, A. (2019). Using statistical analysis software (R, SPSS). In *The Routledge handbook of research methods in applied linguistics* (pp. 385–397). Routledge.

Lazer, D., Pentland, A., Adamic, L., Aral, S., Barabási, A. L., Brewer, D., Christakis, N., Contractor, N., Fowler, J., Gutmann, M., Jebara, T., King, G., Macy, M., Roy, D., & Van Alstyne, M. (2009). Social science: Computational social science. *Science*, *323*(5915), 721–723. DOI: 10.1126/science.1167742 PMID: 19197046

Leavy, P. (2017). *Research design: Quantitative, qualitative, mixed methods, arts-based, and community-based participatory research approaches*. Guilford Publications.

Leech, N. L., & Onwuegbuzie, A. J. (2009). A typology of mixed methods research designs. *Quality & Quantity*, *43*(2), 265–275. DOI: 10.1007/s11135-007-9105-3

Leedy, P., & Ormrod, J. (2005). *Practical research: Planning and design* (7th ed.). Pearson.

Lewin, K. (1945). The Research Centre for Group Dynamics at Massachusetts Institute of Technology. *Sociometry*, *8*(2), 126–136.

Li, N., Lai, S., & Evans, J. (2025). Big data and the computational social science of entrepreneurship and innovation. In O. Sorenson & P. Thornton (Eds.), *De Gruyter handbook of sociology of innovation and entrepreneurship* (pp. 329–352). De Gruyter. DOI: 10.1515/9783111085722-019

Lin-Aung, H., Masumoto, D., Linn, Z., Kobayakawa, Y., Okamura, S., Kurihara, K., Morio, K., Tashiro, Y., Sakurai, H., & Hori, H. (2022). Students' perception of problem-based learning at a Japanese medical school: An exploratory sequential mixed method. *International Journal of Medical Education*, *13*, 322–334. DOI: 10.5116/ijme.636c.c5e1 PMID: 36571846

Lincoln, Y. S., & Guba, E. G. (1985). *Naturalistic inquiry*. SAGE.

Lincoln, Y. S., Lynham, S. A., & Guba, E. G. (2011). Paradigmatic controversies, contradictions, and emerging confluences, revisited. In Denzin, N. K., & Lincoln, Y. S. (Eds.), *The Sage handbook of qualitative research* (4th ed., pp. 97–128). Sage.

Liu, X., Guo, Y., Li, J., & Wang, Y. (2022). Visualization and data analysis of multi-factors for the scientific research training of graduate students. *Applied Sciences (Basel, Switzerland)*, *12*(24), 12845. Advance online publication. DOI: 10.3390/app122412845

Li, Z., & Abramson, C. M. (2025). Ethnography and machine learning: Synergies and new directions. In Borch, C., & Pardo-Guerra, J. P. (Eds.), *The Oxford handbook of the sociology of machine learning* (pp. 245–272). Oxford University Press., DOI: 10.1093/oxfordhb/9780197653609.013.36

Lopes, J. M., Sousa, A., Calçada, E., & Oliveira, J. (2021). A citation and co-citation bibliometric analysis of omnichannel marketing research. *Management Review Quarterly*, 1-34

López-Fernández, O., & Molina-Azorín, J. F. (2011). The use of mixed methods research in interdisciplinary educational journals. *International Journal of Multiple Research Approaches*, *5*(2), 269–283. DOI: 10.5172/mra.2011.5.2.269

Ludviga, I. (2023). Theoretical and conceptual frameworks and models: What are they, when, and how to apply them in teaching research methodology to master's and PhD students?

Luft, , J. AJeong, , SIdsardi, , RGardner, , G. (2022). Literature reviews, theoretical frameworks, and conceptual frameworks: An introduction for new biology education researchers. *CBE Life Sciences Education*, *21*(3). DOI: 10.1187/cbe.21-05-0134

Lukito, J., & Pruden, M. L. (2023). Critical computation: Mixed-methods approaches to big language data analysis. *The Review of Communication*, *23*(1), 62–78. DOI: 10.1080/15358593.2022.2125821

Lynham, S. A. (2002). Theory building: Dubin's method. *Advances in Developing Human Resources*, *4*(3), 242–276.

Macfadyen, L. P., & Dawson, S. (2012). Numbers are not enough. Why e-learning analytics failed to inform an institutional strategic plan. *Journal of Educational Technology & Society*, *15*(3), 149–163.

Mackieson, P., Shlonsky, A., & Connolly, M. (2019). Increasing rigor and reducing bias in qualitative research: A document analysis of parliamentary debates using applied thematic analysis. *Qualitative Social Work: Research and Practice*, *18*(6), 965–980.

Magama, B., & Ngoepe, M. (2025). The proof of the pudding is in the eating: Archival services quality at the national archives of Zimbabwe. *Information Development*, *0*(0), 1–14. DOI: 10.1177/02666669251375096

Malik, S., & Terzidis, O. (2025). Integrating AI and human judgment in systematic literature reviews: A hybrid framework. *Journal of Artificial Intelligence Research*, *12*(1), 88–104. DOI: 10.1016/j.jair.2025.02.005

Mandviwalla, M., Söllner, M., & Tuunanen, T. (Eds.). (2024). *Design science research for a resilient future: 19th International Conference on Design Science Research in Information Systems and Technology, DESRIST 2024, Trollhättan, Sweden, June 3–5, 2024, Proceedings* (Vol. 14621). Springer Nature.

Manion, J. L. (2019). *A mixed methods investigation of student achievement and satisfaction in traditional versus online learning environments* [Doctoral dissertation, Lindenwood University]. Lindenwood University Digital Commons. Retrieved on 15 November 2025 from https://digitalcommons.lindenwood.edu/dissertations/107

Marjaei, S., Yazdi, F. A., & Chandrashekara, M. (2019). MAXQDA and its application to LIS research. *Library Philosophy and Practice*, (e-journal) 2325, 1–10. https://digitalcommons.unl.edu/libphilprac/2325/

Marshall, C., & Rossman, G. B. (2016). *Designing qualitative research* (6th ed.). SAGE.

Mason, P., Augustyn, M. M., & Seakhoa-King, A. (2021). Mixed methods research in tourism: A systematic sequential approach. *Folia Turistica, 56*, 0-0. https://doi .org/DOI: 10.5604/01.3001.0014.8956

Master Academia. (2026). Theoretical vs. conceptual frameworks: Simple definitions and an overview of key differences. Retrieved on 13 February 2026 from https://master -academia.com/theoretical-vs-conceptual-frameworks-definitions-differences/

Matthews, C. R. (2025). *Doing good social science: Lessons from immersion, understanding social life and exploring the in-between.* Routledge.

Maxwell, J. (1997). Designing a qualitative study. In Bickman, L., & Rog, D. J. (Eds.), *Handbook of applied social research methods* (pp. 69–100). Sage.

May, E. M., Hunter, B. A., & Jason, L. A. (2017). Methodological pluralism and mixed methods to strengthen community psychology research: An example from Oxford House. *Journal of Community Psychology, 45*, 100–116. PMID: 28839344

Mayoh, J., & Onwuegbuzie, A. J. (2015). Toward a conceptualization of mixed methods phenomenological research. *Journal of Mixed Methods Research, 9*, 91–107.

McCormick, K., & Salcedo, J. (2017). *SPSS statistics for data analysis and visualization.* John Wiley and Sons.

McKenzie, C. R. (2004). Hypothesis testing and evaluation. In Koehler, D. J., & Harvey, N. (Eds.), *Blackwell handbook of judgment and decision making* (pp. 200–219). Blackwell Publishing.

Mcleod, S. (2023). Sampling methods in research: Types, techniques, and examples. Retrieved on 10 October 2025 from Https://Www. Simplypsychology. Org/ Sampling. Html.

McMillan, J., & Schumacher, S. (2006). *Research in education: Evidence-based inquiry* (6th ed.). Pearson.

McNabb, D. E. (2021). *Research methods for political science: Quantitative and qualitative methods* (3rd ed.). Routledge.

Megheirkouni, M., & Moir, J. (2023). Simple but effective criteria: Rethinking excellent qualitative research. *The Qualitative Report*, *28*(3), 848–864. DOI: 10.46743/2160-3715/2023.5845

Melnikovas, A. (2018). Towards an explicit research methodology: Adapting research onion model for futures studies. *Journal of Futures Studies*, *23*(2), 29–44. DOI: 10.6531/JFS.201812_23(2).0003

Merriam, S. B. (1988). *Case study research in education: A qualitative approach.* Jossey-Bass.

Merriam, S. B., & Tisdell, E. J. (2016). *Qualitative research: A guide to design and implementation* (4th ed.). Jossey Bass.

Mertens, D. M. (2010). Research and evaluation in education and psychology: Integrating diversity with quantitative, qualitative and mixed methods. *Sage (Atlanta, Ga.).*

Mertens, D. M. (2018). *Mixed methods design in evaluation.* SAGE Publications.

Merton, R. K. (1967). *On theoretical sociology*. Free Press.

Mettler, T. (2025). Computational Text Analysis for Qualitative IS Research: A Methodological Reflection. *Communications of the Association for Information Systems*, *56*, 349–378.

Miles, M. B., Huberman, A. M., & Saldaña, J. (2014). *Qualitative data analysis: A methods sourcebook* (3rd ed.). SAGE.

Mingers, J. (2001). Combining IS research methods: Towards a pluralist methodology. *Information Systems Research*, *12*(3), 240–259.

Mirza, H., Mirza, C., & Bellalem, F. (2023). Ethical considerations in qualitative research: Summary guidelines for novice social science researchers., *11*(1), 441–449. Retrieved on 10 October 2025 from https://asjp.cerist.dz/en/article/220704

Mizrahi, M. (2020). Hypothesis testing in scientific practice: An empirical study. *International Studies in the Philosophy of Science*, *33*(1), 1–21.

Moerer-Urdahl, T., & Creswell, J. W. (2004). Using transcendental phenomenology to explore the "ripple effect" in a leadership mentoring program. *International Journal of Qualitative Methods*, *3*(2), 19–35. DOI: 10.1177/160940690400300202

Mogalakwe, M. (2009). The documentary research method: Using documentary sources in social research. *Eastern Africa Social Science Research Review*, *25*(1), 43–58.

Mohammadi, A. (2023). Analytical evaluation of big data applications in e-commerce: A mixed method approach. *Decision Science Letters*, *12*(2), 457–476. DOI: 10.5267/j.dsl.2022.11.003

Molina-Azorín, J. F., & Font, X. (2015). Mixed methods in sustainable tourism research: An analysis of prevalence, designs and application in JOST 2005–2014. *Journal of Sustainable Tourism*, *2015*, 1–25.

Morgan, H. (2022). Conducting a qualitative document analysis. *The Qualitative Report*, *27*(1), 64–77.

Morse, J. M., & Niehaus, L. (2009). *Mixed method design: Principles and procedures.* Retrieved on 20 October from https://www.researchgate.net/publication/290746053_Janice_M_Morse_Linda_Niehaus_2009_mixed_method_design_Principles_and_procedures

Morse, J. M. (1994). Designing qualitative research. In Denzin, N. K., & Lincoln, Y. S. (Eds.), *Handbook of qualitative inquiry* (pp. 220–235). Sage.

Morse, J. M., Barrett, M., Koçer, B., & Omerod, R. (2023). *Qualitative health research: An introduction to methods and practices* (4th ed.). SAGE Publications.

Moschis, G. P. (2024). Conceptual framework and theory development. In *Academic research in business and the social sciences: A guidebook for early career researchers* (pp. 83–99). Springer., DOI: 10.1007/978-3-031-56548-9_5

Moser, A., & Korstjens, I. (2018). Series: Practical guidance to qualitative research. Part 3: Sampling, data collection and analysis. *The European Journal of General Practice*, *24*(1), 9–18. DOI: 10.1080/13814788.2017.1375091 PMID: 29199486

Muganga, L. (2015). The importance of hermeneutic theory in understanding and appreciating interpretive inquiry as a methodology. *Journal of Social Research & Policy*, *6*(1), 65.

Muijeen, K., Kongvattananon, P., & Somprasert, C. (2020). The key success factors in focus group discussions with the elderly for novice researchers: A review. *Journal of Health Research*, *34*(4), 359–371. https://www.emerald.com/insight/content/doi/10.1108/JHR-05-2019-0114/full/html

Mukherjee, D., Lim, W. M., Kumar, S., & Donthu, N. (2022). Guidelines for advancing theory and practice through bibliometric research. *Journal of Business Research, 148*, 101–115. DOI: 10.1016/j.jbusres.2022.04.042

Mullainathan, S., & Spiess, J. (2017). Machine learning: An applied econometric approach. *The Journal of Economic Perspectives, 31*(2), 87–106. DOI: 10.1257/jep.31.2.87

Muskat, M., Blackman, D., & Muskat, B. (2013). *Mixed Methods: Combining expert interviews, cross-impact analysis and scenario development*. University of Canberra.

Mustapha, I., Rattanawiboonsom, V., & Intanon, R. (2023). Data-driven insights in higher education: Exploring the synergy of big data analytics and mobile applications. [iJIM]. *International Journal of Interactive Mobile Technologies, 17*(20), 21–37. DOI: 10.3991/ijim.v17i20.45037

Mwadulo, M. W., & Odoyo, C. O. (2020). Phenomenology approach applicability in information systems and technology field. *Computer Science and Information Technology, 8*(2), 37–42. DOI: 10.13189/csit.2020.080201

Nachmias, C., & Nachmias, D. F. (1996). *Research methods in the social science* (5th ed.). Arnold.

Naeem, M., Ozuem, W., Howell, K., & Ranfagni, S. (2023). A step-by-step process of thematic analysis to develop a conceptual model in qualitative research. *International Journal of Qualitative Methods, 22*, 16094069231205789. DOI: 10.1177/16094069231205789

Nair, S. S., & Prem, S. S. (2020). A framework for mixed-method research. *Shanlax International Journal of Management, 8*(2), 45–53. DOI: 10.34293/management.v8i2.3220

Nanthagopan, Y. (2021). Review and comparison of multi-method and mixed method application in research studies. *Journal of Advanced Research, 2*(3), 55–78.

National Academies of Sciences, Engineering, and Medicine. (2022). *Fostering responsible computing research: Foundations and practices*. The National Academies Press.

Ncube, M. M., & Ngulube, P. (2025). Mapping theoretical approaches: A scoping review of data analytics in higher education. *Discover Education, 4*, 201. DOI: 10.1007/s44217-025-00647-8

Ncube, M. M., & Ngulube, P. (2025). Surge of data analytics in postgraduate education and methodological plurality: A systematic review. *Discover Education, 4*, 29. DOI: 10.1007/s44217-025-00422-9

Ndame, T. (2023). Phenomenological studies. In Okoko, J. M., Tunison, S., & Walker, K. D. (Eds.), *Varieties of qualitative research methods* (pp. 377–382). Springer International Publishing., DOI: 10.1007/978-3-031-04394-9_58

Nelson, L. K. (2020). Computational grounded theory: A methodological framework. *Sociological Methods & Research, 49*(1), 3–42.

Netshakhuma, N. S. (2024). Application of interpretivism theoretical framework on the management of university records: Comparison of selected universities. In A. Kandiero, S. Bigirimana & S. Chizwina. (2024). *Theoretical and conceptual frameworks in ICT research.* (pp. 29–49). IGI Global.

Neubauer, B. E., Witkop, C. T., & Varpio, L. (2019). How phenomenology can help us learn from the experiences of others. *Perspectives on Medical Education, 8*(2), 90–97. DOI: 10.1007/S40037-019-0509-2 PMID: 30953335

Neuman, W. L. (2014). *Social research methods: Qualitative and quantitative approaches* (7th ed.). Pearson.

Newby, P. (2014). *Research methods for education* (2nd ed.). Routledge.

Newman, M. C. (2008). "What exactly are you inferring?" A closer look at hypothesis testing. *Environmental Toxicology and Chemistry, 27*(5), 1013–1019. PMID: 18419185

Neyman, J. (1942). Basic ideas and some recent results of the theory of testing statistical hypotheses. *Journal of the Royal Statistical Society, 105*(4), 292–327.

Ngulube, P. (2015). Qualitative data analysis and interpretation: systematic search for meaning. In Mathipa, E. R., & Gumbo, M. T. (Eds.), *Addressing research challenges: making headway for developing researchers* (pp. 131–156). Mosala-MASEDI Publishers & Booksellers.

Ngulube, P. (2015). Trends in research methodological procedures used in knowledge management studies 2009–2013. *African Journal of Library Archives and Information Science, 25*(2), 1–30.

Ngulube, P. (2018). Overcoming the difficulties associated with using conceptual and theoretical frameworks in heritage studies. In Ngulube, P. (Ed.), *Handbook of research on heritage management and preservation* (pp. 1–23). IGI Global.

Ngulube, P. (2020). Mixed methods research in knowledge management studies (2009-2014): A content analysis of journal articles. *Journal of Information & Knowledge Management, 19*(3), 2040015. Advance online publication. DOI: 10.1142/S0219649220500161

Ngulube, P. (2020). The movement of mixed methods research and the role of information science professionals. In Ngulube, P. (Ed.), *Handbook of research on connecting research methods for information science research* (pp. 425–455). IGI Global., DOI: 10.4018/978-1-7998-1471-9.ch022

Ngulube, P. (2022). Using simple and complex mixed methods research designs to understand research in information science. In Ngulube, P. (Ed.), *Handbook of research on mixed methods research in information science* (pp. 20–46). IGI Global., DOI: 10.4018/978-1-7998-8844-4.ch002

Ngulube, P. (2023). Improving the quality of reporting findings using computer data analysis applications in educational research in context. *Heliyon Education, 9*(9), e19683. DOI: 10.1016/j.heliyon.2023.e19683 PMID: 37810136

Ngulube, P. (2025). Indigenous theories and theorisation in African scholarship in the quest for epistemic freedom. In Ngulube, P. (Ed.), *Harnessing Indigenous epistemologies for sustainable progress* (pp. 139–172). IGI Global., DOI: 10.4018/979-8-3693-8447-3.ch008

Ngulube, P., Mathipa, E. R., & Gumbo, M. T. (2015). Theoretical and conceptual framework in the social sciences. In Mathipa, E. R., & Gumbo, M. T. (Eds.), *Addressing research challenges: making headway in developing researchers* (pp. 43–66). Mosala-MASEDI Publishers & Booksellers., DOI: 10.13140/RG.2.1.3210.7680

Ngulube, P., & Mosha, N. F. V. (2023). Transparency in the application of theoretical frameworks to the advancement of knowledge in selected library and information science journals: A systematic review. *African Journal of Library Archives and Information Science, 33*(2), 129–146.

Ngulube, P., & Ngulube, B. (2015). Mixed methods research in the South African Journal of Economic and Management Sciences: An investigation of trends in the literature. *Suid-Afrikaanse Tydskrif vir Ekonomiese en Bestuurswetenskappe, 18*(1), 1–13.

Ngulube, P., & Ngulube, B. (2017). Application and contribution of hermeneutic and eidetic phenomenology to indigenous knowledge research. In Ngulube, P. (Ed.), *Handbook of research on theoretical perspectives on indigenous knowledge systems in developing countries* (pp. 127–155). IGI Global., DOI: 10.4018/978-1-5225-0833-5.ch006

Ngulube, P., & Ngulube, B. (2022). Are we there yet? Mixed methods research in the *South African Journal of Economic and Management Sciences*. *Acta Commercii*, *22*(1), a1013. Advance online publication. DOI: 10.4102/ac.v22i1.1013

Ngulube, P., & Ukwoma, S. C. (2021). Prevalence of methodological transparency in the use of mixed methods research in library and information science research in South Africa and Nigeria, 2009-2015. *Library & Information Science Research*, *43*(4), 101105. Advance online publication. DOI: 10.1016/j.lisr.2021.101105

Ngulube, P., & Ukwoma, S. C. (2021). Prevalence of methodological transparency in the use of mixed methods research in library and information science research in South Africa and Nigeria, 2009–2015. *Library & Information Science Research*, *43*(4), 101124. DOI: 10.1016/j.lisr.2021.101124

Ngulube, P., & Ukwoma, S. C. (2022). Mixed methods research in open distance learning: A content analysis of the literature. *Open Learning*, •••, 1–17. DOI: 10.1080/02680513.2022.2149321

Nguyen, D. C., & Welch, C. (2025). Generative artificial intelligence in qualitative data analysis: Analyzing—or just chatting? *Organizational Research Methods*, *0*(0). Advance online publication. DOI: 10.1177/10944281251377154

Nguyen, T., Baldassarre, M. T., de Lima, L. F., & de Souza Santos, R. (2024, October). From literature to practice: Exploring fairness testing tools for the software industry's adoption. *In Proceedings of the 18th ACM/IEEE International Symposium on Empirical Software Engineering and Measurement* (pp. 549-555).

Noble, S. U. (2018). *Algorithms of oppression: How search engines reinforce racism*. New York University Press.

Nowell, L. S., Norris, J. M., White, D. E., & Moules, N. J. (2017). Thematic analysis: Striving to meet the trustworthiness criteria. *International Journal of Qualitative Methods*, *16*(1), 160940691773384. DOI: 10.1177/1609406917733847

Nunan, D. (1999). *Research methods in language learning*. Cambridge University Press.

Nurmaidah, N., Kustiningsih, N., & Rahayu, S. (2024). Change management strategy in the implementation of electronic medical record system in the era of digital transformation: Case study at Rahman Rahim Hospital. *Gema Wiralodra*, *15*(1), 228–238.

Nyirenda, L., Kumar, M. B., Theobald, S., Sarker, M., Simwinga, M., Kumwenda, M., Johnson, C., Hatzold, K., Corbett, E. L., Sibanda, E., & Taegtmeyer, M. (2020). Using research networks to generate trustworthy qualitative public health research findings from multiple contexts. *BMC Medical Research Methodology*, *20*(1), 13. DOI: 10.1186/s12874-019-0895-5 PMID: 31964333

Nyumba, T. O., Wilson, K., Derrick, C. J., & Mukherjee, N. (2018). The use of focus group discussion methodology: Insights from two decades of application in conservation. *Methods in Ecology and Evolution*, *9*(1), 20–32. DOI: 10.1111/2041-210X.12860

Nzechukwu, P. O. (2016). *Internal Audit Practice from A to Z*. CRC Press.

O'Cathain, A., Murphy, E., & Nicholl, J. (2007a). Integration and publications as indicators of "yield" from mixed methods studies. *Journal of Mixed Methods Research*, *1*(2), 147–163. DOI: 10.1177/1558689806299094

O'Cathain, A., Murphy, E., & Nicholl, J. (2007b). Why, and how, mixed methods research is undertaken in health services research in England: A mixed methods study. *BMC Health Services Research*, *7*(1). Advance online publication. DOI: 10.1186/1472-6963-7-85 PMID: 17570838

O'Cathain, A., Murphy, E., & Nicholl, J. (2008). The quality of mixed methods studies in health services research. *Journal of Health Services Research & Policy*, *13*(2), 92–98. DOI: 10.1258/jhsrp.2007.007074 PMID: 18416914

O'Cathain, A., Nicholl, J., & Murphy, E. (2009). Structural issues affecting mixed methods studies in health research: A qualitative study. *BMC Medical Research Methodology*, *9*(1). Advance online publication. DOI: 10.1186/1471-2288-9-82 PMID: 20003210

Oates, B. J. (2006). Researching information systems and computing. *Sage (Atlanta, Ga.)*.

Oludapo, S., Akinwale, A., & Ogunleye, O. (2024). Digital transformation failure in African enterprises: A qualitative analysis of 12 critical factors. *International Journal of Information Management*, *76*, 102889. Advance online publication. DOI: 10.1016/j.ijinfomgt.2024.102889

Oluka, A. (2025). Phenomenological research strategy: Descriptive and interpretive approaches. *F1000 Research*, *14*, 725. Retrieved October 10, 2025, from https://f1000research.com/articles/14-725

Oluwadiya, K. S. (2022). Software packages for quantitative and qualitative research. Retrieved on 20 October from https://oluwadiya.com/wp-content/uploads/2022/07/Software-Packages-for-Quantitative-and-Qualitative-Research.pdf

Omar, A. S., & Mwakondo, F. (2024). Evolution of cloud computing: Trends, issues, and future directions: A systematic literature review. *International Journal of Computer Science Trends and Technology*, *12*(3), 102–111.

Onwuegbuzie, A. J., & Collins, K. M. (2007). A typology of mixed methods sampling designs in social science research. *The Qualitative Report*, *12*(2), 281–316. DOI: 10.46743/2160-3715/2007.1638

Onwuegbuzie, A. J., & Collins, K. M. (2017). The role of sampling in mixed methods research. *KZfSS Kölner Zeitschrift Für Soziologie Und Sozialpsychologie*, *69*(2), 133–156.

Onwuegbuzie, A. J., & Collins, K. M. T. (2007). A typology of mixed methods sampling designs in social science research. *Qualitative Report*, *12*(2), 281–316.

Onwuegbuzie, A. J., & Johnson, R. B. (2006). The validity issue in mixed research. *Research in the Schools*, *13*(1), 48–63.

Onwuegbuzie, A. J., & Johnson, R. R. (2021). *The reviewer's guide for mixed methods research analysis*. Routledge.

Onwuegbuzie, A. J., & Leech, N. L. (2006). Linking research questions to mixed methods data analysis procedures. *The Qualitative Report*, *11*, 474–498.

Oxbridge Essays. (2026). Literature review, theoretical framework, and conceptual framework. Retrieved on 13 February 2026 from https://www.oxbridgeessays.com/blog/literature-review-theoretical-framework-conceptual-framework/

Oyewobi, L. O., Okanlawon, T. T., Medayese, S. O., Ogunbode, E. B., & Jimoh, R. A. (2024). Is pursuing a PhD without theoretical and conceptual framework a journey without roadmap? *Environmental Technology and Science Journal*, *15*(2), 138–149. DOI: 10.4314/etsj.v15i2.15

Padilla-Díaz, M. (2015). Phenomenology in educational qualitative research: Philosophy as science or philosophical science. *International Journal of Educational Excellence*, *1*(2), 101–110. DOI: 10.18562/IJEE.2015.0009

Page, M. J., McKenzie, J. E., Bossuyt, P. M., Boutron, I., Hoffmann, T. C., Mulrow, C. D., Shamseer, L., Tetzlaff, J. M., Akl, E. A., Brennan, S. E., Chou, R., Glanville, J., Grimshaw, J. M., Hróbjartsson, A., Lalu, M. M., Li, T., Loder, E. W., Mayo-Wilson, E., McDonald, S., & Moher, D. (2021). The PRISMA 2020 statement: An updated guideline for reporting systematic reviews. *Journal of Clinical Epidemiology*, *134*, 178–189. DOI: 10.1016/j.jclinepi.2021.03.001 PMID: 33789819

Palinkas, L. A., Aarons, G. A., Horwitz, S. M., Chamberlain, P., Hurlburt, M. S., & Landsverk, J. (2010). Mixed method designs in implementation research. *Administration and Policy in Mental Health*, *38*(1), 44–53. DOI: 10.1007/s10488-010-0314-z PMID: 20967495

Palinkas, L. A., Horwitz, S. M., Green, C. A., Wisdom, J. P., Duan, N., & Hoagwood, K. (2013). Purposeful sampling for qualitative data collection and analysis in mixed method implementation research. *Administration and Policy in Mental Health*, *42*(5), 533–544. DOI: 10.1007/s10488-013-0528-y PMID: 24193818

Paperpal. (2024). Conceptual framework vs theoretical framework – Difference and examples. *Paperpal Blog*. Retrieved on 3 November 2025 from https://paperpal.com/blog/researcher/conceptual-framework-vs-theoretical-framework-difference-and-examples

Pardo, A., Jovanovic, J., Dawson, S., Gašević, D., & Mirriahi, N. (2017). Using learning analytics to scale the provision of personalised feedback. *British Journal of Educational Technology*, *50*(1), 128–138. DOI: 10.1111/bjet.12592

Patino, C. M., & Ferreira, J. C. (2018). Inclusion and exclusion criteria in research studies: Definitions and why they matter. *Jornal Brasileiro de Pneumologia : Publicaçao Oficial da Sociedade Brasileira de Pneumologia e Tisilogia*, *44*, 84–84. DOI: 10.1590/S1806-37562018000000088 PMID: 29791550

Patton, M. Q. (2002). *Qualitative evaluation and research methods*. Sage. Retrieved on 10 October 2025 from https://psycnet.apa.org/record/1990-97369-000

Patton, M. (1990). *Qualitative evaluation and research methods* (2nd ed.). Sage.

Patton, M. Q. (2015). *Qualitative research & evaluation methods: integrating theory and practice*. SAGE Publishers.

Paul, J., & Criado, A. R. (2020). The art of writing literature review: What do we know and what do we need to know? *International Business Review*, *29*(4), 101717.

Paul, J., Lim, W. M., O'Cass, A., Hao, A. W., & Bresciani, S. (2021). Scientific Procedures and Rationales for Systematic Literature Reviews (SPAR-4-SLR). *International Journal of Consumer Studies*, *45*(4), O1–O16. DOI: 10.1111/ijcs.12695

Payne, G., & Payne, J. (2004). *Key concepts in social research.* Sage Publishers.

Perez, A. (2023). Practical recommendations for mixed methods sampling in psychological intervention research: A mixed methods case study. Retrieved on 20 October from https://digitalcommons.unl.edu/cehsdiss/427/

Petter, S., & Gallivan, M. (2004). Toward a framework for classifying and guiding mixed method research in information systems. *37th Annual Hawaii International Conference on System Sciences, 2004. Proceedings of the, Big Island, HI, USA,* 2004, pp. 10 pp. https://doi.org/DOI: 10.1109/hicss.2004.1265614

Phillips, D. C. (2004). *Is mixed methods research an epistemological oxymoron?* [Paper presentation]. American Educational Research Association.

Plano Clark, V. L., & Ivankova, N. V. (2016). *Mixed methods research: A guide to the field.* SAGE Publications., DOI: 10.4135/9781483398341

Polit, D. F., & Beck, C. T. (2017). *Nursing research: Generating and assessing evidence for nursing practice* (10th ed.). Wolters Kluwer Health.

Poth, C. N. (Ed.). (2023). *The SAGE handbook of mixed methods research design* (Vol. 1–2). Sage Publications., DOI: 10.4135/9781529682663

Poth, C. N., Bullock, E. P., Eppel, E., Cameron, R., & Golenko, X. (2023). Adaptive mixed methods research design practices to address complexity in business and management research. In *Handbook of mixed methods research in business and management* (pp. 329–347). Edward Elgar Publishing., DOI: 10.4337/9781800887954.00031

Poth, C. N., Wongvorachan, T., Bulut, O., & Otto, S. J. G. (2024). Adaptive case study-mixed methods design practices for researchers studying complex phenomena. *Journal of Mixed Methods Research, 18*(3), 292–303. DOI: 10.1177/15586898241250217

Poth, C., & Munce, S. E. (2020). Commentary—Preparing today's researchers for a yet unknown tomorrow: Promising practices for a synergistic and sustainable mentoring approach to mixed methods research learning. *International Journal of Multiple Research Approaches, 12*(1), 56–64. DOI: 10.29034/ijmra.v12n1commentary

Powell, H., Mihalas, S. T., Onwuegbuzie, A. J., Suldo, S. M., & Daley, C. E. (2008). Mixed methods research in school psychology: A mixed methods investigation of trends in the literature. *Psychology in the Schools, 45*(4), 291–309. DOI: 10.1002/pits.20296

Prem, E. (2023). From ethical AI frameworks to tools: A review of approaches. *AI and Ethics, 3*(3), 699–716. DOI: 10.1007/s43681-023-00258-9

Pretorius, L. (2026). Theoretical and conceptual frameworks in research. Retrieved on 13 February 2026 from https://www.lynettepretorius.com/the_scholars_way_blog/theoretical-and-conceptual-frameworks-in-research/

Prieto, L. P., Pishtari, G., Dimitriadis, Y., Rodríguez-Triana, M. J., Ley, T., & Odriozola-González, P. (2023). Single-case learning analytics: Feasibility of a human-centered analytics approach to support doctoral education. *Journal of Universal Computer Science*, *29*(9), 1033–1068.

Prior, L. (2012). The role of documents in social research. In *Handbook of qualitative research in education*. Edward Elgar Publishing.

Prus, R. (1996). *Symbolic interaction and ethnographic research: Intersubjectivity and the study of human experience*. State University of New York Press.

Python Software Foundation. (2024). Python official documentation. Retrieved on 10 October 2025 from https://docs.python.org/

Quinney, L., Dwyer, T., & Chapman, Y. (2016). Who, where, and how of interviewing peers: Implications for a phenomenological study. *SAGE Open*, *6*(3), 2158244016659688. DOI: 10.1177/2158244016659688

Rabiee, F. (2004). Focus-group interview and data analysis. *The Proceedings of the Nutrition Society*, *63*(4), 655–660. DOI: 10.1079/pns2004399 PMID: 15831139

Radhakrishnan, G. (2014). Sampling in mixed methods research. *International Journal of Advances in Nursing Management*, *2*(1), 24–27.

Raftery, A. E., Gilks, W. R., Richardson, S., & Spiegelhalter, D. (1995). Hypothesis testing and model. In W. R. Gilks, S. Richardson, & D. Spiegelhalter (Eds.), *Markov Chain Monte Carlo in practice* (pp. 165–187). Chapman and Hall.

Rahman, S. (2017). The advantages and disadvantages of using qualitative and quantitative approaches and methods in language "testing and assessment" research: A literature review. *Journal of Education and Learning*, *6*(1), 102–112.

Ramanujan, P., Bhattacharjea, S., & Alcott, B. (2022). A multi-stage approach to qualitative sampling within a mixed methods evaluation: Some reflections on purpose and process. *The Canadian Journal of Program Evaluation*, *36*(3), 355–364. DOI: 10.3138/cjpe.71237

Raschka, S. (2020). [*Machine learning lecture notes*. University of Wisconsin–Madison.]. *Stat*, •••, 451.

Ravi, M., Tyagi, S., Gopikumar, V., de Wit, E. E., Bunders, J., Padgett, D., & Regeer, B. (2025). Participatory research in low-resource settings: Endeavours in epistemic justice at The Banyan, India. *Journal of Responsible Technology, 23*, 100123. Advance online publication. DOI: 10.1016/j.jrt.2025.100123

Ravitch, S. M., & Riggan, M. (2017). *Reason & rigor: How conceptual frameworks guide research* (2nd ed.). SAGE.

Ravn, S. (2023). Integrating qualitative research methodologies and phenomenology—Using dancers' and athletes' experiences for phenomenological analysis. *Phenomenology and the Cognitive Sciences, 22*(1), 107–127. DOI: 10.1007/s11097-021-09735-0

Reason, P., & Bradbury, H. (2008). *The SAGE handbook of action research: Participative inquiry and practice*. SAGE Publications.

Resnik, D. B. (2020). *What is ethics in research and why is it important* (Vol. 1). December. https://www.niehs.nih.gov/research/resources/bioethics/whatis/index.cfm

Reynolds, P. D. (2007). *A primer in theory construction: An A&B classics edition*. Routledge.

Ringelstein, D., & Patel, K. (2023). A conceptual framework for informatics. [files.eric.ed.gov]. *E-Journal of Business Education & Scholarship of Teaching, 17*(1), 9–14. https://files.eric.ed.gov/fulltext/EJ1376059.pdf

Ritchie, J., Lewis, J., Nicholls, C. M., & Ormston, R. (2013). Qualitative research practice: A guide for social science students and researchers. *Sage (Atlanta, Ga.)*.

Ritzer, G. (1991). *Metatheorizing in sociology*. Lexington Books.

Robinson, O. C. (2023). Probing in qualitative research interviews: Theory and practice. *Qualitative Research in Psychology, 20*(3), 382–397. DOI: 10.1080/14780887.2023.2238625

Ruane, J. M. (2016). *Introducing social research methods*. Wiley.

Salawu, R. O., Shamsuddin, A.-O. S., & Masibo, S. (2023). Theoretical and conceptual frameworks in research: Conceptual clarification. *European Chemical Bulletin, 12*(12), 2103–2117.

Salmona, M., Lieber, E., & Kaczynski, D. (2019). Qualitative and mixed methods data analysis using Dedoose: A practical approach for research across the social sciences. *Sage (Atlanta, Ga.)*.

Sanscartier, M. D. (2018). The craft attitude: Navigating mess in mixed methods research. *Journal of Mixed Methods Research, 14*(1), 47–62. DOI: 10.1177/1558689818816248

Saraswati, P., & Devy, A. (2023). Mixed methods-research methodology: An overview. *Mathews Journal of Nursing and Health Care, 5*(4), Article 24. Retrieved on 15 November 2025 from https://www.mathewsopenaccess.com/scholarly-articles/mixed-methods-research-methodology-an-overview.pdf

Sardana, D., Kahu, S. Y., Gračanin, D., & Matković, K. (2021). Multi-modal data exploration in a mixed reality environment using coordinated multiple views. In Yamamoto, S., & Mori, H. (Eds.), *Human interface and the management of information. information presentation and visualization.*

Sarfo, J. O., Debrah, T., Gbordzoe, N. I., Afful, W. T., & Obeng, P. (2021). Qualitative research designs, sample size and saturation: Is enough always enough. *Journal of Advocacy. Research in Education, 8*(3), 60–65. Retrieved October 10, 2025, from https://www.academia.edu/download/101557437/jare.2021.3.pdf

Sarker, S., Sarker, S., Sahaym, A., & Bjørn-Andersen, N. (2012). Exploring value co-creation in relationships between an ERP vendor and its partners: A revelatory case study. *Management Information Systems Quarterly, 36*(1), 317–338. DOI: 10.2307/41410419

Saunders, M. N. K., Gray, D. E., Tosey, P., & Sadler-Smith, E. (2015). Concepts and theory building. In Anderson, L., Gold, J., Stewart, J., & Thorpe, R. (Eds.), *A guide to professional doctorates in business and management* (pp. 35–56). Sage., DOI: 10.4135/9781473921412.n3

Scerbo, M. W., Calhoun, A. W., & Hui, J. (2019). Research and hypothesis testing: Moving from theory to experiment. In *Healthcare simulation research: A practical guide* (pp. 161–167). Springer International Publishing.

Schlegel, D., Müller, J., & Wagner, C. (2024). Why AI projects fail: A qualitative study of 12 critical factors. *AI & Society, 39*(2), 301–318. DOI: 10.1007/s00146-023-01788-2

Schneider, M. A. (2006). *The theory primer: A sociological guide.* Rowman and Littlefield.

Schoonenboom, J. (2024). Design patterns in mixed methods research: Embedding mixed methods matched comparisons sampling in previous design decisions. *Journal of Mixed Methods Research, 18*(3), 383–392. DOI: 10.1177/15586898241257284

Schumpeter, J. A. (1942). *Capitalism, socialism and democracy.* Harper & Row.

SciPy Community. (2024). SciPy user guide and tutorials. Retrieved on 10 October 2025 from https://docs.scipy.org/

Scott, J. (2014). *A matter of record: documentary sources in social research*. John Wiley & Sons.

Sedgwick, P. (2013). Sample size: How many participants are needed in a trial? *BMJ (Clinical Research Ed.)*, •••, 346. PMID: 23418280

Sedgwick, P. (2014). Pitfalls of statistical hypothesis testing: Multiple testing. *BMJ (Clinical Research Ed.)*, *349*, g6288. PMID: 25172184

Sharma, L. R., Bidari, S., Bidari, D., Neupane, S., & Sapkota, R. (2023). Exploring the mixed methods research design: Types, purposes, strengths, challenges, and criticisms. *Global Academic Journal of Linguistics and Literature*, *5*(1), 3–12. DOI: 10.36348/gajll.2023.v05i01.002

Sharp, J. L., Mobley, C., Hammond, C., Withington, C., Drew, S. F., Stringfield, S., & Stipanovic, N. (2011). A mixed methods sampling methodology for a multisite case study. *Journal of Mixed Methods Research*, *6*(1), 34–54. DOI: 10.1177/1558689811417133

Shin, D.-H. (2016). Cross-platform users' experiences toward designing interusable systems. *International Journal of Human-Computer Interaction*, *32*(7), 503–514. DOI: 10.1080/10447318.2016.1177277

Shmueli, G. (2010). To explain or to predict? *Statistical Science*, *25*(3), 289–310. DOI: 10.1214/10-STS330

Shoko, J., Muboko, N., & Gandiwa, E. (2024). *Focus group discussion method as a tool for eliciting local community views for decision-making: A case study of the Save Valley Conservancy in the Southeast Lowveld of Zimbabwe*. Retrieved on 10 October 2025 from https://www.preprints.org/frontend/manuscript/c7945ac3f5ccefd910faef27424c5943/download_pub

Silvola, A., Sjöblom, A., Näykki, P., Gedrimiene, E., & Muukkonen, H. (2023). Learning analytics for academic paths: Student evaluations of two dashboards for study planning and monitoring. *Frontline Learning Research*, *11*(2), 78–98. Retrieved November 15, 2025, from https://files.eric.ed.gov/fulltext/EJ1414726.pdf

Sivarajah, U., Kamal, M. M., Irani, Z., & Weerakkody, V. (2017). Critical analysis of big data challenges and analytical methods. *Journal of Business Research*, *70*, 263–286. DOI: 10.1016/j.jbusres.2016.08.001

Slade, S., & Prinsloo, P. (2013). Learning analytics: Ethical issues and dilemmas. *The American Behavioral Scientist*, *57*(10), 1510–1529. DOI: 10.1177/0002764213479366

Sloan, A., & Bowe, B. (2014). Phenomenology and hermeneutic phenomenology: The philosophy, the methodologies, and using hermeneutic phenomenology to investigate lecturers' experiences of curriculum design. *Quality & Quantity*, *48*(3), 1291–1303. DOI: 10.1007/s11135-013-9835-3

Smith, D. W. (2018). Phenomenology. In *The Stanford encyclopedia of philosophy*. Metaphysics Research Lab, Stanford University. Retrieved on 10 October 2025 from https://plato.stanford.edu/archives/sum2018/entries/phenomenology

Smith, B., Paton, C., & Ramaraj, P. (2023). Teaching basic surgical skills using a more frugal, near-peer, and environmentally sustainable way: Mixed methods study. *JMIR Perioperative Medicine*, *6*(e50212), 1–10. DOI: 10.2196/50212 PMID: 37966886

Smith, J. A. (2011). Evaluating the contribution of interpretive phenomenological analysis. *Health Psychology Review*, *5*, 9–27.

Smith, J. A., Flowers, P., & Larkin, M. (2009). Interpretive phenomenological analysis: Theory, method and research. *Sage (Atlanta, Ga.)*.

Sobhiyeh, S., & Naraghi-Pour, M. (2016). Hypothesis testing with dependent observations. *IEEE Transactions on Signal Processing*, *65*(5), 1183–1195.

Song, D., Shin, Y., & Hong, H. (2022). *Learning analytics as an integrated approach in mixed methods research*. Routledge.

Stahl, N. A., & King, J. R. (2020). Expanding approaches for research: Understanding and using trustworthiness in qualitative research. *Journal of Developmental Education*, *44*(1), 26–28. Retrieved October 10, 2025, from https://files.eric.ed.gov/fulltext/EJ1320570.pdf

Stake, R. E. (1995). *The art of case study research*. Sage Publishers.

Statsmodels Developers. (2024). Statsmodels documentation and examples. Retrieved on 10 October 2025 from https://www.statsmodels.org/

Steffen, E. (2016). Ethical considerations in qualitative research. *Analysing Qualitative Data in Psychology*, *2*, 31–44.

Stenfors, T., Kajamaa, A., & Bennett, D. (2020). How to assess the quality of qualitative research. *The Clinical Teacher*, *17*(6), 596–599. DOI: 10.1111/tct.13242 PMID: 32790137

Stojanov, A., & Daniel, B. K. (2023). A decade of research into the application of big data and analytics in higher education: A systematic review of the literature. *Education and Information Technologies*, *29*(5), 5807–5831. DOI: 10.1007/s10639-023-12033-8

Straits, B. C., & Singleton, R. A. (2018). *Social research: Approaches and fundamentals*. Oxford University Press.

Strang, K. D. (2017). Beyond engagement analytics: Which online mixed-data factors predict student learning outcomes? *Education and Information Technologies*, *22*(2), 917–937. DOI: 10.1007/s10639-016-9464-2

Sukhera, J. (2022). Narrative reviews: Flexible, rigorous, and practical. *Journal of Graduate Medical Education*, *14*(4), 414–417. PMID: 35991099

Sundler, A. J., Lindberg, E., Nilsson, C., & Palmér, L. (2019). Qualitative thematic analysis based on descriptive phenomenology. *Nursing Open*, *6*(3), 733–739. DOI: 10.1002/nop2.275 PMID: 31367394

Sutton, R. I., & Staw, B. M. (1995). What theory is not. *Administrative Science Quarterly*, *40*(3), 371–384. DOI: 10.2307/2393788

Taherdoost, H. (2020). Different types of data analysis; data analysis methods and techniques in research projects. *International Journal of Academic Research in Management*, *9*(1), 1–9.

Taherdoost, H. (2022). What are different research approaches? Comprehensive review of qualitative, quantitative, and mixed method research, their applications, types, and limitations. *Journal of Management Science & Engineering Research*, *5*(1), 53–63. https://doi.org/https://doi.org/10.30564/jmser.v5i1.4538

Tani-Yıldız, E., & Tam, M. S. (2023). Using qualitative data analysis software (QDAS) in communication studies: A systematic review. *Akdeniz İletişim, (20.Yıl Özel Sayısı - 42)*, 51–68. https://doi.org/DOI: 10.31123/akil.1330172

Tashakkori, A., & Creswell, J. W. (2007). Editorial: The new era of mixed methods. *Journal of Mixed Methods Research*, *1*(1), 3–7. DOI: 10.1177/2345678906293042

Tashakkori, A., Johnson, R. B., & Teddlie, C. (2020). *Foundations of mixed methods research: Integrating quantitative and qualitative approaches in the social and behavioral sciences* (2nd ed.). SAGE Publications.

Tashakkori, A., & Teddlie, C. (1998). Mixed methodology: Combining qualitative and quantitative approaches. *Sage (Atlanta, Ga.)*.

Tashakkori, A., & Teddlie, C. (2003a). Major issues and controversies in the use of mixed methods in social and behavioural sciences. In Tashakkori, A., & Teddlie, C. (Eds.), *Mixed methods in social and behavioural research*. Sage.

Tashakkori, A., & Teddlie, C. (2003b). *Handbook of mixed methods in social and behavioural research*. Sage.

Tavakol, M., & Sandars, J. (2025). Twelve tips for using phenomenology as a qualitative research approach in health professions education. *Medical Teacher*, •••, 1–6. DOI: 10.1080/0142159X.2025.2478871 PMID: 40089895

TechTarget. (n.d.). Data analytics definition. Retrieved on 15 November 2025 from https://www.techtarget.com/searchbusinessanalytics/resources/Data-science-and-analytics

Teddlie, C., & Tashakkori, A. (2010). *Overview of contemporary issues in mixed methods research* (Vol. 2). Sage. https://doi.org/DOI:10.4135/9781506335193

Teddlie, C., Johnson, R. B., & Tashakkori, A. (2021). *Foundations of mixed methods research: Integrating quantitative and qualitative techniques in the social and behavioral sciences* (2nd ed.). Sage.

Teddlie, C., & Tashakkori, A. (2009). Foundations of mixed methods research: Integrating quantitative and qualitative approaches in the social and behavioral sciences. *Sage (Atlanta, Ga.).*

Teddlie, C., & Tashakkori, A. (2009). Foundations of mixed methods research: Integrating quantitative and qualitative approaches in the social and behavioural sciences. *Sage (Atlanta, Ga.).*

Teddlie, C., & Yu, F. (2007). Mixed methods sampling. *Journal of Mixed Methods Research*, *1*(1), 77–100. DOI: 10.1177/1558689806292430

Teddlie, C., & Yu, F. (2007). Mixed methods sampling: A typology with examples. *Journal of Mixed Methods Research*, *1*(1), 77–100.

Tegtmeyer, J. (2023). Building a dissertation conceptual and theoretical framework: A recent doctoral graduate narrates behind the curtain development. *Perspectives on Urban Education*, 20(1). Retrieved on 14 February 2026 from https://urbanedjournal.gse.upenn.edu/archive/volume-20-issue-1-fall-2022

Thelwall, M., & Sud, P. (2022). Scopus 1900–2020: Growth in articles, abstracts, countries, fields, and journals. *Quantitative Science Studies*, *3*(1), 37–50.

Thomas, S. P., & Sohn, B. K. (2023). From Uncomfortable squirm to self-discovery: A phenomenological analysis of the bracketing experience. *International Journal of Qualitative Methods, 22*, 16094069231191635. DOI: 10.1177/16094069231191635

Tisdell, E. J., Merriam, S. B., & Stuckey-Peyrot, H. L. (2025). *Qualitative research: A guide to design and implementation*. John Wiley & Sons. Williams, B. (n.d). *Role of document analysis in qualitative research methods. Insight*. Retrieved on 16 October, 2025 from https://insight7.io/role-of-document-analysis-in-qualitative-research-methods/

Tiwari, A. B., Rai, D. A., Singh, A. M., & Jaiswal, K. J. (2024). Ethical and security considerations in the era of quantum computing: A framework for responsible implementation strategies. *International Journal of Research Publication and Reviews, 5*(5), 7266–7271. DOI: 10.55248/gengpi.5.0524.1295

Tiwari, A., & Darbari, M. (2025). *Emerging trends in computer science and its application*. CRC Press., DOI: 10.1201/9781003606635

Togia, A., & Malliari, A. (2017). Research methods in library and information science. In Oflazoglu, S. (Ed.), *Qualitative versus quantitative research*. InTech., DOI: 10.5772/intechopen.68749

Tomaszewski, L. E., Zarestky, J., & Gonzalez, E. (2020). Planning qualitative research: Design and decision making for new researchers. *International Journal of Qualitative Methods, 19*, 160940692096717. DOI: 10.1177/1609406920967174

Toyon, M. A. S. (2021). Explanatory sequential design of mixed methods research: Phases and challenges. *International Journal of Research in Business & Social Science, 10*(5), 253–260. DOI: 10.20525/ijrbs.v10i5.1262

Trent, A., & Cho, J. (2020). *Interpretation in qualitative research: What, why, how*. Retrieved on 10 October 2025 from https://academic.oup.com/edited-volume/34283/chapter-abstract/290650730

Trieu, V.-H., Burton-Jones, A., Green, P., & Cockcroft, S. (2022). Applying and extending the theory of effective use in a business intelligence context. *Management Information Systems Quarterly, 46*(1), 645–678. DOI: 10.25300/MISQ/2022/14880

Tunarosa, A., & Glynn, M. A. (2016). Strategies of integration in mixed methods research. *Organizational Research Methods, 20*(2), 224–242. DOI: 10.1177/1094428116637197

Tuncel, E. (2005). On error exponents in hypothesis testing. *IEEE Transactions on Information Theory, 51*(8), 2945–2950.

Turner, J. H. (1991). *The structure of sociological theory* (5th ed.). Wadsworth.

Turner, S. F., Cardinal, L. B., & Burton, R. M. (2016). Research design for mixed methods. *Organizational Research Methods, 20*(2), 243–267. DOI: 10.1177/1094428115610808

Ukwoma, S. C., & Ngulube, P. (2021). (in press). To borrow or not to borrow is the question? Theory borrowing in library information science postgraduate research in Nigeria and South Africa. *The International Information & Library Review, 53*(1). Advance online publication. DOI: 10.1080/10572317.2020.1790261

UNESCO. (2022). UNESCO recommendation on open science. Available from: https://doi.org/DOI: 10.54677/MNMH8546

Urcia, I. A. (2021). Comparisons of adaptations in grounded theory and phenomenology: Selecting the specific qualitative research methodology. *International Journal of Qualitative Methods, 20*, 16094069211045474. DOI: 10.1177/16094069211045474

Ustun, A. B., Zhang, K., Karaoğlan-Yilmaz, F. G., & Yilmaz, R. (2022). Learning analytics-based feedback and recommendations in flipped classrooms: An experimental study in higher education. *Journal of Research on Technology in Education, 55*(5), 841–857. DOI: 10.1080/15391523.2022.2040401

van Ameijde, J., Cheng, S., & Wang, H. (2023). Integrating urban analytics into postgraduate urban design pedagogy: A mixed-methods teaching approach to addressing urban liveability. *Journal of Planning Education and Research*. Advance online publication. DOI: 10.1177/0739456X231162842

Van de Ven, A. H. (2007). *Engaged scholarship: A guide for organizational and social research*. Oxford University Press., DOI: 10.1093/oso/9780199226290.001.0001

Van der Waldt, G. (2020). Constructing conceptual frameworks in social science research. *The Journal for Transdisciplinary Research in Southern Africa, 16*(1), 758. DOI: 10.4102/td.v16i1.758

Van Eck, N., &Waltman, L. (2010). Software survey: VOSviewer, a computer program for bibliometric mapping. *scientometrics, 84*(2), 523-538.

Van House. N. A. (1991). Assessing the quantity, quality, and impact of LIS research. In C. R. McClure and P. Hernon (Eds.), *Library and information science research: Perspectives and strategies for improvement* (pp. 85-100). Ablex.

Varpio, L., Paradis, E., Uijtdehaage, S., & Young, M. (2020). The distinctions between theory, theoretical framework, and conceptual framework. *Academic Medicine, 95*(7), 989–994. DOI: 10.1097/ACM.0000000000003075 PMID: 31725464

Venkatesh, V., Brown, S., & Sullivan, Y. (2023). *Conducting mixed-methods research: From classical social sciences to the age of big data and analytics.* Virginia Tech Publishing. https://hdl.handle.net/10919/117223

Venkatesh, V., Brown, S. A., & Bala, H. (2013). Bridging the qualitative–quantitative divide: Guidelines for conducting mixed methods research in information systems. *Management Information Systems Quarterly*, *37*(1), 21–54. DOI: 10.25300/MISQ/2013/37.1.02

Viberg, O., Baars, M., Mello, R. F., Weerheim, N., Spikol, D., Bogdan, C., Gasevic, D., & Paas, F. (2024). Exploring the nature of peer feedback: An epistemic network analysis approach. *Journal of Computer Assisted Learning*, *40*(4), 1567–1579. DOI: 10.1111/jcal.13035

Viberg, O., Hatakka, M., Bälter, O., & Mavroudi, A. (2018). The current landscape of learning analytics in higher education. *Computers in Human Behavior*, *89*, 98–110. DOI: 10.1016/j.chb.2018.07.027

Wang, X., Chen, L., & Li, Z. (2023). Information phenomenology: An informational interpretation of Husserl's phenomenology. *Computer Sciences & Mathematics Forum, 8*(1), 86. https://www.mdpi.com/2813-0324/8/1/86

Wang, F., King, R. B., Zeng, L. M., Zhu, Y., & Leung, S. O. (2022). The research experience of postgraduate students: A mixed-methods study. *Studies in Higher Education*, *48*(4), 616–629. DOI: 10.1080/03075079.2022.2155344

Wang, H. I., Manolas, C., & Xanthidis, D. (2022). Statistical analysis with Python. In Hamada, T. M. (Ed.), *Handbook of computer programming with Python* (pp. 373–408). Chapman and Hall/CRC.

Weber, R. (2012). Evaluating and developing theories in the information systems discipline. *Journal of the Association for Information Systems*, *13*(1), 1–30. DOI: 10.17705/1jais.00284

Weick, K. E. (1995). What theory is not, theorizing is. *Administrative Science Quarterly*, *40*, 385–390.

West, D., Luzeckyj, A., Searle, B., Toohey, D., Vanderlelie, J., & Bell, K. R. (2020). Perspectives from the stakeholder: Students' views regarding learning analytics and data collection. *Australasian Journal of Educational Technology*, *36*(6), 72–88. DOI: 10.14742/ajet.5957

Westein, M. P. D., Koster, A. S., Daelmans, H. E. M., & Kusurkar, R. A. (2023). How progress evaluations are used in postgraduate education with longitudinal supervisor-trainee relationships: A mixed method study. *Advances in Health Sciences Education : Theory and Practice*, *28*(1), 205–222. DOI: 10.1007/s10459-022-10153-3 PMID: 36094680

Westland, H., Vervoort, S., Kars, M., & Jaarsma, T. (2025). Interviewing people on sensitive topics: Challenges and strategies. *European Journal of Cardiovascular Nursing*, *24*(3), 488–493. https://academic.oup.com/eurjcn/article-abstract/24/3/488/7761136 PMID: 39295513

Whetten, D. A. (1989). What constitutes a theoretical contribution? *Academy of Management Review*, *14*(4), 490–495.

Whitney, C., & Evered, J. A. (2022). The qualitative research distress protocol: A participant-centered tool for navigating distress during data collection. *International Journal of Qualitative Methods*, *21*, 16094069221110317. DOI: 10.1177/16094069221110317

William, F. K. A. (2024). Unveiling the significance and influence of theoretical and conceptual frameworks in social science research. *International Journal of Research Publications*, *142*(1), 137–142. DOI: 10.47119/ijrp1001421220246059

Williamson, K., & Johanson, G. (2018). *Research methods: Information, systems, and contexts*. Chandos Publishing.

Willig, C. (2017). *Interpretation in qualitative research. The SAGE Handbook of Qualitative Research in Psychology, 2, 274–288*. Sage.

Woods, M., Paulus, T., Atkins, D. P., & Macklin, R. (2016). Advancing qualitative research using qualitative data analysis software (QDAS): Reviewing potential versus practice in published studies using Atlas.ti and NVivo, 1994–2013. *Social Science Computer Review*, *34*(5), 597–617. DOI: 10.1177/0894439315519631

Wu, S., Diao, E., Elkhalil, K., Ding, J., & Tarokh, V. (2022). Score-based hypothesis testing for unnormalized models. *IEEE Access : Practical Innovations, Open Solutions*, *10*, 71936–71950.

Xiao, M., Wei, P., & Tai, H. M. (2012). Estimation of the number of sources based on hypothesis testing. *Journal of Communications and Networks (Seoul)*, *14*(5), 481–486.

Yaakop, N. (2023). A content validation of focus group discussions based on need analysis in a physical education training module for primary school teachers. *Retos: Nuevas Tendencias En Educación Física. Deporte y Recreación*, *50*, 1115–1122. DOI: 10.1177/0739456X17723971

Yadav, D. (2022). Criteria for good qualitative research: A comprehensive review. *The Asia-Pacific Education Researcher*, *31*(6), 679–689.

Yahav, I., Shmueli, G., & Mani, D. (2016). A tree-based approach for addressing self-selection in impact studies with big data. *Management Information Systems Quarterly*, *40*(4), 819–848. DOI: 10.25300/MISQ/2016/40.4.02

Yatsko, V. A. (2018). Informatics, Information Science, and Computer Science. *Scientific and Technical Information Processing*, *45*, 235–240. DOI: 10.3103/S0147688218040081

Yin, R. K. (1994). Case study research: Design and methods. *Sage (Atlanta, Ga.)*.

Yin, R. K. (2009). *Case study research: Design and methods* (4th ed.). Sage.

Yin, R. K. (2018). *Case study research and applications: Design and methods* (6th ed.). SAGE.

Younas, A., Pedersen, M., & Tayaben, J. L. (2019). Review of mixed-methods research in nursing. *Nursing Research*, *68*(6), 464–472. DOI: 10.1097/nnr.0000000000000372 PMID: 31693552

Zeitouni, O., & Gutman, M. (2002). On universal hypothesis testing via large deviations. *IEEE Transactions on Information Theory*, *37*(2), 285–290.

Zhao, X., Sriwarnasinghe, S. M., Tang, J., Wang, H., Wang, S., & Morikawa, S. (2024). Collaborative participatory research with LLM agents in South Asia: Field evidence from Sri Lanka. *arXiv*. Retrieved on 19 October 2025 from https://arxiv.org/abs/2411.08294

Zhao, G., Xie, X., Wang, Y., Liu, S., Jones, P., & López, C. (2024). Barrier analysis to improve big data analytics capability of the maritime industry: A mixed-method approach. *Technological Forecasting and Social Change*, *203*, 123345–123345. DOI: 10.1016/j.techfore.2024.123345

Zhou, Y., Yuchun, Z., & Machtmes, K. (2023). Mixed methods integration strategies used in education: A systematic review. *Methodological Innovations*, *17*(1), 41–49. DOI: 10.1177/20597991231217937

Zohrabi, M. (2013). Mixed method research: Instruments, validity, reliability and reporting findings. *Theory and Practice in Language Studies*, *3*(2), 254–262.

Zupic, I., & Čater, T. (2015). Bibliometric methods in management and organization. *Organizational Research Methods*, *18*(3), 429–472.

About the Contributors

Patrick Ngulube (PhD) is one of the leading researchers in the field of information science in Africa. He is a professor of information science and interdisciplinary research at the University of South Africa, a National Research Foundation rated researcher. As a teacher and a mentor, Prof Ngulube, has supervised more than sixty masters and PhD studies to completion. He has taught in many Universities in South Africa and outside South Africa. Above all he has mentored many postdoctoral fellows, researchers, students and Librarians who have excelled in different fields. He has published over a hundred articles in international and local journals, over 15 book chapters and edited more than five books. He has vast experience in facilitating workshops for postgraduate students and supervisors since 2010 with a focus on research methodologies (mixed methods research) and use of theories and theorizing in research (application of indigenous theories and epistemologies). He is a Guest Editor of an issue Frontiers of Research Metrics and Analytics on the research topic "Research Ethics and Integrity in the Artificial Intelligence Era" the Guest Editor of Sustainability (Special Issue) "Sustainable Digitalization: Access to Information and Information Preservation". He is a former Chairperson of the Social Media Interest Group of the Association for Information Science & Technology (ASIS&T). Additionally, he is the Editor of the Scopus Journal New Review of Information Networking and Journal of Electronic Resources Librarianship.

***

Berraouna Abdelkader is an Assistant Professor in the Department of Computer Science at University of Souk-Ahras, Algeria and Visiting Researcher at IRISA Laboratory of Bretagne Sud University Vannes, France. He is a PhD student at the University of Badji Mokhtar-Annaba, Algeria. He received his Magister in the Department of Information Technology at same university, in 2014. He has published several papers on the business processes modelling and workflow evolution in several international conferences. His current research interests include workflow evolution, MDA technology and business process development.

Collence Takaingenhamo Chisita holds a PhD in Library and Information Science. He is a Lecturer at the Durban University of Technology (DUT) and research fellow at the University of Johannesburg Methods lab South Africa. He has co-authored and co-edited numerous books, book chapters and more than 28 scholarly articles. He is the recipient of the South African National Research Foundation (NRF) for recognized researchers with a constant record of productivity in LIS. His studies have been presented as keynotes in many countries across the globe. He is a trailblazing knowledge activist with a voracious appetite for reading.

Pradeep Kumar Gupta is an accomplished Professor of Accounting and Finance at LM Thapar School of Management, TIET, with over 20 years of academic and administrative experience. Holding a PhD in Finance from Thapar University, he specialises in corporate governance, CSR, and investment strategies. With 25+ peer-reviewed publications (Scopus/WoS indexed), 353 Google Scholar citations, and active roles in international conferences (EURAM, AoM), his research impacts both academia and industry. As Associate Dean (Accreditations & Rankings), he drives institutional excellence in business education, supported by an ICSSR-funded project (INR 1.5M). With UGC-NET qualifier and several professional certifications, he blends teaching expertise in financial management with global collaborations across the US, Europe, and Asia.

Sharneet Singh Jagirdar, FRM, is a distinguished financial professional with 20 years of diverse experience in Banking, Telecom and Education sector. Holding a PhD in Finance and certified as a Financial Risk Manager (FRM) by US GARP, Dr. Jagirdar excels in driving performance enhancement, implementing best practices, and fostering talent development. His extensive industry expertise is complemented by his academic prowess, with notable research publications and presentations at prestigious global conferences like Academy of Management, USA and European Academy of Management. As an expert in financial modelling and valuation, Dr. Jagirdar has a proven track record of success in both industry and academia. An avid runner, reader, and traveller, Dr. Jagirdar brings a well-rounded perspective to his professional endeavours.

Martha Lyaka (PhD) is a postdoctoral fellow in the Department of Interdisciplinary Research and Postgraduate Studies at the University of South Africa (UNISA). Her research spans research design and methodology, artificial intelligence in the library environment, indigenous knowledge systems, knowledge management, the use of information and communication technologies, and the digitisation and preservation of archival materials. She is particularly committed to advancing sustainable access to information. She has authored several peer-reviewed publications addressing themes such as knowledge management, knowledge retention, and broader information science issues.

Blessed Magama is a seasoned archives and records management practitioner for over sixteen years, and a recent PhD graduate in information Science from the University of South Africa (UNISA). He is the Records Manager for the Zimbabwe Revenue Authority (ZIMRA), with research interests in records management and preservation systems; records management policy, standards, and procedures; quality service delivery within the records and information management discipline.

Tlou Maggie Masenya holds PhD in Information Science from the University of South Africa (UNISA) and a Masters in Information Technology from the University of Pretoria (UP). She has 11 years of working experience as a senior lecture in the field of Information Science, Information Systems and Technology at University of South Africa, University of Zululand and Durban University of Technology. She is currently Associate Professor in the Department of Information Systems at the Durban University of Technology. She is a distinguished researcher with extensive, impressive national and international research standing and has published numerous book chapters and articles in peer-reviewed journals within the broad niche area of Digital transformation, ICT4D, Digital preservation and Data Curation, Records and Knowledge Management,Technopreneurship, Digital Entrepreneurship, Cloud Computing, Innovative Pedagogies, Disruptive Technologies and Indigenous Knowledge Systems, which had a positive impact on her scholarly growth. She recently edited four books, published by IGI- Global International Academic Publisher in 2022, 2023 and 2025. Her research publications have consistently pushed the frontiers in Information Science and Technology and related fields and have become highly cited by academics and scholars globally. she is constantly reviewing journal articles and book chapters for Springer, SAJIM, Mousaion, JSASA, IFLA Journals and IGI-Global book chapters, just to name a few. She has also participated and contributed to numerous community engagement research projects throughout her academic career.

Isaac Mukungu An academic at higher education institutions in Uganda. An academic librarian at Busitema University and a lecturer in information science at the Islamic University in Uganda. Research interest in information behaviour, community library provisions, Digital library services, user studies and information literacy

Mthokozisi Ncube Driven by a commitment to excellence in the field of information and knowledge management, I strive to grow as a well-rounded academic through continuous professional and scholarly development. I value integrity, honesty, and diligence in all professional and personal advancement activities. I am currently a Post-Doctoral Fellow at the University of South Africa and holds a PhD from the Zimbabwe Open University. My academic training spans both library and information science and management. I hold a Master of Science in Library and Information Science, awarded with Distinction, and a Master of Business Administration, obtained with Merit, both from the National University of Science and Technology. I also hold a Postgraduate Diploma in Tertiary Education from Bindura University of Science Education, earned with Merit. My foundational qualifications include a Bachelor of Science in Library and Information Science from the National University of Science and Technology, where I graduated with an Upper Second Class, and a Diploma in Information Processing (IT) from City and Guilds of London, obtained with Credit. Through this diverse academic background, I am committed to contributing meaningfully to the advancement of knowledge in information and knowledge management.

Krishna Kumari is a distinguished academic and career development expert at SRM Institute of Science and Technology (SRM IST), Kattankulathur, where she serves as a key faculty member at the Department of Mathematics. With a strong background in career counseling and professional development, Dr. Kumari has become a vital resource for students and professionals seeking to advance their careers. Dr. Kumari holds a Ph.D. in Mathematics and has a rich academic and professional background. Her research and publications focus on [Formal languages and Automata Theory, Graph based machine learning, Educational Pedagogy], and she has authored several papers in peer-reviewed journals that contribute to her field. In her role at the Career Development Centre, Dr. Kumari is actively involved in designing and conducting various workshops and training programs aimed at enhancing students' career readiness. She has organized numerous workshops on essential skills such as resume writing, interview techniques, and job search strategies. These workshops are tailored to address the needs of students across different disciplines, preparing them for successful transitions into the professional world. Dr. Kumari has also been a key facilitator in several Faculty Development Programs (FDPs) and training sessions. These programs are designed to equip faculty members with the latest tools and techniques in teaching and career guidance. Her contributions to these FDPs have been instrumental in promoting best practices in career development and educational excellence. Her involvement in these areas extends to professional organizations and networks related to career development and higher education. Dr. Kumari is an active member of various professional bodies, where she collaborates with peers to stay updated on industry trends and advancements. Dr. Krishna Kumari's commitment to career development and her extensive experience in organizing workshops and FDPs underscore her significant contributions to SRM IST and the broader academic community. Her work not only enhances the career prospects of students but also supports the ongoing professional development of faculty members, reflecting her dedication to excellence in education and career guidance.

Nina Olivia Rugambwa is a lecturer in the Department of Library and Information Science, School of Computing and Information Science at Kyambogo University in Kampala Uganda. She is both an educationist and Information scientist and has 15 years of teaching experience in university settings. Nina obtained her PhD in 2020 from Makerere University. She is passionate about teaching and research. Her research interests are; information seeking behavior, Indigenous Knowledge Management, Health informatics and child centered research methodogies in Information science research. Nina is also a student mentor at Makerere University MasterCard scholars program. She coaches young people to excel and succeed in their academic endeavors. This is a voluntary service she has offered to her community since 2016 to date. She is also a mentor to child mothers and out of school teenage girls at Tusitukirewamu women group in Bwaise.

Aniket Sachan is a PhD Candidate in Information Systems and Analytics at the Indian Institute of Management Tiruchirappalli (IIM Trichy). He earned his Master of Science in Computational Social Science, with a specialisation in Computational Sociology, from the Indian Institute of Technology Jodhpur (IIT Jodhpur). During his master's program, he also pursued research as a summer intern at the Indian Institute of Technology Guwahati (IIT Guwahati). Aniket has qualified for the University Grants Commission (UGC) National Eligibility Test (NET) for Assistant Professorship. He holds a Bachelor's degree from Chhatrapati Sahu Ji Maharaj University, Kanpur. His research interests lie at the intersection of society and technology. He received recognition for his model presentation at the Research and Industrial Conclave (RIC), the flagship event of IIT Guwahati, jointly organised by the Students' Academic Board and the IIT Guwahati Research Park. RIC serves as a multidisciplinary platform uniting academia, industry, and policymakers to drive innovation for a sustainable future. Aniket's recent research was published in the PSIT International Journal of Multidisciplinary Research (ISSN: 2456-2556, Volume 5). Aniket presented his work at the 2nd India Conference on Information Systems (InCIS 2026), the 19th ISDSI-Global International Conference, and the 10th Pan IIM World Management Conference. In addition to his academic pursuits, he enjoys writing non-fiction poems, which have also been published in the IIM Trichy

Hindi magazine "Sangam. His research was also recognized in various conferences in Information Systems.

Index

www.ingramcontent.com/pod-product-compliance
Lightning Source LLC
LaVergne TN
LVHW080908300926
849162LV00003B/597
* 9 7 9 8 3 6 9 3 4 0 1 8 9 *